The Illustrated Encyclopedia of Palynology: Volume 1

The Illustrated Encyclopedia of Palynology: Volume 1

Kristen McKenzie

www.callistoreference.com

Callisto Reference,
118-35 Queens Blvd., Suite 400,
Forest Hills, NY 11375, USA

Visit us on the World Wide Web at:
www.callistoreference.com

ISBN: 978-1-64116-749-9 (Hardback)

Cataloging-in-Publication Data

The illustrated encyclopedia of palynology: Volume 1 / Kristen McKenzie.
p. cm.
Includes bibliographical references and index.
ISBN 978-1-64116-749-9
1. Palynology. 2. Palynology--Encyclopedias. 3. Palynology--Pictorial works. I. McKenzie, Kristen.
QK658 .I45 2023
582.046 3--dc23

Table of Contents

Preface

Palynology is a scientific discipline that involves the study of plant pollen, spores and certain microscopic planktonic organisms, in both living and fossil form. Palynology has several applications in areas such as aerobiology, archaeology, forensic science and crime scene investigation, and allergy studies. Botany makes use of living pollen and spores to study plant relationships and their evolutions. The study of pollen in honey, with the purpose of identifying the source plants used by bees in the production of honey is known as melissopalynology. This study is beneficial to the production of honey as the honey produced by pollen and nectar from certain plants such as mesquite, buckwheat, tupelo or citrus trees is priced higher than that produced by other plant sources. It also studies the plants that produce nectar and pollen that is dangerous to human health. This book is compiled in such a manner that it will provide in-depth knowledge about the theory and practice of palynology. It is an essential guide for both botanists and students who wish to pursue this discipline further.

This book is a comprehensive compilation of works of different researchers from varied parts of the world. It includes valuable experiences of the researchers with the sole objective of providing the readers (learners) with a proper knowledge of the concerned field. This book will be beneficial in evoking inspiration and enhancing the knowledge of the interested readers.

In the end, I would like to extend my heartiest thanks to the authors who worked with great determination on their chapters. I also appreciate the publisher's support in the course of the book. I would also like to deeply acknowledge my family who stood by me as a source of inspiration during the project.

Kristen McKenzie

Part I
Introduction to Palynology

1

History of Palynological Research and the Development of Classification System

Palynology is the science of palynomorphs, a general term for all entities found in palynological preparations (e.g., pollen, spores, cysts, diatoms). A dominating object of the palynomorph spectrum is the pollen grain. The term palynology was coined by Hyde and Williams (1955; Fig. 1). It is a combination of the Greek verb paluno (παλύνω, "I strew or sprinkle"), palunein (παλύνειν, "to strew or sprinkle"), the Greek noun pale (παλη, in the sense of "dust, fine meal," and very close to the Latin word pollen, meaning "fine flour, dust"), and the Greek noun logos (λογος, "word, speech").

The History of Palynology

Assyrians are said to have known the principles of pollination (they practiced hand pollination of date palms), but it is unclear if they recognized the nature of pollen itself. The invention of the first microscopes and especially the compound microscope in the late sixteenth century represents the starting point of a new fascinating era. Some of the most important findings and scientists within the long tradition of light microscopy are mentioned here. For a more comprehensive overview, see Wodehouse (1935) and Ducker and Knox (1985).

Following the invention of the simple microscope by J. Janssen and Z. Janssen in 1590, the first compound microscope was developed by Hooke (1665). This was an important contribution to the study of pollen morphology. Malpighi in his "Anatomia Plantarum" was the first to describe pollen grains as having germination furrows while Grew noted in his famous work "The Anatomy of Plants" the constancy of pollen characters within the same species (Fig. 2; Grew 1682; Malpighi 1901). They are both considered the founders of pollen morphology. Camerarius described several pollination experiments and communicated the results in his letters about plant sexuality to Valentini (Camerarius 1694). He stated that male "seed dust" is necessary for seed development. Von Linné (also known before his ennoblement as Carl Nilsson Linnæus) first used the term pollen (in 1750). In the 18th and the early nineteenth centuries, there was considerable progress in pollen research and the understanding of pollination. In 1749, Gleditsch demonstrated in a spectacular experiment (Experimentum Berolinense) the central role of pollen in double fertilization. He organized the transport of an inflorescence from a male fan palm in Leipzig to a hitherto "sterile" female fan palm growing in a greenhouse in Berlin. After pollination, the female flowers produced fertile seeds for the first time in the palms lifetime (Gleditsch 1751, 1765). Placing male inflorescences within groups of female date palms for pollination was according to Theophrast already

of assistance ... suggestions that you might care to offer." (William W. Rubey, Chairman, Division of Geology and Geography, National Research Council, August 30, 1944)

THE RIGHT WORD. - "The question raised by Dr. Antevs: 'Is pollen analysis the proper name for the study of pollen and its applications?' and his suggestion to replace it by 'pollen science' interest us very much. We entirely agree that a new term is needed but in view of the fact that pollen analysts normally include in their counts the spores of such plants as ferns and mosses we think that some word carrying a wider connotation than pollen seems to be called for. We would therefore suggest palynology (from Greek παλύνω (paluno), to strew or sprinkle; cf. παλη (palē), fine meal; cognate with Latin pollen, flour, dust): the study of pollen and other spores and their dispersal, and applications thereof. We venture to hope that the sequence of consonants p-l-n, (suggesting pollen, but with a difference) and the general euphony of the new word may commend it to our fellow workers in this field. We have been assisted in the coining of this new word by Mr. L. J. D. Richardson, M.A., University College, Cardiff." (H.A. Hyde and D. A. Williams, July 15, 1944. Wales)

"I have been toying with the idea of 'micro-paleobotany' as including most of the work on pollen and spores and also all minor constituents of peat and humus layers of vegetative remains which

Fig. 1 The right word. Excerpt from Hyde and Williams (1955). Pollen Analysis Circular no. 8, p. 6

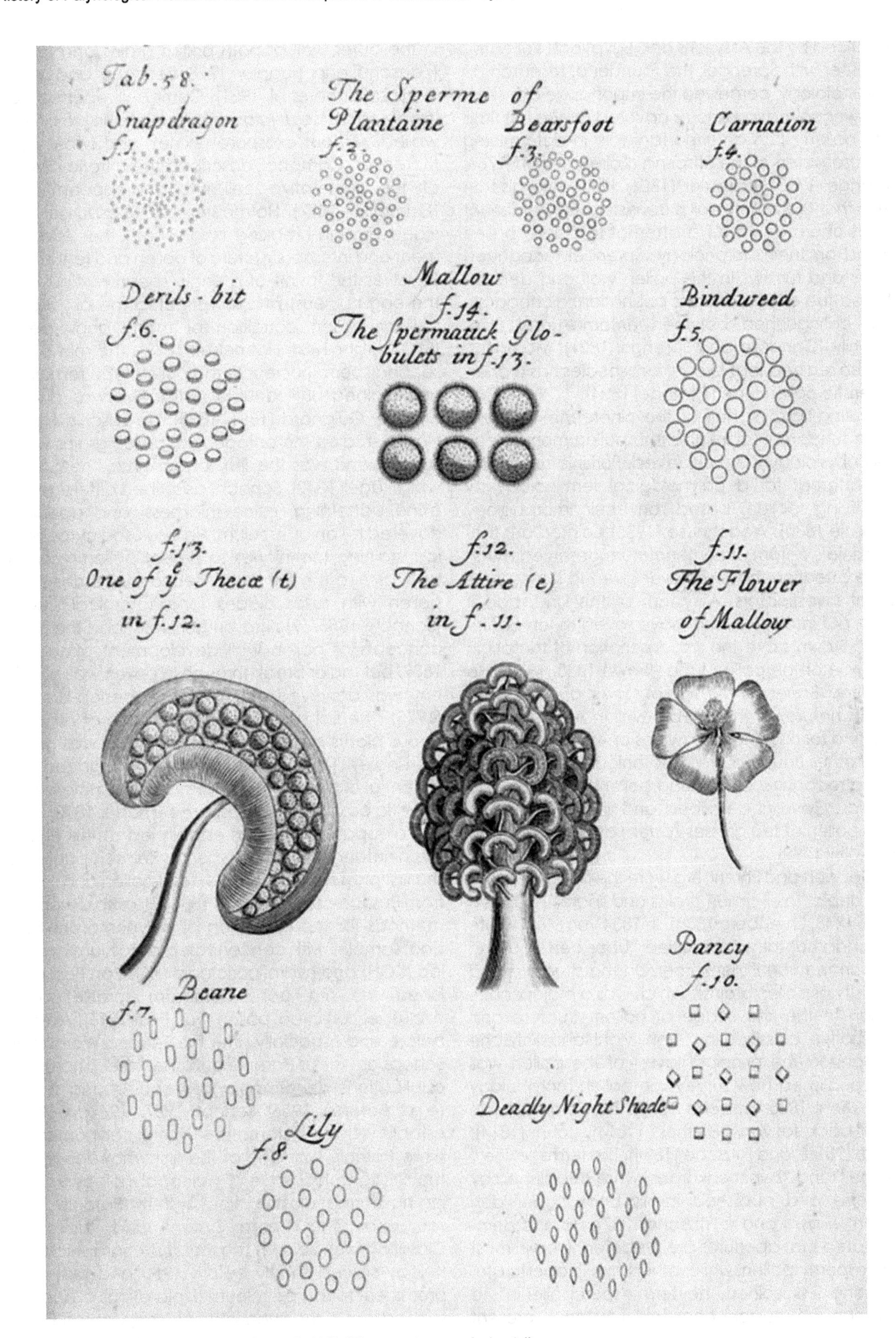

Fig. 2 First drawings of pollen. Grew (1682) "The anatomy of plants"

practiced by the Assyrians and Egyptians. Kölreuter, together with Sprengel, the founder of research on flower biology, perceived the importance of insects in flower pollination and discovered for the first time that pollen plays an important role in determining the characters of the offspring (Kölreuter 1761-1766; Sprengel 1793). Kölreuter (1806, 1811) also discovered that the pollen wall is consisting of two distinct layers and made the first attempt to classify pollen based on their morphology. Sprengel recognized pores and furrows in the pollen wall and demonstrated the effects of cross pollination, dichogamy, and distinguished between entomo- and anemophily (Candolle and Sprengel 1821). Moreover, he also realized that every plant species has a characteristic pollen type (Sprengel 1804).

During the first half of the nineteenth century, some fundamental insights into pollen morphology and physiology were achieved. Purkinje made the first attempt for a palynological terminology by classifying pollen based on their morphology (Purkinje 1830). Wodehouse (1935) pointed out that "Purkinje's system of nomenclature deserved much more attention than was ever given to it by subsequent investigators. A system of this kind, had it been put into use, would have saved much confusion." Brown gave the first description of the origin and role of the pollen tube (Brown 1828, 1833). He credited Bauer as the first observer of the pollen tube's nature, of the double wall in *Asclepias* pollen, and for his minute drawings of *Asclepias* pollen. His brother Bauer, a great botanical artist, was the first to recognize compound pollen in *Acacia* and orchids. Cavolini described and illustrated the filiform pollen of sea grasses *Zostera* and *Cymodocea* (Cavolini 1792).

Göppert and Ehrenberg were the first to describe and depict fossil pollen grains and spores (Göppert 1837, 1848; Ehrenberg 1838). In 1834 von Mohl wrote his fundamental work entitled "Über den Bau und die Formen von Pollenkörner/On the structure and diversity of pollen grains," which was a major contribution to the knowledge of pollen structure and descriptive classification. von Mohl and Fritzsche recognized the principal layers of the pollen wall and published new surveys on pollen morphology (von Mohl 1835; Fritzsche 1837). The term pollenin goes back to von Grotthuss (1814), John (1814), Stolze (1816), and Fritzsche (1834). The terms "exine," "intine," and "Zwischenkörper" were established by Fritzsche and published in his book "Über den Pollen" (Figs. 3 and 4; Fritzsche 1837). He also demonstrated that apertures are predetermined in most angiosperm pollen while others are inaperturate. Zetzsche first coined the term "sporopollenin" to describe the resistant chemical substance present in the outer wall of both pollen grains and spores (Zetzsche and Huggler 1928; Zetzsche and Vicari 1931; Zetzsche et al. 1931). Campbell reported pollen of seagrasses (*Naias* and *Zannichellia*) to be thin walled, without exospore (exine), and two-celled. Moreover, Campbell described the mitotic division of the generative cell into two (sperm) cells (Campbell 1897). Hofmeister and Strasburger provided ground-breaking insights into the development and internal structure of pollen and fertilization (involves the fusion of a single sperm nucleus and the egg nucleus) and investigated the bi- and tricellular pollen condition of many angiosperms (Strasburger 1884; Hofmeister 1849). The role of the second sperm nucleus in the pollen tube remained unexplained until double fertilization was discovered by Guignard (1891, 1899); Nawaschin (1898). Nägeli studied the ontogeny of pollen grains within anthers and was the first to recognize the callose wall (Nägeli 1842). Schacht described differences in exine patterning, exine thickness, and apertures covered by an operculum. He also used cytochemical staining techniques to detect pollen reserves. He was also the first to cut sections of embedded pollen with razor blades for anatomical studies (Schacht 1856/59). Strasburger described the basic concepts of pollen wall development already in 1889, but major break-through in pollen wall ontogeny was achieved much later by Heslop-Harrison (1975). The first successful classification of orchidaceous plants based on pollen features was made by Lindley (1836). Later, Fischer recognized the potential of pollen morphology in aiding the phylogenetic position of angiosperms (Fischer 1890).

Paleopalynology was established at the end of the nineteenth century, when P. Reinsch published the first photomicrographs of fossil pollen and spores from Russian coals (Reinsch 1884). He also described methods for the extraction of palynomorphs from coal samples with concentrated potassium hydroxide (KOH) and hydrofluoric acid (HF). Von Post published the first pollen diagram (profile) using exclusively arboreal pollen (von Post 1916). Already before and especially after the Second World War, Schopf as well as Potonié published their impressive publications devoted to fossil spores and pollen (e.g., Potonié 1956; Schopf 1957, 1964). Schopf established the systematic study of palynomorphs, while Potonié was one of the first who recognized the stratigraphic value of paleopalynology, applying his "turmal classification" system (Potonié 1934; see also "The Treme System and the NPC-Classification" below). The rise of stratigraphic palynology started shortly before 1950 and played a prominent role in petroleum explorations during the second half of the twentieth century (Manten 1966).

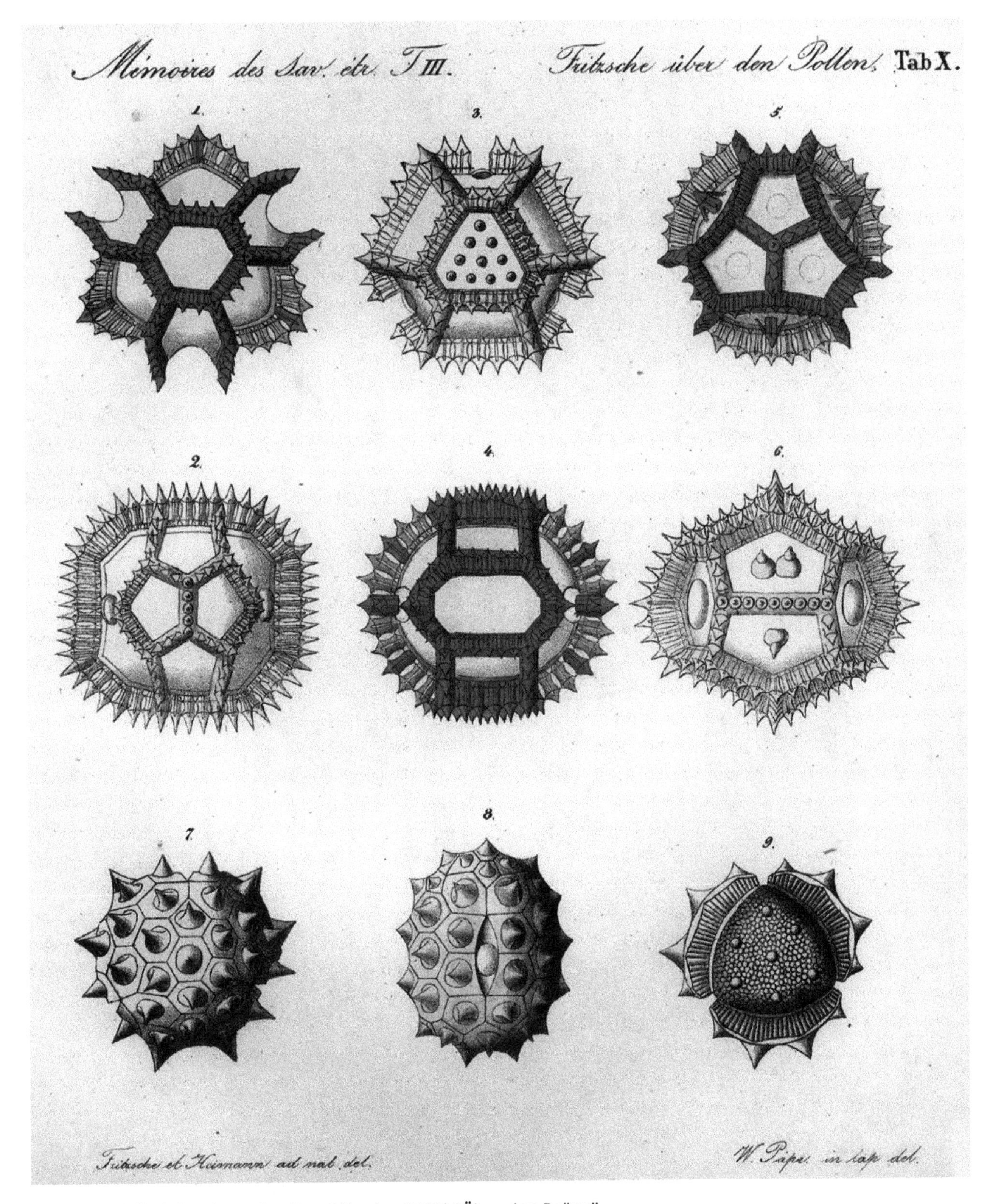

Fig. 3 Detailed drawings of pollen. Fritzsche (1837) "Über den Pollen"

The key role of palynology in stratigraphy depends upon the fact that the natural biopolymer sporopollenin in the spore/pollen walls is extremely resistant; thus, pollen/spores are often abundantly preserved in sedimentary rocks.

The twentieth century up to ca 1960 was dominated by the skillful use of the LM, with many new findings; for example, the LO-analysis, a method for analyzing patterns of exine organization by light microscopy, focusing at different levels distinct

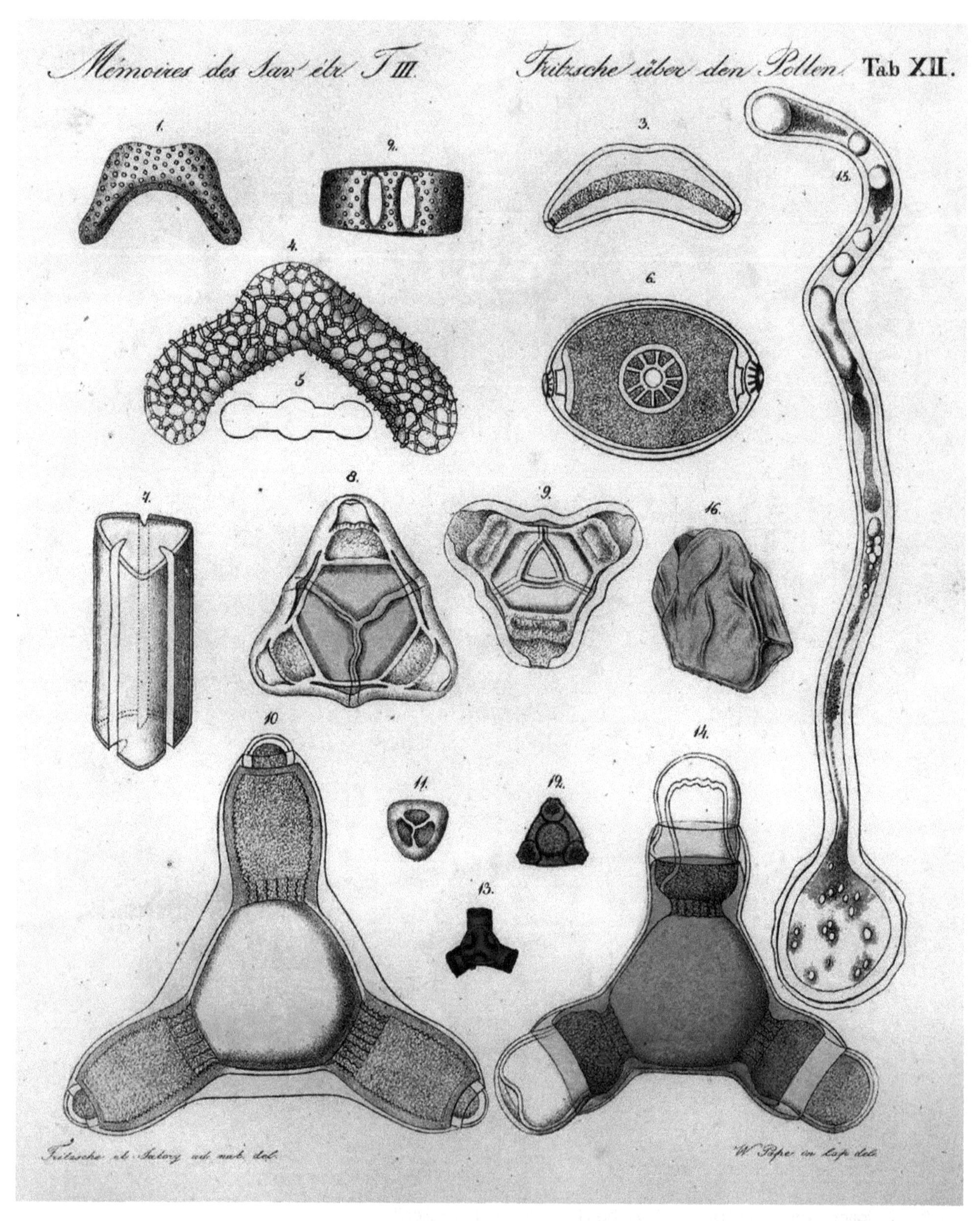

Fig. 4 First detailed drawings of pollen. Fritzsche (1837) "Über den Pollen"

features appear bright (*L* = Lux) or dark (*O* = Obscuritas). Textbooks by Wodehouse, Erdtman, or Fægri and Iversen summarized the knowledge of palynology from that time, but are still in good use (Erdtman 1943, 1952, 1957, 1969; Fægri and Iversen 1950, 1989; Wodehouse 1935). During this time palynology also became more diverse and applied in numerous fields among others: aeropalynology, biostratigraphy, copropalynology, cryopalynology, forensic palynology, iatropalynology, melissopalynology, paleopalynology, archeology, paleoclimatology, and palynotaxonomy.

Electron Microscopy with its two most important instrument types, the Transmission Electron Microscope (TEM) and the Scanning Electron Microscope (SEM), facilitated major breakthroughs in palynology. The TEM revealed new and stunning insights into pollen wall development and stratification. This prompted authors to publish new descriptions and create new terms. As pointed out by Knox: "The terminology applied to the pollen wall is daunting, especially as it has been developed from early light microscopy work, and then transposed to the images seen in the transmission and scanning electron microscopes" (Knox 1984, p. 204).

One of the first reports on the ultrastructure of recent pollen using TEM were published by H. Fernandez-Moran and A. O. Dahl (1952), and by K. Mühlethaler (1953). The first reports on the ultrastructure of fossil pollen were published by Ehrlich and Hall (1959; Pettitt and Chaloner (1964). During the 1950s and early 1960s considerable progress in TEM preparation methods (from fixation to microtome sectioning and staining) took place. EM-based information on ornamentation details of pollen grains was rare up to the mid-1960s. Only TEM-based casts or replica methods were available, all of them with limited resolution and depth of focus (e.g., the single-stage carbon replica technique; Mühlethaler 1955; Bradley 1958; Rowley and Flynn 1966). The time-consuming and laborious TEM replica procedures were an obstacle to extensive surveys of pollen morphology and later replaced by SEM (Harley and Ferguson 1990). The introduction of SEM in palynology in the mid of the 1960s was a key innovation in the study of the fine relief (sculpture) of pollen and spore surfaces. Advantages of SEM include the relatively simple and rapid preparation methods and the supreme depth of focus. SEM was considered from the very first moment as the quantum leap in EM (Hay and Sandberg 1967). The first SEM micrographs of pollen grains were published by Thornhill et al. (1965) and Erdtman and Dunbar (1966). Since then palynologists have been provided with a plethora of beautiful micrographs. Like Blackmore noted "The scanning electron microscope has provided a greater impetus to palynology than any other technical development during the history of the subject." Blackmore (1992). The LM with basic and advanced equipment, such as the fluorescent super-resolution microscopy, is overcoming the Abbe limit of LM resolution (especially STED microscopy, Hell 2009). The super-resolution LM and the two main types of EM form an expedient combination of imaging techniques. The LM remains the "workhorse method" (Traverse 2007; see the compendia by Reille 1992, 1995, 1998), but is limited regarding various morphological and structural features. Therefore, the role of SEM as an essential part in illustrating exine sculpture and ornamentation cannot be overrated (Harley and Ferguson 1990). The TEM still plays an important role, for example, in elucidating the complex steps of exine formation and development (e.g., Blackmore et al. 2007, 2010; Gabarayeva and Grigorjeva 2010; Gabarayeva et al. 2010).

The first and especially the second half of the twentieth century saw palynology at its peak, combining light microscopy with electron microscopy techniques. In addition to the above-mentioned scientists, other great palynologists have also promoted our science toward its present multifaceted appearance. These include among others: B. Albert, H.-J. Beug, G. El Ghazali, F. Firbas, M. Harley, J. Jansonius, W. Klaus, G. O. W. Kremp, B. Lugardon, S. Nadot, A. Maurizio, J. Muller, S. Nilsson, J. R. Rowley, J. J. Skvarla, H. Straka, G. Thanikaimoni, R. H. Tschudy, M. van Campo, T. van der Hammen, and A. Le Thomas.

Categories, Classification Systems and Systematic Value of Pollen Features

For the scientist, categories are essential for classifying natural characters in their diversity, defining their range and placing them in a systematic order. In addition to the theoretical concept, categorization always depends on the manner in which a feature is perceived: i.e. on the **visibility** of a feature, and/or their specific value. Categorization also greatly depends on the technical equipment and method(s) used, as well as on the **subjective interpretation** of character(s) (see "Methods in Palynology"). Thus, categorization of features is difficult to standardize. An example is the category **pollen size**: there is not just a natural size variation within a single anther/flower/taxon, dimensions may also vary depending on the preparation method(s) used, and the observer's evaluation. Moreover, sometimes the size of a pollen grain is just at the boundary between two adjacent pollen size categories (for size categories: see "Pollen Morphology and Ultrastructure").

When describing and categorizing pollen, two basic groupings are known from the literature: pollen type and pollen class. **Pollen type** is a general term categorizing pollen grains by a distinct combination of characters and is used in connection with systematics, affiliating the pollen type with a distinct

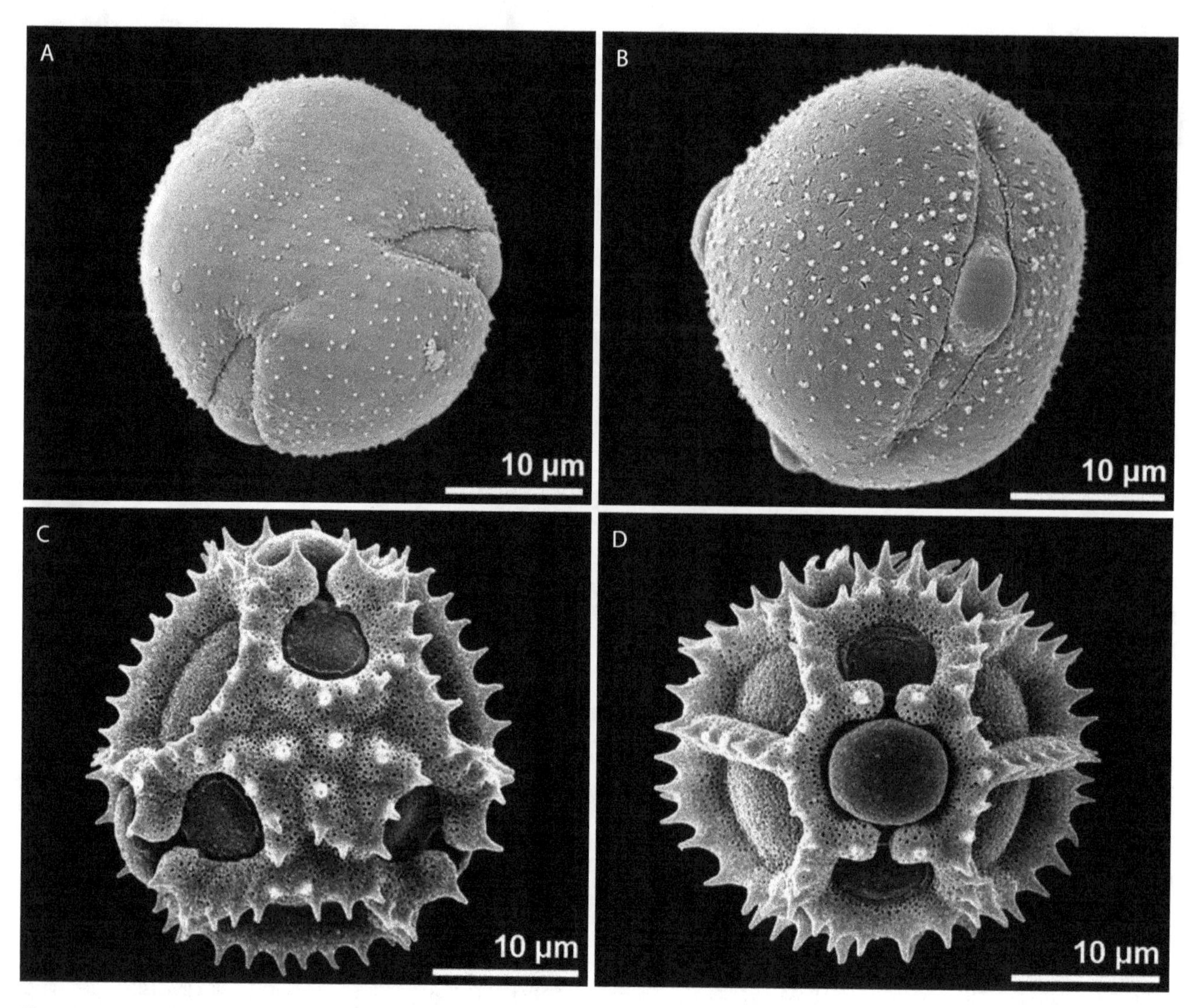

Fig. 5 Pollen type vs pollen class. A-B. *Polygonum aviculare*, Polygonaceae; all *Polygonum* pollen sharing the combined features observed here belong to the *Polygonum* aviculare type. This pollen can also be included in the pollen class "tricolporate". **C-D.** *Leontodon saxatilis*, Asteraceae; all Asteraceae pollen sharing the combined features observed here (lophate, tricolporate, echinate) belong to the *Leontodon* type, characteristic for the "Liguliflorae" group within the Asteraceae. This pollen can also be included in the pollen class "tricolporate"

taxon/a (e.g., *Polygonum aviculare* type/*Leontodon* type, Fig. 5). The term "pollen type" is sometimes (colloquially) misused: for example, *Croton* type, which is a distinct feature of ornamentation and is correctly termed *Croton* pattern.

Pollen class is an artificial grouping of pollen grains that share a single or more, distinctive characters (see "Illustrated Pollen Terms"). Pollen classes can refer to pollen units (e.g., polyads, tetrads), to shape (e.g., saccate, polygonal, heteropolar, arcus), to aperture type and location (e.g., inaperturate, sulcate, ulcerate, colpate, colporate, porate, synaperturate, spiraperturate), or to an extremely distinctive ornamentation character (e.g., lophate, clypeate). These classes can be useful in identification keys as they have a good diagnostic, although mostly no systematic, value. In general, a pollen grain may belong to more than one pollen class; in such cases, the more significant feature should be ranked first (e.g., *Pistia*: plicate-inaperturate, *Hemigraphis*: plicate-colporate, *Typha*: tetrads-ulcerate, *Rhododendron*: tetrads-colporate).

Many terms in palynology were coined at a time when only LM observations were available. Mainly for historical reasons, inconsequent nomenclatural applications, enumerations of synonyms, and even differing definitions have been found for one and the same term. During the twentieth century, questions of terminology became more and more problematic. The main reasons were the increasing numbers of publications in palynology, dealing with sometimes insufficiently described or "uncommon" pollen features, and simultaneously the advent of manifold applied fields of palynology. For various

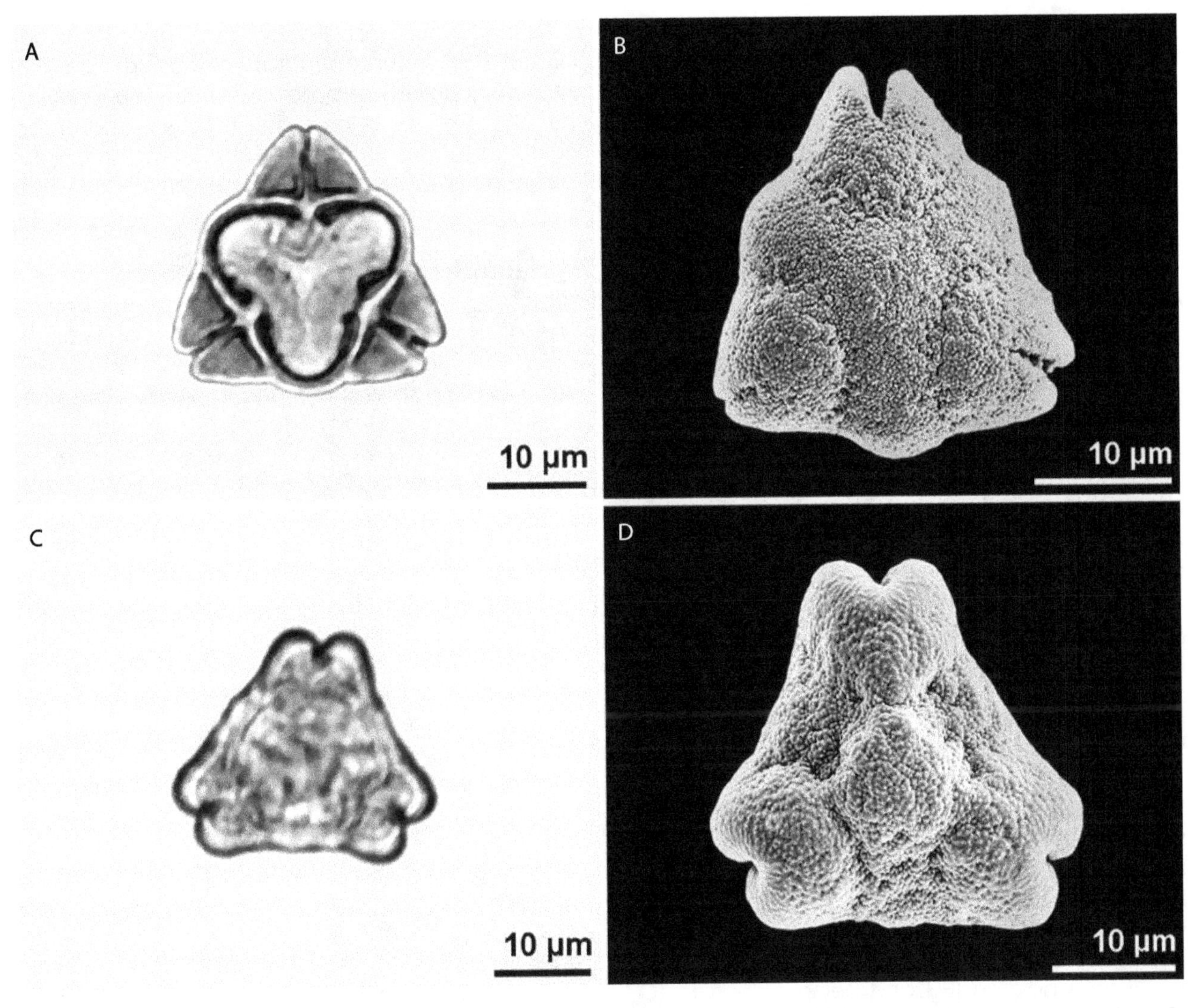

Fig. 6 Nomenclature in Paleopalynology. A. *Oculopollis* sp., fossil, Upper Cretaceous, Hungary, polar view. **B.** *Oculopollis* sp., fossil, Upper Cretaceous, Hungary, polar view. **C.** *Trudopollis* sp., fossil, Upper Cretaceous, Hungary, polar view. **D.** *Trudopollis* sp., fossil, Upper Cretaceous, Hungary, polar view

reasons, nearly all authors used their own terminology. Nonetheless, in the 1950s attempts were made to restrict the wording and to state the definitions of terms more precisely (Erdtman 1947; Erdtman and Vishnu-Mittre 1956). A limited list of pollen morphological terms and definitions was published as early as 1950 by Iversen and Troels-Smith. Later, Kremp (1968), in his famous encyclopedia, provided a monumental enumeration of all known terms. Reitsma (1970) took the first resolute step to overcome the problem of synonyms in palynological terminology, though unfortunately not taking into account the variation range of palynological features. Fægri and Iversen (1989, 4th ed.) restricted their glossary of terms exclusively used in their book. Moore et al. (1991, 2nd ed.) provided a glossary of selected terms used in their pollen and spore keys. Standardization came with the glossary by Punt et al. (1994, 2007). The main advance of their concise and comprehensive terminology is the consistent use of drawings and the critical comments on terms.

A complex category issue in (Paleo-) Palynology is the nomenclature question. In Paleopalynology, for morphotaxa often form-generic names are used. The nomenclature of form-genera is either artificial when the relationship is not known at all (e.g., *Oculopollis* and *Trudopollis* from the Normapolles group, Fig. 6), or semi-biological, when reference to an extant taxon is suspected but not proven (e.g., *Liliacidites*). However, if reference to extant taxa is certain, then a biological nomenclature is possible (e.g., *Quercus* sp.).

The Turmal System

A quite different classification and nomenclature is **Potonié's turmal system.** This is an artificial, informal, neutral suprageneric classification scheme for fossil (especially Carboniferous or Permian) pollen and spores. It is subdivided into a hierarchy of progressively finer units (ranks): anteturma, turma, subturma, infraturma, subinfraturma, and corresponds mostly to morphological features (for details see Traverse 2007).

The Treme System and the NPC-Classification

The "**-treme system**" of aperture configuration as an alternative or an addition to the traditional nomenclature was introduced by Erdtman and Straka (1961). The suffix -treme is derived from trema (pl. tremata) and is synonymous with aperture. In combination with prefixes such as cata-, ana-, zono-, and panto-, the position of germination sites in relation to pollen polarity can be designated. Catatreme indicates the proximal, anatreme the distal, zonotreme the equatorial, and pantotreme the global position of apertures. Other prefixes such as mono-, di-, tri-, tetra- indicate the number of apertures irrespective of their position.

The **NPC-classification** by Erdtman and Straka (1961), resting upon the -treme system, is a morphological system for classifying pollen and spores. This system is based on the aperture features: their number (N), position (P), and character (C). Their NPC-system for spore/pollen classification was used as a diagnostic tool in systematics. As an example, the three apertures (N_3) of pollen grains, having a zonotreme position (P_4) and being colporate (C_5) have the NPC-formula 345 (Fig. 7; Erdtman and Straka 1961). Taxa with the same general NPC-formula are grouped together, those showing a different formula, separately. This system does not work in, e.g., heteroaperturate or inaperturate (formula 000) pollen, or pollen tetrads. Unfortunately, the NPC-system ignores other pollen characters including shape and ornamentation that are indispensable for a complete description.

Systematic Value of Pollen Features

One of the main research interests in palynology focuses on the taxon-specific patterns of the pollen wall, how they developed and evolved. Moreover, pollen can provide phylogenetic evidence important to plant systematics (Hesse and Blackmore 2013). The reconstruction of phylogenies has continuously developed. The advances in modern phylogenetic approaches are resulting in constant changes in plant systematics, even whole genomes are being used together with multiple DNA analyses for a better insight into relationships (Stuessy and Funk 2013). Critically evaluated pollen features may be a useful tool for systematics with a significant diagnostic value, supporting or contradicting the results of molecular studies ("The palynological compass" sensu Blackmore 2000; Hesse and Blackmore 2013). Palynological features are very valuable, especially in delimiting taxa (Ulrich et al. 2012). Regarding multiple-gene tree studies with conflicting results, pollen data combined with other morphological evidence (e.g., floral characters) have more recently become an important indicator of which tree may be the best representative (Stuessy and Funk 2013; Ulrich et al. 2012, 2013). Furthermore, pollen morphological studies proved to be indispensable for the understanding of evolutionary

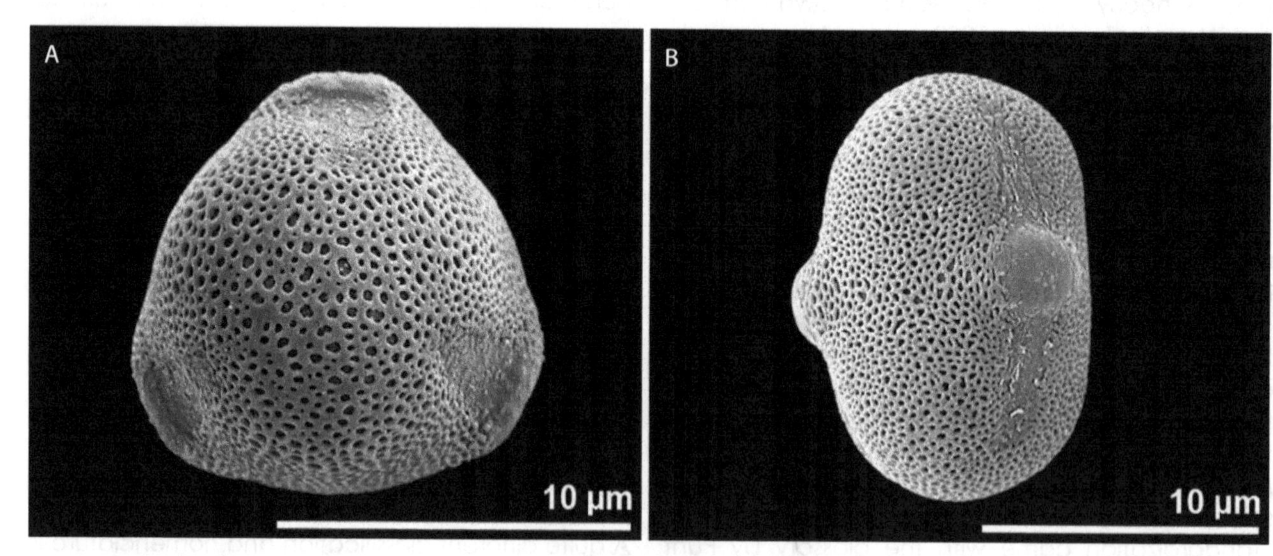

Fig. 7 NPC-classification of pollen. A-B. *Androsace chamaejasme*, Primulaceae, a tricolporate pollen with the formula $N_3P_4C_5$

processes and systematics. For taxonomic studies, pollen features that have value for the lower and higher taxonomic levels should be obtained by a combined study using LM, SEM, and TEM (Stuessy 1979).

Alternation of generations is a unique feature of plants that occurs in green algae, mosses, ferns, gymnosperms, and angiosperms. Pollen grains develop in anthers as the result of meiosis and mitoses (two in angiosperms, three to five in gymnosperms) and represent an extra generation, the highly reduced male gametophyte. Therefore, pollen grains are not simply small parts of a plant like leaves or seeds; they are the complete (hidden) haploid counterpart to the more dominant plant, which represents the diploid generation (Kessler and Harley 2004). During dispersal, pollen grains are completely separated from the parent plant and perfectly adapted for their role — the transfer of male genetic material — and are able to resist hostile environmental stresses on their way to the female flower parts. Usually, pollen does not suffer to the same extent from the various and harsh selective pressures to which the diploid plant is subjected. Because selective pressures (e.g., temperature, precipitation) upon pollen characters are predominantly absent or low, compared to those on the diploid plant, pollen features may remain constant for millions of years, meaning pollen features can be conservative and of taxonomic value (Wodehouse 1928, 1935; Hao et al. 2001; Grímsson et al. 2014, 2016, 2017a, b). Therefore, identical and rare conditions in fossil vs recent pollen probably belong to only one group and were not invented independently in distant groups (e.g., fossil *Spinizonocolpites* pollen and recent *Nypa* pollen, Arecaceae; Zetter and Hofmann 2001; Gee 2001). Selective pressures might concern especially the pollen aperture number, but also the pollen sculpture and the mode of pollination ecology (Furness and Rudall 2004). Pollen features are, if used for a systematic purpose, at least as important as any other morphological character of the diploid generation. For this reason pollen morphology claims a crucial role in, e.g., systematics and palynostratigraphy, for example in elucidating the early history of angiosperms. Angiosperm pollen from the Early Cretaceous are usually sulcate (typical for basal angiosperms) with a columellate infratectum (which is restricted to angiosperms). The first appearance of dispersed tricolpate pollen, typical for eudicots, is not known before the latest Barremian, is rare in the Aptian of Southern Laurasia and Northern Gondwana, but is ubiquitous in the Albian of both provinces. Tricolporate pollen appears first in the late Albian, and triporate pollen in the middle Cenomanian (Doyle and Endress 2010; Friis et al. 2011; Doyle 2012). For a detailed overview of structural pollen diversification and of the stratigraphic appearance of major angiosperm pollen types during the Cretaceous, see Friis et al. (2011) and Mendes et al. (2014).

Palynological data may be helpful at all levels of systematics, especially in angiosperms (c.f. Stuessy 2009). When pollen of a taxon (representing family/ies or genus/era) is characteristic and similar among species they are termed **stenopalynous** (Fig. 8), and occur, for example, in Poaceae, Lamiaceae, Asclepiadaceae, Brassicaceae, Asteroideae, and Cichorioideae. On the contrary, **eurypalynous** (Fig. 9) taxa are heterogeneous and pollen can vary among others in size, aperture, and in exine stratification. Examples for eurypalynous groups are Acanthaceae (Sarawichit 2012) and Araceae (Harley and Baker 2001; Ulrich et al. 2017).

At the highest taxonomic level (e.g., angiosperms vs gymnosperms, dicots vs monocots), a columellate exine condition occurs exclusively in angiosperms. A lamellate endexine is typical for gymnosperms, whereas the angiosperm endexine is usually not lamellate, except in immature stages (*Orobanche hederae*). But in very few cases there is a continuously lamellate endexine present, like in *Ambrosia* (Furness and Rudall 1999a, b, Weber and Ulrich 2010). In inaperturate pollen of Araceae the endexine is exceptionally thick and spongy, which may be a functional benefit and of systematic value. A strong phylogenetic signal comes from the aperture arrangement: the "tricolpate" condition is a synapomorphy for eudicots, tricolporate pollen occurs only in core eudicots, while sulcate pollen is a plesiomorphic condition in basal angiosperms (Nadot et al. 2006). Palynologists have long wondered about the two fundamental evolutionary shifts occurring at the base of the eudicot clade, both in aperture position (from distal to equatorial) and number (from one to three or more). Most probably, these changes in pollen morphology have a systematic and simultaneously a functional background. The shift from a distal, single aperture to equatorially or globally situated apertures, increases the number of possible germination sites (Furness and Rudall 2004). Pollen morphology does not support sharp delimitation between dicots and monocots, as dicotyledonous pollen characters also occur in some monocots and conversely. In early-diverging angiosperms the formation of pollen features appears to be more plastic than in dicots (especially in eudicots). Manifold combinations of pollen features are typical for basal angiosperms and even for the most basal eudicots, the Ranunculales. All of them are more or less eurypalynous. In contrast, late-divergent eudicots are often stenopalynous and

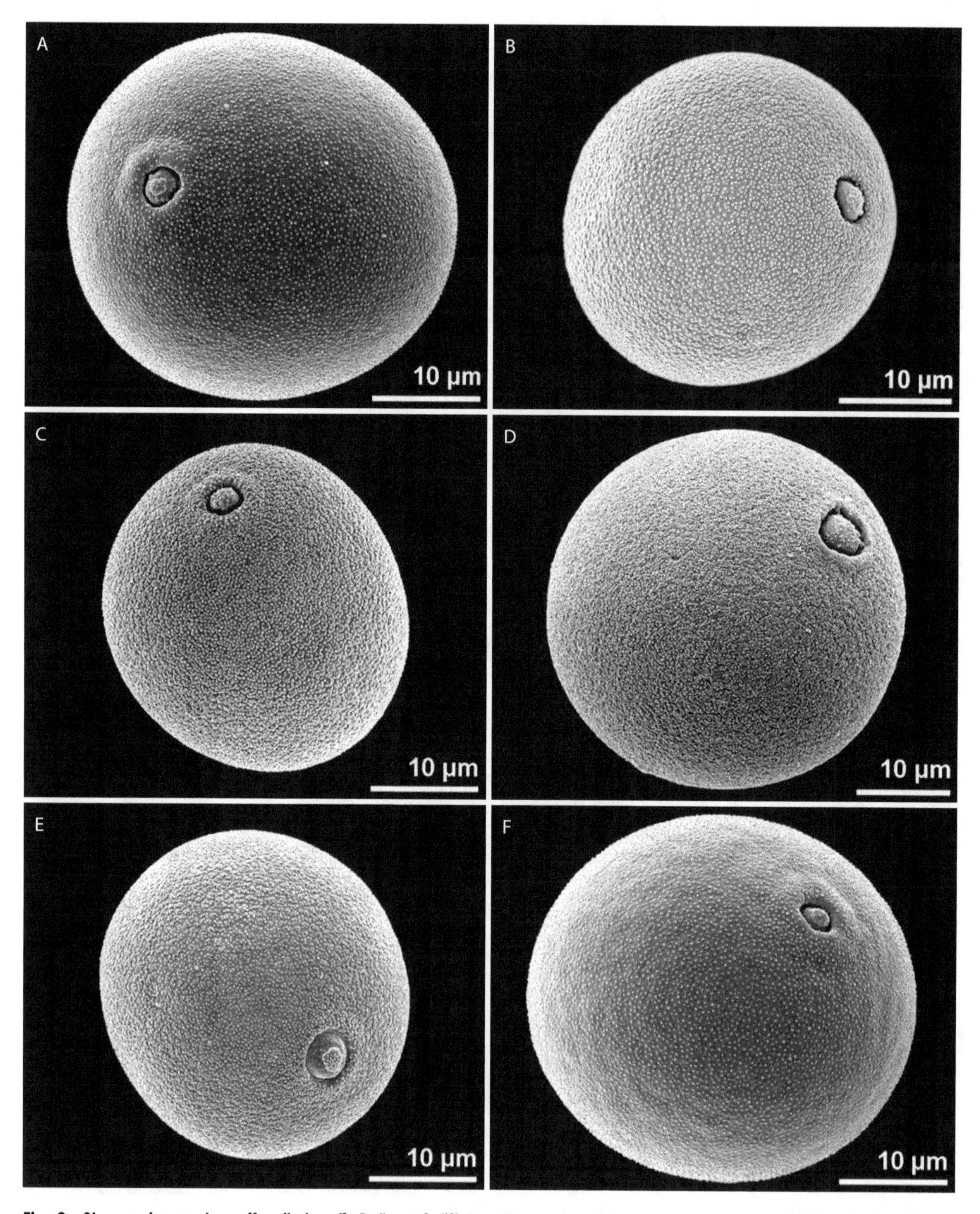

Fig. 8 Stenopalynous taxa (family level). Pollen of different Poaceae all look very similar, for example in *Alopecurus* (**A**), *Cutandia* (**B**), *Dactylis* (**C**), *Fargesia* (**D**), *Poa* (**E**), *Sesleria* (**F**) the pollen is spherical, ulcerate with nano-sized sculpture elements

Fig. 9 Eurypalynous taxa (family level). **A-F.** Pollen of different Araceae genera look very different. **A.** *Ambrosina* pollen is plicate and inaperturate. **B.** *Dracunculus* pollen is verrucate and inaperturate. **C.** *Pinellia* pollen is echinate and inaperturate. **D.** *Cyrtosperma* pollen is reticulate and ulcerate. **E.** *Anthurium* pollen is reticulate-microechinate and diporate. **F.** *Monstera* pollen is psilate, with ring-like aperture

appear somewhat "poor" regarding the diversity of pollen features (Hesse et al. 2000). In general, the richness and variation of morphological features in pollen decreases in eudicots (Furness and Rudall 1999a). In Alismatales, many pollen features are adaptive and related to their aquatic/semiaquatic habitat, e.g., thin-walled, inaperturate pollen have evolved iteratively, even filamentous pollen is not rare (Furness and Banks 2010).

Fine example for adaptive and simultaneously systematic values is the ring-like aperture found especially in monocots, while only few occur in dicots. A ring-like aperture was probably the best way to a target-oriented harmomegathic movement, to contract or expand a large area adapted for pollen tube formation. This type of aperture might be relict of early angiosperms, before the advent of the "eudicot-tricolpates".

Examples for diagnostic features at lower taxonomic levels (family) are saccate pollen, typical for Pinaceae and Podocarpaceae. A small papilla is characteristic for Taxodioideae pollen (see "Illustrated Pollen Terms"). Another example for a strong phylogenetic signal comes from an aroid subfamily, the aperigoniate Aroideae (Araceae). They are characterized by several synapomorphies: inaperturate pollen, often with an outermost non-sporopollenin layer (exine absent) and a thick spongy endexine. The absence of callose in pollen development is the reason for this uncommon wall structure, that differs from all other currently known angiosperms (Anger and Weber 2006; Hesse 2006a, b).

At the lowest taxonomic level (genus, species) a combination of distinct morphological and structural features usually refers to a particular genus or species. Even very inconspicuous features can represent an example of systematic value, like the *Pinus* subgenus *Strobus* (Haploxylon) type and the *Pinus* subgenus *Pinus* (Diploxylon) type (see "Pollen Morphology and Ultrastructure"). Another example is the large genus *Amorphophallus* (Araceae), showing high diversity in ornamentation (e.g., Ulrich et al. 2017). As a result of the harmomegathic effect, the shape of pollen may change, which is enabled by the elasticity of the exine and infoldings of the apertures. The aperture type and arrangement may lead to characteristic infoldings. Therefore, the shape of pollen in dry state can be typical for a family or genus (e.g., Halbritter and Hesse 2004). For example, tricolporate pollen of the genus *Chaenarrhinum* (Plantaginaceae) is heteropolar. The heteropolarity is only apparent in dry condition. Also, tricolpate pollen of Lamiaceae is highly characteristic in dry condition: it is prolate, extremely flattened, and with apertures arranged in a very distinct manner (Fig. 10; see also "harmomegathic effect" in "Pollen Morphology and Ultrastructure").

Future Perspective

Nowadays, palynology serves as an indispensable tool for various applied sciences such as systematics (Doyle and Endress 2010; Dransfield et al. 2008), melissopalynology (Jones and Bryant 1996), and forensics (e.g., Mildenhall et al. 2006; Bryant 2013; Weber and Ulrich 2016), but should also stand alone as a basic field in science. In general, compared to the sporophyte the male gametophyte in seed plants is poorly investigated. From ca. 260.000 to 422.000 plant species (e.g., Thorne 2002; Govaerts 2003; Scotland and Wortley 2003; *The Plant List* currently accepts 350.699 species) only about 10% have been studied with respect to pollen grain morphology, and regarding pollen ultrastructure it is even much less. Therefore, it is important to continue classical and more advanced palynological studies.

Despite the long tradition of palynology and its application in many fields, it should be considered why it is important and where it is heading in the near future. In the twenty-first century, no matter what role palynology will play, being a basic field of science or more probably a bundle of applied fields, a vital issue will be the increase of our knowledge of pollen grains and in this context the enhancement of pollen terminology. Online pollen databases (efficient in data storage, data transmitting and dissemination) will get more and more important for the exchange of pollen and spore information (for example, *PalDat;* Weber and Ulrich 2017). Journals are nowadays published simultaneously in print as well as in electronic format, both have manifold advantages and disadvantages. Nevertheless, illustrated monographs, like this one, will retain their role of detailed information and long-living documentation.

Fig. 10 Characteristic shape of pollen in dry condition. A-B. *Lamium maculatum*, Lamiaceae, pollen in hydrated and dry condition. **C-D.** *Microrrhinum minus*, Plantaginaceae, pollen in hydrated and dry condition. **E-F.** *Scutellaria baicalensis*, Lamiaceae, pollen in hydrated and dry condition

References

Anger E, Weber M (2006) Pollen wall formation in *Arum alpinum*. Ann Bot 97: 239–244
Blackmore S (1992) Scanning electron microscopy in palynology. In: Nilsson S, Praglowski J (eds) Erdtman's Handbook of Palynology. 2nd edition, Munksgaard, Copenhagen, p. 403–431
Blackmore S (2000) The palynological compass: the contribution of palynology to systematics. In: Nordenstam B, El-Ghazaly G, Kassas M (eds) Plant Systematics for the 21st Century. Portland Press, London, p. 161–177
Blackmore S, Wortley A, Skvarla JJ, Rowley JR (2007) Pollen wall development in flowering plants. New Phytol 174: 483–498
Blackmore S, Wortley AH, Skvarla JJ, Gabarayeva NI, Rowley JR (2010) Developmental origins of structural diversity in pollen walls of Compositae. Plant Syst Evol 284: 17–32
Bradley DE (1958) The study of pollen grain surfaces in the Electron Microscope. New Phytol 57: 226–229
Brown R (1828) A brief account of microscopical observations made in the months of June, July, and August, 1827, on the particles contained in the pollen of plants; and on the general existence of active molecules in organic and inorganic bodies. Richard Taylor, London
Brown R (1833) On the organs and mode of fecundation in Orchideae and Asclepiadeae. In: The miscellaneous botanical works by Robert Brown. The Ray Society, London (1866)
Bryant VM (2013) Pollen and spore use in forensics. In: Jamieson J, Moenssens A (eds) Wiley Encyclopedia of Forensic Science, 2nd edition, John Wiley & Sons Ltd, Chichester, U.K.
Camerarius R J (1694) Ueber das Geschlecht der Pflanzen (De sexu plantarum epistola). Uebersetzt und herausgegeben von M. Mobius. Ostwald's Klassiker der exakten Wissenschaften 105
Campbell DH (1897) A morphological study of *Naias* and *Zannichellia*. Proc Calif Acad, Bot 3(1): 1–70
Candolle AP, Sprengel K (1821) Elements of the philosophy of plants. Edinburgh, printed for William Blackwood
Cavolini F (1792) *Zosterae oceanicae* Linnei ΑΝΗΣΙΣ. Contemplatus est Philippus Caulinus Neapolitanus. Annis 1787 et 1791. Neapoli
Doyle JA (2012) Molecular and fossil evidence on the origin of Angiosperms. Ann Rev Earth Planet Sci 40: 301–326
Doyle JA, Endress PK (2010) Integrating Early Cretaceous fossils into the phylogeny of living angiosperms: Magnoliidae and eudicots. J Syst Evol 48: 1–35
Dransfield J, Uhl NW, Asmussen CB, Baker WJ, Harley MM, Lewis CE (2008) Genera Palmarum. The Evolution and Classification of Palms. Kew Publishing, Kew
Ducker S, Knox B (1985) Pollen and pollination: a historical review. Taxon 34: 401–419
Ehrenberg CG (1838) Über die Bildung der Kreidefelsen und des Kreidemergels durch unsichtbare Organismen. Abh Kgl Akademie Wiss Berlin 1838: 59–147
Ehrlich HG, Hall JW (1959) The ultrastructure of eocene pollen. Grana Palynol 2: 32–35
Erdtman G (1943) An introduction to pollen analysis. Chronica Botanica, Waltham, Mass
Erdtman G (1947) Suggestions for the classification of fossil and recent pollen grains and spores. Svensk Bot Tidskr 41: 104–114
Erdtman G (1952) Pollen Morphology and Plant Taxonomy. Angiosperms. Almqvist & Wiksell, Stockholm
Erdtman G (1957) Pollen and Spore Morphology. Plant Taxonomy. Gymnospermae, Pteridophyta, Bryophyta. Almqvist & Wiksell, Stockholm
Erdtman G (1969) Handbook of Palynology – An Introduction to the Study of Polllen Grains and Spores. Munksgaard, Copenhagen
Erdtman G, Dunbar A (1966) Notes on electron micrographs illustrating the pollen morphology in *Armeria maritima* and *Armeria sibirica*. Grana Palynol 6: 338–354
Erdtman G, Straka H (1961) Cormophyte spore classification. Geol Fören Förenhandl 83: 65–78
Erdtman G, Vishnu-Mittre (1956) On terminology in pollen and spore morphology. The Palaeobotanist 5: 109–111
Fægri K, Iversen J (1950) Textbook of modern pollen analysis. Munksgaard, Copenhagen
Fægri K, Iversen J (1989) Textbook of Pollen analysis. 4th edition, John Wiley & Sons, Chichester
Fernandez-Moran H, Dahl AO (1952) Electron microscopy of ultrathin frozen sections of pollen grains. Science 116: 465–467
Fischer H (1890) Beiträge zur vergleichenden Morphologie der Pollenkörner. Thesis, Breslau
Friis EM, Crane PR, Pedersen KR (2011) Early Flowers and Angiosperm Evolution. Cambridge University Press, Cambridge
Fritzsche J (1834) Ueber den Pollen der Pflanzen und das Pollenin. Ann Phys 108: 481–492
Fritzsche J (1837) Über den Pollen. Mém Sav Étrang Acad Sci Pétersbourg 3: 649–672
Furness CA, Banks H (2010) Pollen evolution in the early-divergent monocot order Alismatales. Int J Plant Sci 171: 713–739
Furness CA, Rudall PJ (1999a) Microsporogenesis in Monocotyledons. Ann Bot 84: 475–499
Furness CA, Rudall PJ (1999b) Inaperturate pollen in monocotyledons. Int J Pl Sci 160: 395–414
Furness CA, Rudall PJ (2004) Pollen aperture evolution – a crucial factor for eudicot success? Trends Plant Sci 9: 154–158
Gabarayeva NI, Grigorjeva VV (2010) Sporoderm ontogeny in *Chamaedorea microspadix* (Arecaceae): self-assembly as the underlying cause of development. Grana 49: 91–114
Gabarayeva NI, Grigorjeva VV, Rowley JR (2010) A new look at sporoderm ontogeny in *Persea americana* and the hidden side of development. Ann Bot 105: 939–955
Gee CT (2001) The mangrove palm Nypa in the geologic past of the New World. Wetl Ecol Manag 9: 181–194
Gleditsch JG (1751) Essai d'une Fécondation artificielle, fait sur l'espèce de Palmier qu'on nomme, Palma dactylifera folio flabelliformi. – Histoire de l'Académie Royale des Sciences et Belles Lettres de Berlin année 1749: 103–108
Gleditsch JG (1765) Kurze Nachricht von einer künstlichen wohlgelungenen Befruchtung eines Palmbaumes im Königlichen Kräutergarten zu Berlin. – Vermischte Physikalisch-Botanisch-Ökonomische Abh 1: 94–104
Göppert HR (1837) De floribus in statu fossili, commentatio botanica. Thesis, Breslau
Göppert HR (1848) Über das Vorkommen von Pollen im fossilen Zustande. Neues Jahrbuch für Mineralogie, Geognosie, Geologie und Petrefaktenkunde 11: 338–340
Govaerts R (2003) How many species of seed plants are there? – a response. Taxon 52: 583–584

Grew N (1682) The Anatomy of plants, with an idea of a philosophical history of plants, and several other lectures, read before the Royal Society. W. Rawlins, London
Grímsson F, Grimm GW, Zetter R, Denk T (2016) Cretaceous and Paleogene Fagaceae from North America and Greenland: evidence for a Late Cretaceous split between *Fagus* and the remaining Fagaceae. Acta Palaeobotanica 56: 247–305
Grímsson F, Grimm GW, Zetter R (2017a) Tiny pollen grains: first evidence of Saururaceae from the Late Cretaceous of western North America. Peer J 5:e3434 https://doi.org/10.7717/peerj.3434
Grímsson F, Kapli P, Hofmann C, Zetter R, Grimm GW (2017b) Eocene Loranthaceae pollen pushes back divergence ages for major splits in the family. PeerJ 5:e3373 https://doi.org/10.7717/peerj.3373
Grímsson F, Zetter R, Halbritter H, Grimm GW (2014) *Aponogeton* pollen from the Cretaceous and Paleogene of North America and West Greenland: Implications for the origin and palaeobiogeography of the genus. Rev Palaeobot Palynol 200: 161–187
Guignard L (1891) Nouvelles études sur la fécondation. Ann Sc Nat, Bot ser 7, 14: 163–296
Guignard L (1899) Sur les antherozoides et la double copulation sexuelle chez les végétaux angiosperms. Rev Gen Bot 11: 129–135
Halbritter H, Hesse M (2004) Principal modes of infoldings in tricolp(or)ate Angiosperm pollen. Grana 43: 1–14
Hao G, Chye M-L, Saunders RMK (2001) A phylogenetic analysis of the Schisandraceae based on morphology and nuclear ribosomal ITS sequences. Bot J Linn Soc 135: 401–411
Harley MM, Baker WJ (2001) Pollen aperture morphology in Arecaceae: application within phylogenetic analyses, and a summary of the fossil record of palm-like pollen. Grana 40: 45–77
Harley MM, Ferguson IK (1990) The role of the SEM in pollen morphology and plant systematics. In: Claugher D (ed) Scanning Electron Microscopy in Taxonomy and Functional Morphology. Syst Ass Special Volume 41: 45–68. Clarendon Press, Oxford
Hay WW, Sandberg PA (1967) The Scanning Electron Microscope, a major break-through for micropaleontology. Micropaleontology 13: 407–418
Hell SW (2009) Microscopy and its focal switch. Nature Methods 6: 24–32
Heslop-Harrison J (1975) The physiology of the pollen grain surface. Proc R Soc, London B 190: 275–299
Hesse M (2006a) Reason and consequences of the lack of a sporopollenin ektexine in Aroideae (Araceae). Flora 201: 421–428
Hesse M (2006b) Conventional and novel modes of exine patterning in members of the Araceae – the consequence of ecological paradigm shifts? Protoplasma 228: 145–149
Hesse M, Blackmore S (2013) Editorial: Preface to the Special Focus manuscripts. Plant Syst Evol 299: 1011–1012
Hesse M, Weber M, Halbritter H (2000) A comparative study of the polyplicate pollen types in Arales, Laurales, Zingiberales and Gnetales. In: Harley MM, Morton CM, Blackmore S (eds) Pollen and spores: morphology and biology. Royal Botanic Gardens, Kew, p. 227–239
Hofmeister W (1849) Die Entstehung des Embryo der Phanerogamen. Friedrich Hofmeister, Leipzig
Hooke R (1665) Micrographia, or, Some physiological descriptions of minute bodies made by magnifying glasses, with observations and inquiries thereupon. Printed by Jo. Martyn and Ja. Allestry, London
Hyde HA (1955) Oncus, a new term in pollen morphology. New Phytol 54: 255
Iversen J, Troels-Smith J (1950) Pollenmorfologiske definitioner og typer. Pollenmorphologische Definitionen und Typen. Danm Geol Unders, ser 4, 3: 1–54
John JF (1814) Ueber den Befruchtungsstaub, nebst einer Analyse des Tulpenpollens. J Chem Phys 12: 244–252
Jones GD, Bryant VM Jr. (1996) Melissopalynology. In: Jansonius J, McGregor DC (eds) Palynology: principles and applications. American Association of Stratigraphic Palynologists Foundation, vol 3, AASP Foundation, Dallas, p. 933–938
Kesseler R, Harley MM (2004) Pollen. The hidden sexuality of flowers. Papadakis Publisher, London
Knox RB (1984) The pollen grain. In: Johri BM (ed) Embryology of Angiosperms. Springer, Berlin
Kölreuter JG (1761-1766) Vorläufige Nachricht von einigen das Geschlecht der Pflanzen betreffenden Versuchen und Beobachtungen. 4 Vol., Gleditsch, Leipzig
Kölreuter JG (1806) De antherarum pulvere. Nova acta Academiae Scientiarum Imperialis Petropolitanae 15: 359–398
Kölreuter JG (1811) Dissertationis de antherarum pulvere continuato. Mem Acad Sci Petersbourg 3: 159–199
Kremp GOW (1968) Morphologic Encyclopedia of Palynology. 2nd edition, Arizona Press, Tucson
Lindley J (1836) A natural system of botany; or, A systematic view of the organization, natural affinities, and geographical distribution of the whole vegetable kingdom: together with the uses of the most important species in medicine, the arts, and rural or domestic economy (2nd edition), Longman, London
Linnaeus C (1750) Sponsalia plantarum. J. G. Wahlbom, Stockholm. Facs. edition, Rediviva No. 19, Stockholm 1971
Malpighi M (1901) Die Anatomie der Pflanzen. I und II Theil, London 1675 und 1679. Bearbeitet von M. Möbius. Ostwald's Klassiker der exakten Wissenschaften Nr. 120, pp. 163
Manten AA (1966) Half a century of modern palynology. Earth-Sci Rev 2: 277–316
Mendes MM, Dinis J, Pais J, Friis EM (2014) Vegetational composition of the Early Cretaceous Chicalhão flora (Lusitanian Basin, western Portugal) based on palynological and mesofossil assemblages. Rev Palaeobot Palynol 200: 65–81
Mildenhall DC, Wiltshire PEJ, Bryant VM (2006) Forensic palynology: Why do it and how it works. Forensic Sci Int 163: 163–172
Moore PD, Webb JA, Collinson ME (1991) Pollen analysis. 2nd edition. Blackwell Scientific Publication, Oxford
Mühlethaler K (1953) Untersuchungen über die Struktur der Pollenmembran. Mikroskopie 8: 103–110
Mühlethaler K (1955) Die Struktur einiger Pollenmembranen. Planta 46: 1–13
Nadot S, Forchioni A, Penet L, Sannier J, Ressayre A (2006) Links between early pollen development and aperture pattern in monocots. Protoplasma 228: 55–64
Nägeli K (1842) Zur Entwicklungsgeschichte des Pollens bei den Phanerogamen. Orell, Füssli & Comp., Zürich
Nawaschin S (1898) Resultate einer Revision der Befruchtungsvorgänge bei *Lilium martagon* und *Fritillaria tenella*. Bull Acad Imp Sci St. Petersbourg, ser. 5, 9: 377–382
Pettitt JM, Chaloner WG (1964) The ultrastructure of the Mesozoic pollen Classopollis. Pollen Spores 6: 611–620
Potonié R (1934) I. Zur Morphologie der fossilen Pollen und Sporen. Arb Inst Paläobotanik Petrographie Brennsteine 4: 5–24

Potonié R (1956) Synopsis der Gattungen der Sporae dispersae, I. Teil: Sporites. Beih Geol Jahrb 23: 1–103

Punt W, Blackmore S, Nilsson S, Le Thomas A (1994) Glossary of Pollen and Spore Terminology. LPP Foundation, Laboratory of Palaeobotany and Palynology, University of Utrecht, Utrecht. LPP Contributions Series 1

Punt W, Hoen PP, Blackmore S, Nilsson S, Le Thomas A (2007) Glossary of pollen and spore terminology. Rev Palaeobot Palynol 143: 1–81

Purkinje JE (1830) De Cellulis antherarum fibrosis nec non de granorum pollinarium formis: Commentatio phytotomica. Grueson, Breslau

Reille M (1992) Pollen et Spores d'Europe et d'Afrique du Nord. Laboratoire de Botanique Historique et Palynologie, Marseille

Reille M (1995) Pollen et Spores d'Europe et d'Afrique du Nord, Supplement 1. Laboratoire de Botanique Historique et Palynologie, Marseille

Reille M (1998) Pollen et Spores d'Europe et d'Afrique du Nord, Supplement 2. Laboratoire de Botanique Historique et Palynologie, Marseille

Reinsch P (1884) Micro-Palaeophytologia formationis carboniferae. Krische, Erlangen

Reitsma TJ (1970) Suggestions towards unification of descriptive terminology of angiosperm pollen grains. Rev Palaeobot Palynol 10: 39–60

Rowley JR, Flynn JJ (1966) Single–stage carbon replicas of microspores. Stain Technol 41: 287–290

Sarawichit P (2012) Pollen and orbicular walls of selected species of *Justicieae* (Acanthaceae) and their systematic significance. Thesis, University of Vienna

Schacht H (1856/59) Lehrbuch der Anatomie und Physiologie der Gewächse. 2 Vol., Müller, Berlin

Schopf JM (1957) Spores and related plant microfossils – Paleozoic. In: Ladd HS (ed) Treatise on marine ecology and paleoecology 2, Paleoecology: Geological Scociety of America Memoir 67/2, p. 703–707

Schopf JM (1964) Practical problems and principles in study of plant microfossils. In: Cross, AT (ed) Palynology in oil exploration – A symposium: Society of Economic Paleontologists and Mineralogists Special Publication 11, p. 29–57

Scotland RW, Wortley AH (2003) How many species of seed plants are there? Taxon 52: 101–104

Sprengel CK (1793) Das entdeckte Geheimnis der Natur im Bau und in der Befruchtung der Blumen. Vieweg, Berlin

Sprengel K (1804) Anleitung zur Kenntniß der Gewächse. In Briefen. Erste Sammlung, Kimmel, Halle

Stolze GH (1816) Der Pollen der Pflanzen in chemischer Hinsicht; nebst einer Analyse des Pollens der Haselnusstaude (*Corylus avellana* Linn.). Jahrb Pharm 17: 159–187

Strasburger E (1884) Neue Untersuchungen über den Befruchtungsvorgang bei den Phanerogamen als Grundlage für eine Theorie der Zeugung. Fischer, Jena

Stuessy TF (1979) Ultrastructural data for the practicing plant systematist. Am Zool 19: 621–635

Stuessy TF (2009) Plant Taxonomy: The Systematic Evaluation of Comparative Data. 2nd edition, Columbia University Press, New York

Stuessy TF, Funk VA (2013) New trends in plant systematics – Introduction. Taxon 62: 873–875

Thorne RF (2002) How many species of seed plants are there? Taxon 51: 511–522

Thornhill JW, Matta RK, Wood WH (1965) Examining three–dimensional microstructures with the scanning electron microscope. Grana Palynol 6: 3–6

Traverse A (2007) Paleopalynology. 2nd ed, Springer, Dordrecht

Ulrich S, Hesse M, Bröderbauer D, Bogner J, Weber M, Halbritter H (2013) *Calla palustris* (Araceae): New insights with special regard to its controversial systematic position and to closely related genera. Taxon 62: 701–712

Ulrich S, Hesse M, Bröderbauer D, Wong SY, Boyce PC (2012) *Schismatoglottis* and *Apoballis* (Araceae: Schismatoglottideae): A new example for the significance of pollen morphology in Araceae systematics. Taxon 61: 281–292

Ulrich S, Hesse M, Weber M, Halbritter H (2017) *Amorphophallus*: New insights into pollen morphology and the chemical nature of the pollen wall. Grana 56: 1–36

Von Grotthuss T (1814) Analysis des Tulpensamenstaubs. J Chem Phys 11: 281–380

Von Mohl H (1835) Sur la structure et les formes des grains de pollen. Ann Sci nat 2. Ser., 3: 148–180, 220–236, 304–346

Von Post L (1916) Einige südschwedische Quellmoore. Bull Geol Inst Univ Uppsala 15: 219–278

Weber M, Ulrich S (2010) The endexine: a frequently overlooked pollen wall layer and a simple method for detection. Grana 49: 83–90

Weber M, Ulrich S (2016) Forensic Palynology: How pollen in dry grass can link to a crime scene. In: Kars H, van den Eijkel L (eds) Soil in Criminal and Environmental Forensics: Proceedings of the Soil Forensics Special, 6th European Academy of Forensic Science Conference. The Hague, Springer, p. 15–23

Weber M, Ulrich S (2017) *PalDat* 3.0 – second revision of the database, including a free online publication tool. Grana 56: 257–262

Wodehouse RP (1928) The phylogenetic value of pollen grain characters. Ann Bot 42: 891–934

Wodehouse RP (1935) Pollen grains. Their structure, identification and significance in science and medicine. McGraw–Hill, New York

Zetter R, Hofmann C (2001) New aspects of the palynoflora of the lowermost Eocene (Krappfeld Area, Carinthia). In: Piller WE, Rasser MW (eds) Paleogene of the Eastern Alps. Österreichische Akademie der Wissenschaften, Schriftenreihe der Erdwissenschaftlichen Kommissionen 14, p. 473–507

Zetzsche F, Huggler K (1928) Untersuchungen über die Membran der Sporen und Pollen. I. 1. *Lycopodium clavatum* L. Ann Chem 461: 89–108

Zetzsche F, Kalt P, Leichti J, Ziegler E (1931) Zur Konstitution des Lycopodiumsporonins, des Tasmanins und des Lange–Sporonins. J Prakt Chem 148: 67–84

Zetzsche F, Vicari H (1931) Untersuchungen über die Membran der Sporen und Pollen II, *Lycopodium clavatum* L., Untersuchungen über die Membran der Sporen und Pollen III. *Picea orientalis*, *Pinus silvestris* L., *Corylus avellana* L. Helv Chim Acta 14: 58–67

2

Formation and Development of Pollen Grains

Microsporogenesis and Microgametogenesis

Pollen is source and transport unit for the male gametes (or their progenitor cell). The unicellular pollen grain represents the microspore of seed plants, the multicellular pollen grain the male gametophytic generation. The development of a pollen grain includes **microsporogenesis** and **microgametogenesis** (Figs. 1 and 2, Gomez et al. 2015; Keijzer and Willemse 1988). Microsporogenesis starts with the differentiation of microspore mother cells (MMC) respectively **pollen mother cells** (PMC). These diploid cells become enclosed by a thick **callose** wall and undergo meiosis, forming a tetrad of four haploid **microspores**, each encased in another callose wall insulating them from each other and from the surrounding diploid tapetal cells (Figs. 1 C-E, and 2). Cytokinesis following meiotic nuclear divisions is accompanied by the formation of cleavage planes determined by the configuration and orientation of the meiotic spindle axes. In the case of **successive cytokinesis**, planes are formed after the first and second meiotic divisions leading to the formation of various microspore tetrad types (see "Pollen Morphology and Ultrastructure"). During **simultaneous cytokinesis** the cleavage planes are formed simultaneously after the second meiotic division and microspores become arranged in a **tetrahedral tetrad** (Furness and Rudall 1999, 2001).

Pollen wall formation starts while the microspores are arranged in tetrads, encased by callose. The first step starts with the deposition of **primexine**, a fibrillar polysaccharidic material, on the surface of the microspores. The primexine forms a template where sporopollenin precursors and subsequently **sporopollenin** are deposited, building the final pollen wall (Fig. 1 E). Apertures are formed where the endoplasmic reticulum has prevented the deposition of primexine.

During pollen formation and maturation the **tapetum** plays an important role, usually forming a single layer of cells circumscribing the loculus. Tapetal cells are specialized and have a short lifespan. They finally lose their cellular organization and are reabsorbed. Two types of tapetum are known: the **secretory** (or glandular or parietal) and the **amoeboid** (or periplasmodial). In the secretory type (e.g., in Apiaceae) the tapetal cells remain stationary until they finish their physiological functions. In the amoeboid type (e.g., in Araceae) cells lose their individuality at an early developmental stage by degeneration of the cell walls (Furness and Rudall 1999, 2001). The protoplasts then fuse and intrude into the locule where they enclose the pollen grains (Fig. 3). The tapetum plays an important role during several stages of pollen development (Pacini 1997). Its main function is the nourishment of the microspores, but it also synthesizes enzymes (e.g., callase), exine precursors, pollen coatings, forms Ubisch bodies (orbicules) and viscin threads (both equivalents to the ektexine). The most striking material produced by the tapetum is **pollenkitt** (and **tryphine** in Brassicaceae), a sticky, heterogeneous material composed of neutral lipids, flavonoids, carotenoids, proteins and polysaccharides. Pollenkitt serves numerous functions: keeping pollen grains together during transport, protecting pollen (from water loss, ultraviolet radiation, hydrolysis and exocellular enzymes), and maintaining sporophytic proteins inside exine cavities.

Microgametogenesis (Fig. 1 G-K) in angiosperms includes first and second pollen mitosis, leading to the formation of the male gametes, the **sperm cells** (Mccormick 1993; Cresti et al. 1992). Microgametogenesis starts with formation of a central vacuole within the uninucleate microspore, pushing the nucleus towards the pollen wall. As long as the nucleus is in a central position within the cytoplasm, the cell is called a **microspore** (Fig. 1 F). With the dislocation of the microspore nucleus the cell becomes the young **pollen grain** (Fig. 1 G).

The **first pollen mitosis** is followed by an asymmetric cell division, leading to the formation of a smaller generative cell and a larger **vegetative cell** with a **vegetative nucleus** (Figs. 4 and 5). Subsequently, the generative cell detaches from the pollen wall and is finally located within the cytoplasm of the vegetative cell (Fig. 1 I). The **generative cell**, sparse in organelles, becomes **spindle-shaped** and the shape of the generative nucleus changes correspondingly (Figs. 6 and 7).

The **second pollen mitosis** includes a symmetric cell division, and divides the generative cell into two **sperm cells** (Figs. 8 and 9), the final stage of gametophytic development (Fig. 1 J-K). Angiosperm pollen is either **two-celled** (75%) or **three-celled** (25% of investigated taxa) at the time of anthesis (Brewbaker 1967; Edlund et al. 2004; Williams et al. 2014). In the latter case the second pollen mitosis takes place in the **pollen tube** (Fig. 1 K), after **germination** of the pollen grain on a stigma or on a corresponding structure (Figs. 10 and 11, Edlund et al. 2004, Mascarenhas 1993). In some families, genera with three- as well as two-celled/nuclear pollen grains occur (e.g., Araceae, Brewbaker 1967).

Microgametogenesis in gymnosperms includes several mitotic divisions. Normally, pollen grains of conifers, cycads and allies are multicelled at anthesis, and comprise prothallial cell(s), a large tube cell and a small antheridial cell. The tube cell becomes a pollen tube; the antheridial cell undergoes division into the stalk cell and the spermatogenous cell, the latter finally dividing into the male gametes (sperm cells or spermatozoids).

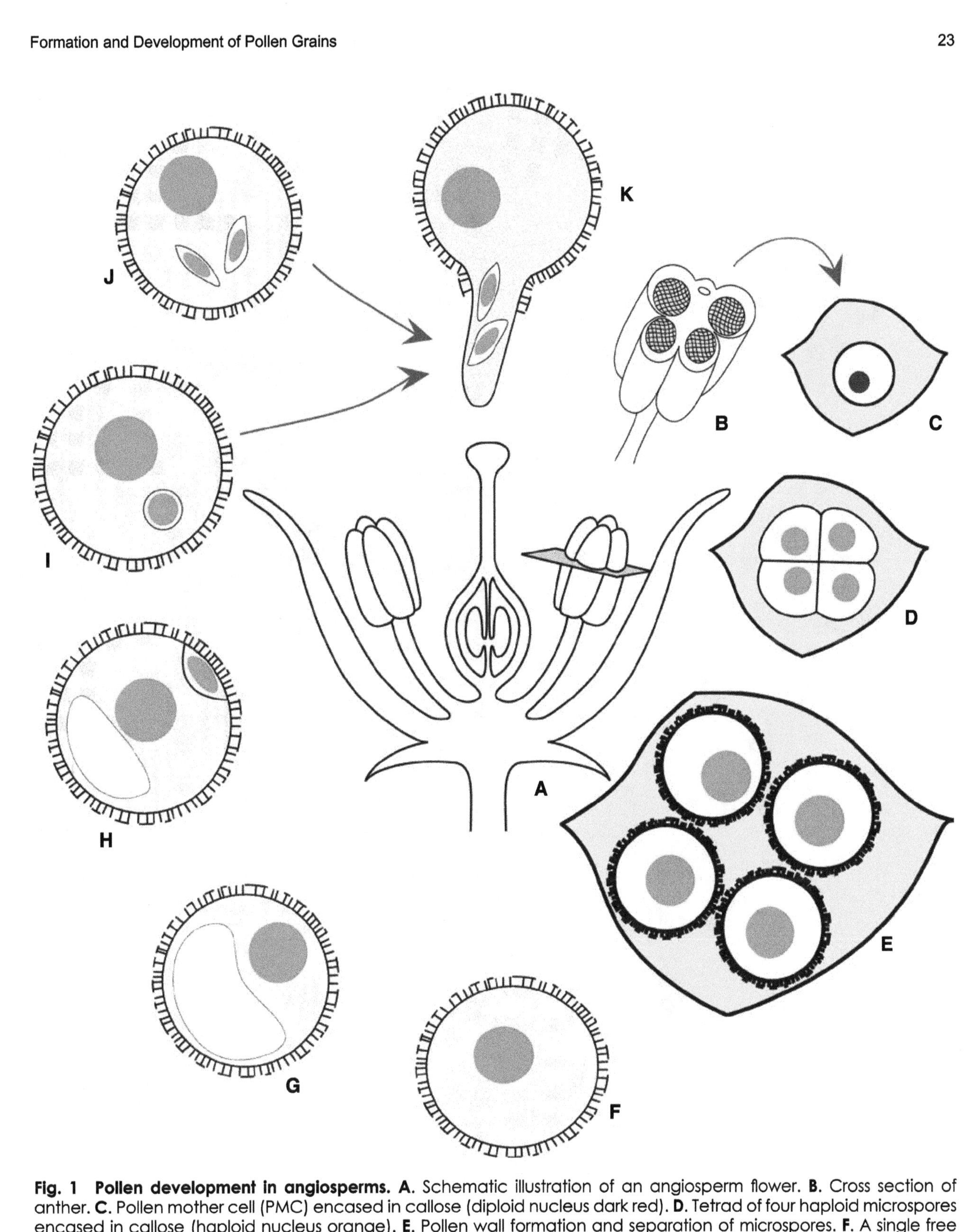

Fig. 1 Pollen development in angiosperms. A. Schematic illustration of an angiosperm flower. **B.** Cross section of anther. **C.** Pollen mother cell (PMC) encased in callose (diploid nucleus dark red). **D.** Tetrad of four haploid microspores encased in callose (haploid nucleus orange). **E.** Pollen wall formation and separation of microspores. **F.** A single free microspore with central haploid nucleus. **G.** Beginning of gametogenesis, formation of a central vacuole (white). **H.** First pollen mitosis, lens-shaped generative cell with generative nucleus attached to pollen wall. **I.** Two-celled pollen grain, generative cell detached from pollen wall. **J.** Three-celled pollen grain after second pollen mitosis, note two sperm cells with sperm nuclei. **K.** Germination can occur from either a two-celled pollen grain, followed by the formation of sperm cells, or from a three-celled pollen grain (pathways indicated by green arrows)

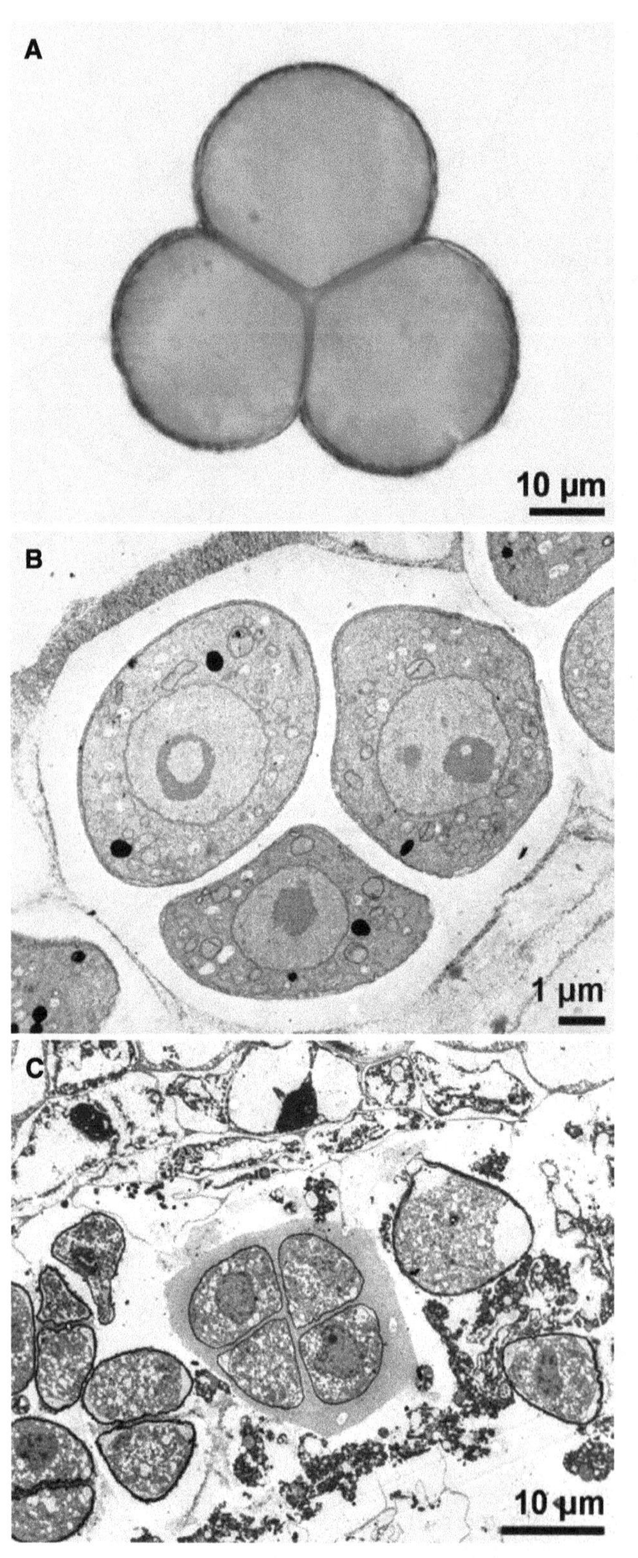

Fig. 2 Microsporogenesis. A. *Scrophularia nodosa*, Scrophulariaceae, tetrad tetrahedral, iodine. **B.** *Spirea* sp., Rosaceae, tetrad tetrahedral, Thiéry test. **C.** *Orobanche hederae*, Orobanchaceae, tetrad planar, potassium iodine

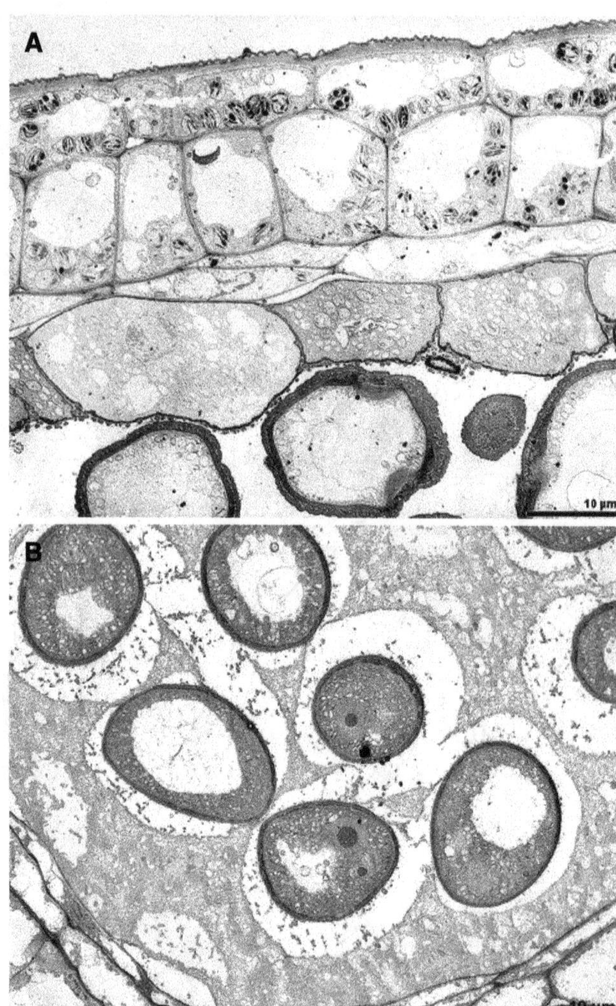

Fig. 3 Tapetum types. A. *Hacquetia epipactis*, Apiaceae, secretory tapetum in young anther, Thiéry test. **B.** *Zantedeschia aethiopica*, Araceae, amoeboid tapetum, U+Pb

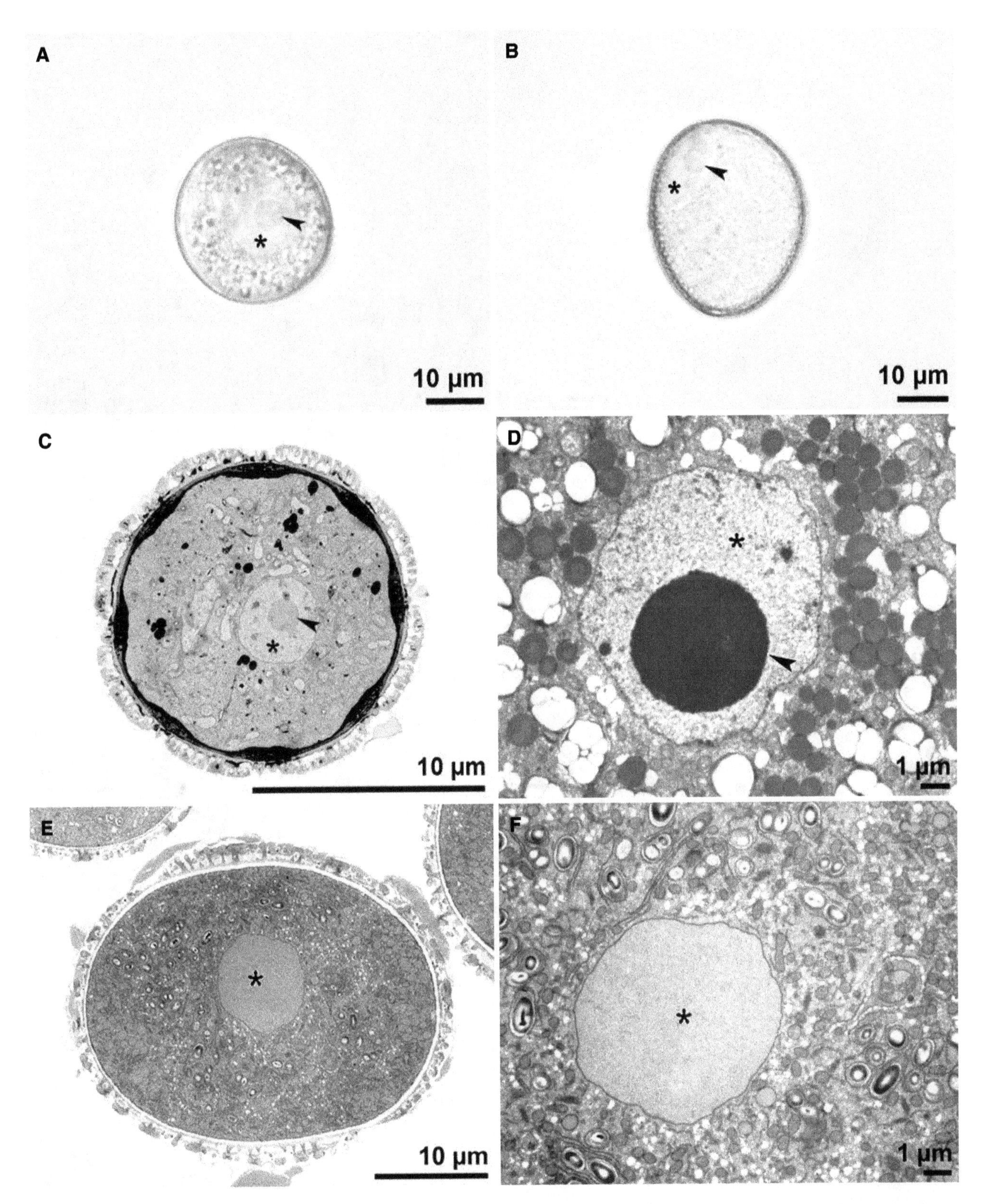

Fig. 4 Variability of the vegetative nucleus in LM and TEM (cross sections). A-B. *Dracontium asperum*, Araceae, pollen hydrated in water, vegetative cell with vegetative nucleus (asterisk) and nucleolus (arrowhead). **C.** *Galium odoratum*, Rubiaceae, vegetative cytoplasm and nucleus (asterisk) with nucleolus (arrowhead), U+Pb. **D.** *Salvia nemorosa*, Lamiaceae, vegetative nucleus (asterisk) with nucleolus (arrowhead) surrounded by cytoplasm, U+Pb. **E.** *Thymus glabrescens*, Lamiaceae, vegetative cytoplasm and nucleus (asterisk), modified Thiéry test. **F.** *Thymus glabrescens*, Lamiaceae, vegetative nucleus (asterisk) surrounded by cytoplasm, modified Thiéry test

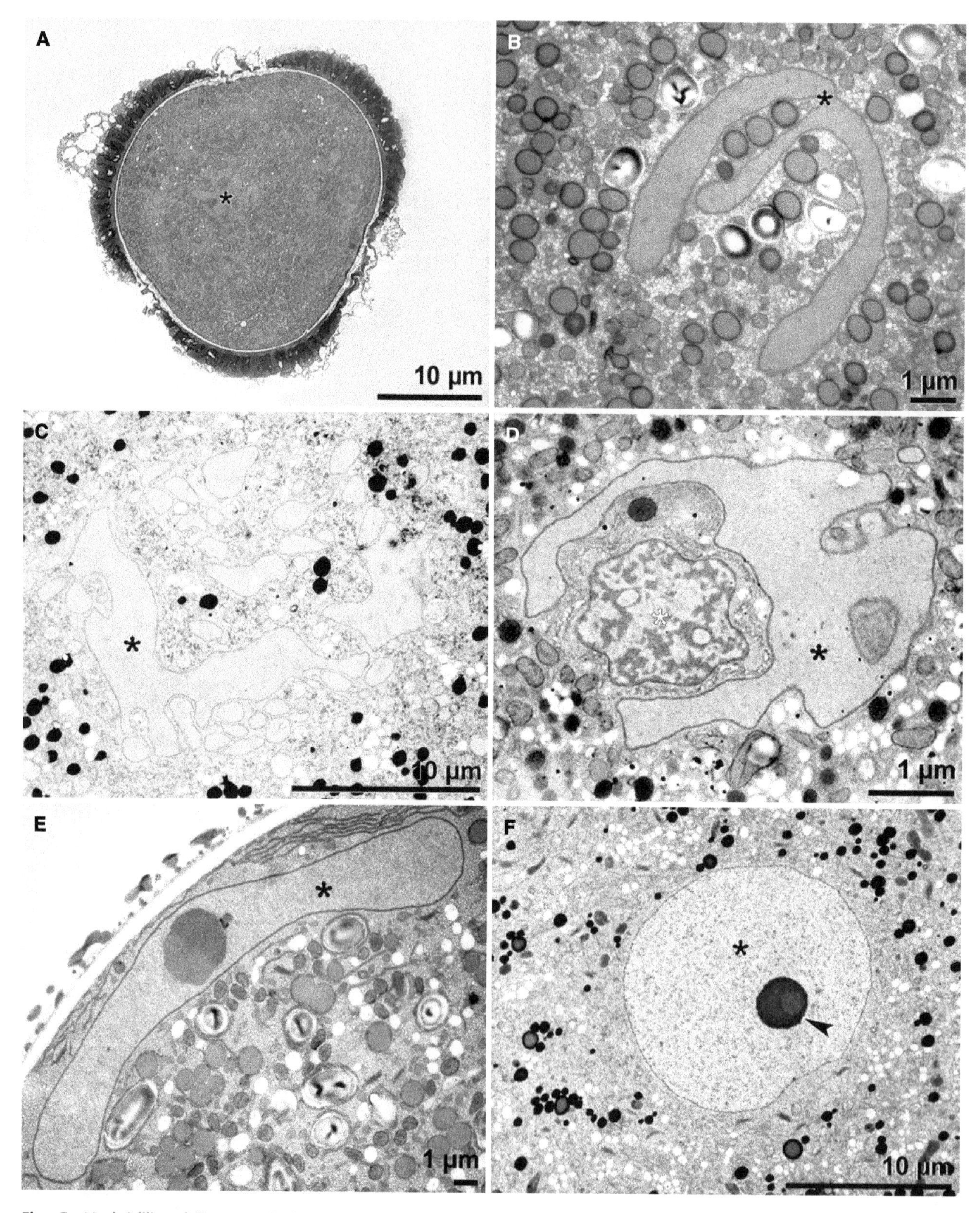

Fig. 5 Variability of the vegetative nucleus in TEM (cross sections). A. *Brassica napus*, Brassicaceae, vegetative cytoplasm and nucleus (asterisk), modified Thiéry test. **B.** *Salvia verticillata*, Lamiaceae, vegetative cytoplasm and nucleus (asterisk), modified Thiéry test. **C.** *Iris pumila*, Iridaceae, vegetative cytoplasm and nucleus (asterisk), modified Thiéry test. **D.** *Consolida regalis*, Ranunculaceae, vegetative nucleus (black asterisk) and generative cell (white asterisk), modified Thiéry test. **E.** *Acinos alpinus*, Lamiaceae, vegetative nucleus (asterisk) with nucleolus (arrowhead) surrounded by cytoplasm, modified Thiéry test. **F.** *Stachys palustris*, Lamiaceae, vegetative nucleus (asterisk) with nucleolus (arrowhead) surrounded by cytoplasm, U+Pb

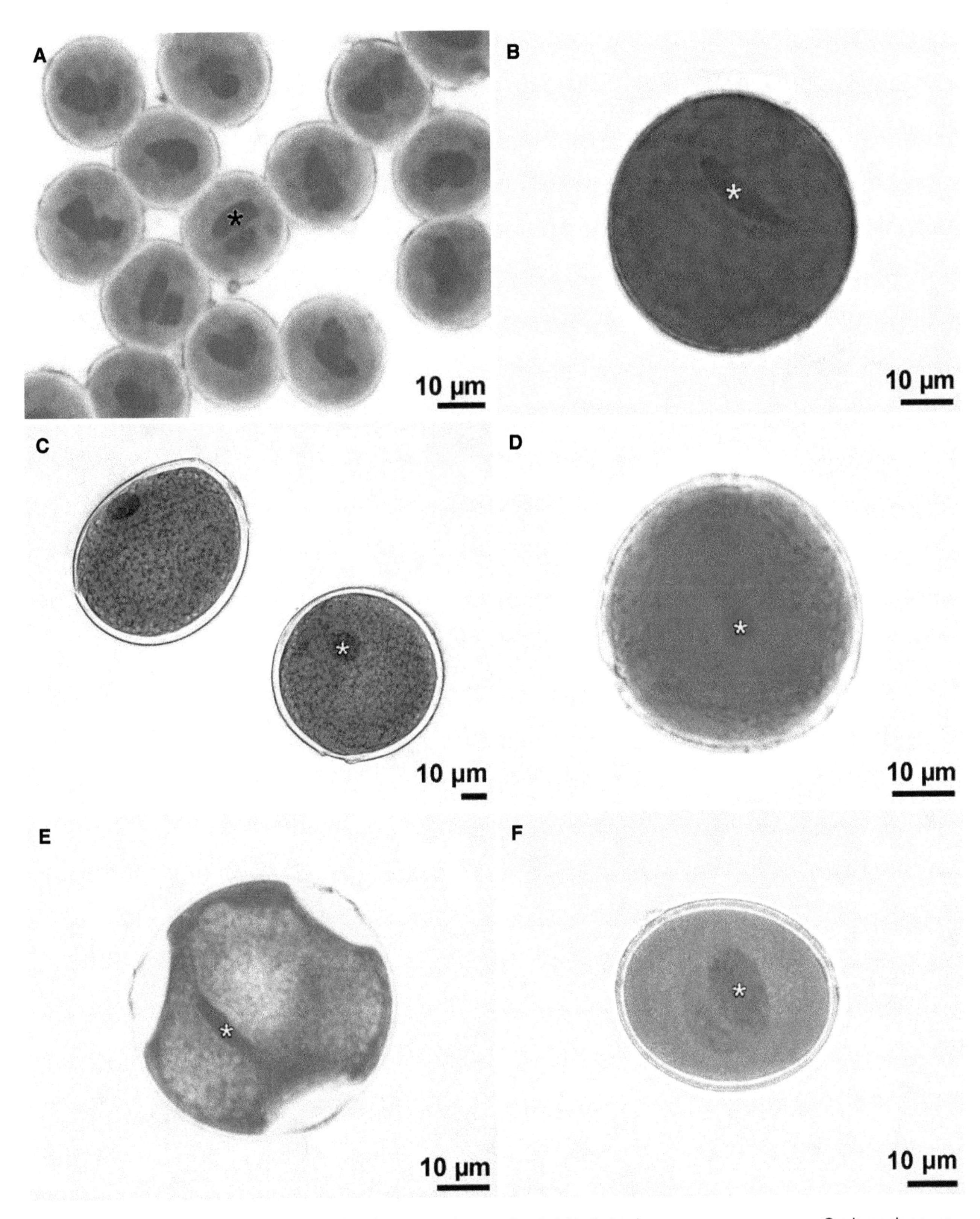

Fig. 6 Generative cell and nucleus stained with acetocarmine in LM. A. *Melampyrum nemorosum*, Orobanchaceae, spindle-shaped generative cell/nucleus (asterisk) and vegetative nucleus. **B.** *Betonica officinalis*, Lamiaceae, spindle-shaped generative cell/nucleus (asterisk). **C.** *Anchomanes welwitschii*, Araceae generative cell/nucleus (asterisk). **D.** *Quercus robur*, Fagaceae, generative cell/nucleus (asterisk). **E.** *Carpinus betulus*, Betulaceae, spindle-shaped generative cell/nucleus (asterisk). **F.** *Asterostigma lividum*, Araceae, generative cell/nucleus (asterisk) and vegetative nucleus

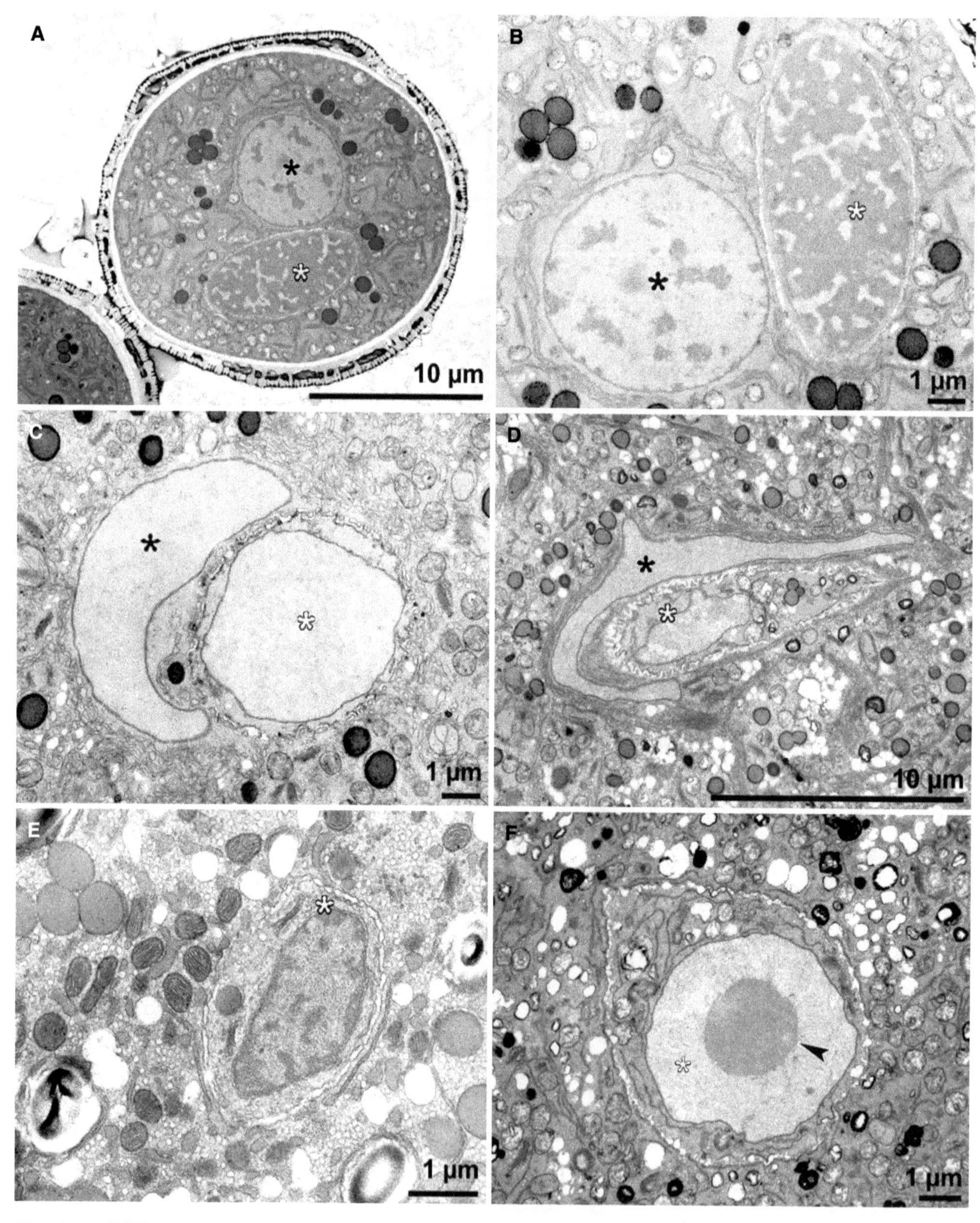

Fig. 7 Variability of the generative cell and nucleus in TEM (cross sections). A. *Melampyrum nemorosum*, Orobanchaceae, pollen in overview, vegetative nucleus (black asterisk), generative cell/nucleus (white asterisk), modified Thiéry test. **B.** *Melampyrum nemorosum*, Orobanchaceae, vegetative nucleus (black asterisk), generative cell/nucleus (white asterisk), modified Thiéry test. **C.** *Betonica officinalis*, Lamiaceae, vegetative nucleus (black asterisk) and generative cell/nucleus (white asterisk) surrounded by cytoplasm, modified Thiéry test. **D.** *Ajuga reptans*, Lamiaceae, vegetative nucleus (black asterisk) and generative cell/nucleus (white asterisk) surrounded by cytoplasm, modified Thiéry test. **E.** *Acinos alpinus*, Lamiaceae, generative cell/nucleus (white asterisk) surrounded by cytoplasm, modified Thiéry test. **F.** *Stachys palustris*, Lamiaceae, generative cell/nucleus (white asterisk) with nucleolus (arrowhead) surrounded by cytoplasm, modified Thiéry test

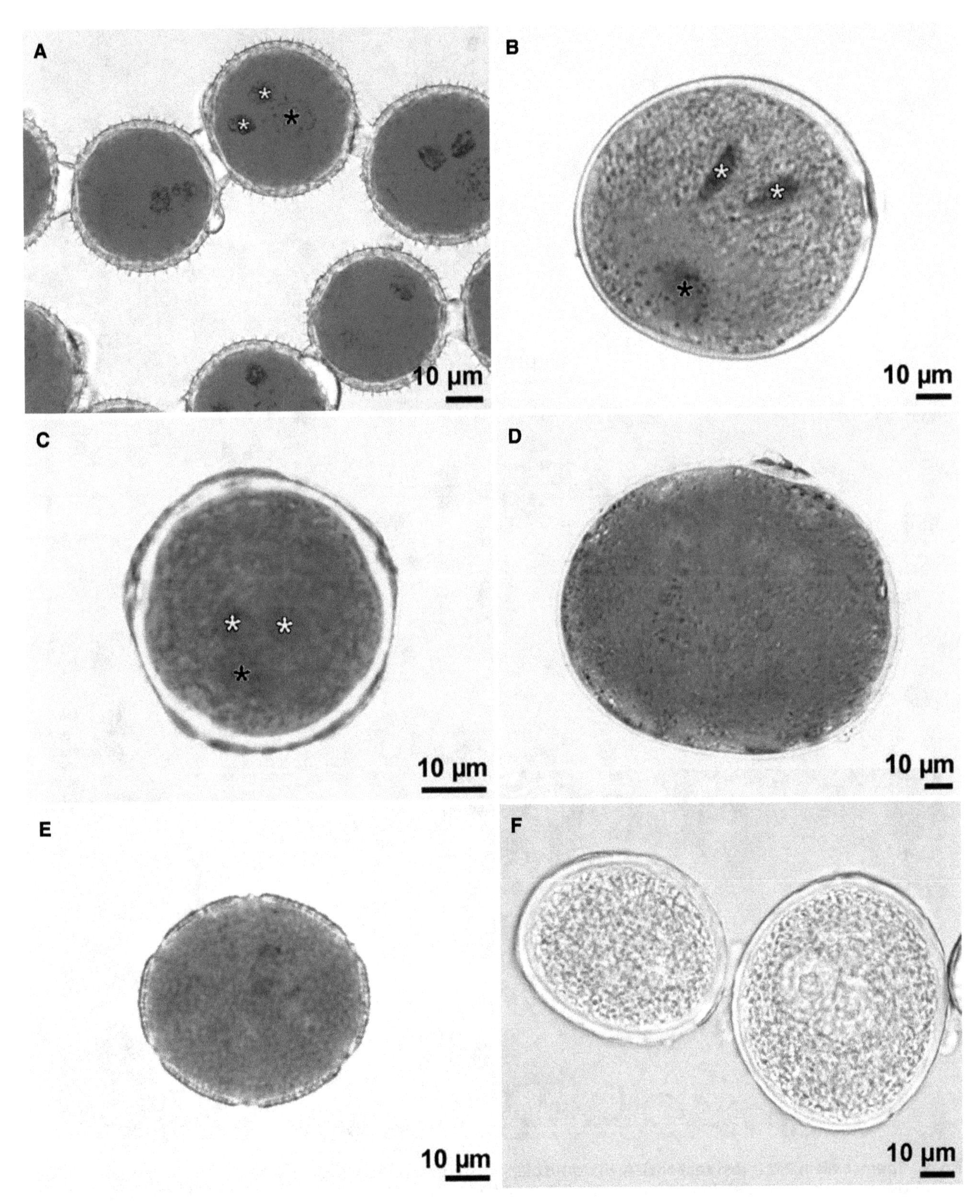

Fig. 8 Sperm cells of different species in LM. A. *Filarum manserichense*, Araceae, stained pollen showing two sperm cells (white asterisks) and vegetative nucleus (black asterisk), acetocarmine. **B.** *Triticum aestivum*, Poaceae, stained pollen showing two sperm cells (white asterisks) and vegetative nucleus (black asterisk), acetocarmine. **C.** *Ulmus minor*, Ulmaceae, stained pollen showing two sperm cells (white asterisks) and vegetative nucleus (black asterisk), acetocarmine. **D.** *Zea mays*, Poaceae, stained pollen showing two sperm cells, acetocarmine. **E.** *Thymus odoratissimus*, Lamiaceae, stained pollen showing two sperm cells, acetocarmine. **F.** *Amorphophallus taurostigma*, Araceae, pollen showing two sperm cells with nuclei, glycerine

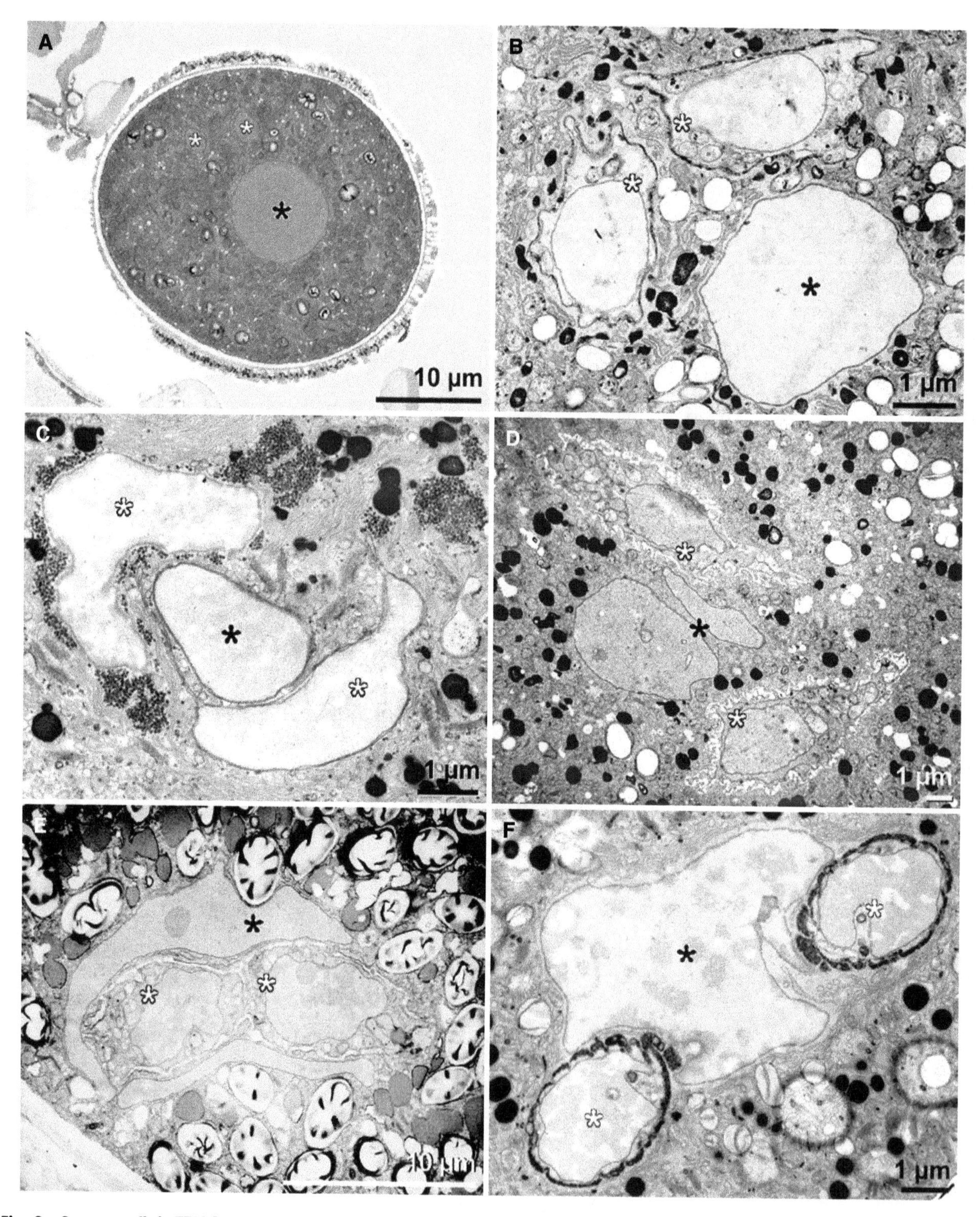

Fig. 9 Sperm cells in TEM (cross sections). A. *Hyssopus officinalis*, Lamiaceae, vegetative cytoplasm, vegetative nucleus (black asterisk), two sperm cells/nuclei (white asterisk), modified Thiéry test. **B.** *Galium odoratum*, Rubiaceae, vegetative nucleus (black asterisk) and two sperm cells/nuclei (white asterisk) surrounded by cytoplasm, modified Thiéry test. **C.** *Smyrnium perfoliatum*, Apiaceae, vegetative nucleus (black asterisk) and two sperm cells/nuclei (white asterisk) surrounded by cytoplasm, Thiéry test. **D.** *Jasminum nudiflorum*, Oleaceae, vegetative nucleus (black asterisk) and two sperm cells/nuclei (white asterisk) surrounded by cytoplasm, Lipid-test. **E.** *Zantedeschia aetiopica*, Araceae, vegetative nucleus (black asterisk) and two sperm cells/nuclei (white asterisk) surrounded by cytoplasm; sperm cells still in contact with each other and enclosed by the vegetative nucleus, modified Thiéry test. **F.** *Melampyrum pratense*, Orobanchaceae, vegetative nucleus (black asterisk) and two sperm cells/nuclei (white asterisk) surrounded by cytoplasm, modified Thiéry test

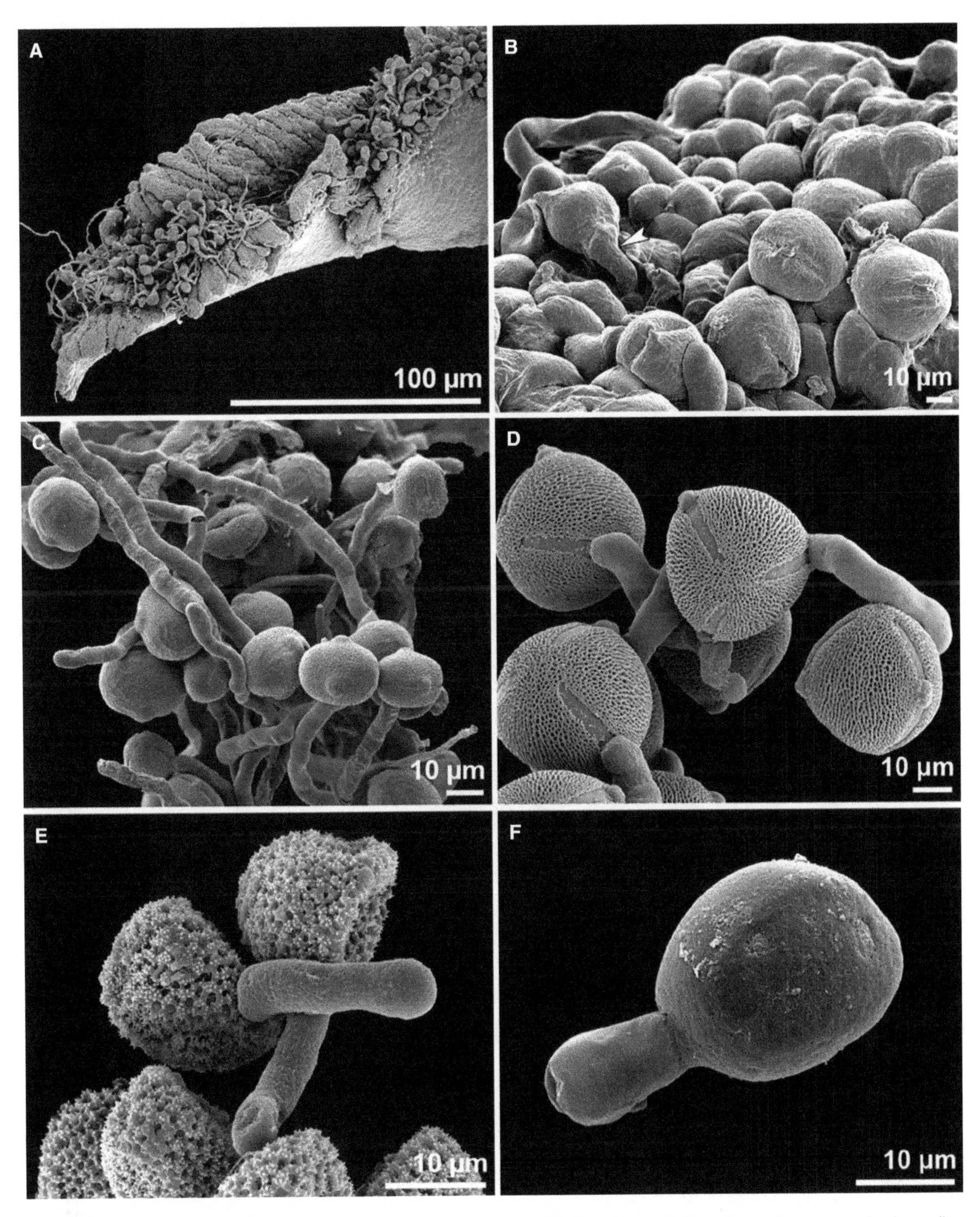

Fig. 10 Pollen germination and pollen tubes in SEM. A. *Cryptanthus bromelioides*, Bromeliaceae, sulcate pollen germinating on stigma. **B.** *Prunus* sp., Rosaceae, tricolporate pollen, note germinating pollen on stigma (left side). **C.** *Oxytropis jacquinii*, Fabaceae, tricolporate pollen. **D.** *Tuberaria guttata*, Cistaceae, tricolporate pollen. **E.** *Anthurium gracile*, Araceae, inaperturate pollen. **F.** *Vanilla pompona*, Orchidaceae, porate pollen

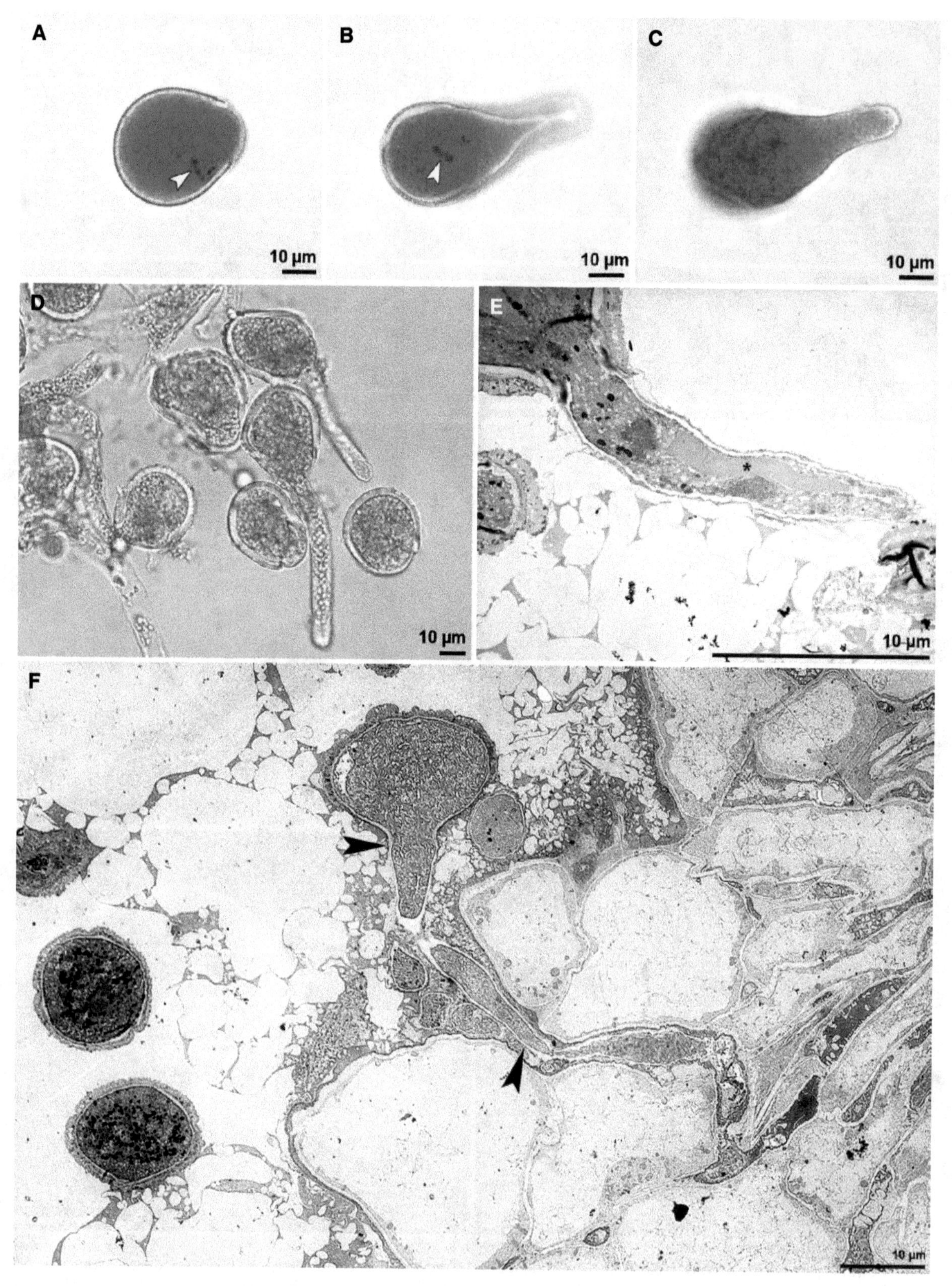

Fig. 11 Pollen germination and pollen tubes in LM and TEM. A-C. *Arum cylindraceum*, Araceae, three-celled, inaperturate pollen, germination can occur anywhere on the pollen surface, staining with acetocarmine, note the two sperm nuclei (arrowhead) staining dark red with acetocarmine, pictures. **B-C** showing optical section and upper focus. **D.** *Colocasia antiquorum*, Araceae, pollen grains germinating in water. **E-F.** *Smyrnium perfoliatum*, Apiaceae, TEM sections of germinating pollen (arrowheads), Thiéry test; detail of pollen tube with sperm nucleus (**E**, asterisk)

References

Brewbaker JL (1967) The distribution and phylogenetic significance of binucleate and trinucleate pollen grains in the angiosperms. Am J Bot 54: 1069–1083

Cresti M, Blackmore S, Van Went JL (1992) Atlas of sexual reproduction in flowering plants. Springer, Berlin, Heidelberg

Edlund AF, Swanson R, Preuss D (2004) Pollen and Stigma Structure and Function: The Role of Diversity in Pollination. Plant Cell 16: 84–97

Furness CA, Rudall PJ (2001) Pollen and anther characters in monocot systematics. Grana, 40: 17–25

Furness CA, Rudall PJ (1999) Microsporogenesis in Monocotyledons. Ann Bot 84: 475–499

Gomez JF, Talle B, Wilson ZA (2015) Anther and pollen development: A conserved developmental pathway. J Integr Plant Biol 57: 876–891

Keijzer CJ, Willemse MTM (1988) Tissue interactions in the developing locule of *Gasteria verrucosa* during microsporogenesis. Acta Bot Neerl 37: 493–508

Mascarenhas JP (1993) Molecular mechanisms of pollen tube growth and differentiation. Plant Cell 5: 1303–1314

McCormick S (1993) Male gametophyte development. Plant Cell 5: 1265–1275

Pacini E (1997) Tapetum character states: analytical keys for tapetum types and activities. Can J Bot 75: 1448–1459

Williams JH, Taylor ML, O'Meara BC (2014) Related evolution of tricellular (and bicellular) pollen. Am J Bot 101: 559–571

3

Characteristics of Pollen: Structure and Sculpture

The study of pollen should encompass all structural and ornamental aspects of the grain. Pollen morphology is studied using LM and SEM and is important to visualize the general features of a pollen grain, including, e.g., symmetry, shape, size, aperture number and location, as well as ornamentation. TEM investigations are used to highlight the stratification and the uniqueness of pollen wall layers as well as cytoplasmic features. The following sections explain the most important structural and sculptural pollen features a palynologist should observe.

Polarity and Shape

Mature pollen is shed in **dispersal units.** When the post-meiotic products become separated the dispersal unit is a single pollen grain, a **monad.** Post-meiotic products also become partly separated or remain permanently united, resulting in **dyads** (a rare combination), **tetrads** or **polyads. Pollinaria** are dispersal units of two pollinia including a sterile, interconnecting appendage (see "Glossary of Palynological Terms").

Pollen shape and aperture location relate directly to pollen **polarity.** The polarity is determined by the spatial orientation of the microspore in the meiotic tetrad and can be examined in the **tetrad stage** (Fig. 1). The **polar axis** of each microspore/pollen runs from the **proximal pole,** orientated towards the tetrad center, to the **distal pole** of the microspore/pollen (Fig. 2). The **equatorial plane** is located at the microspore's center, perpendicular to the polar axis (Fig. 2). Therefore, the **equatorial plane** divides the microspore/pollen into a proximal and a distal half, comparable to the northern and southern hemisphere of our planet Earth.

The polarity gives rise to the polar and to the equatorial view. In dicots there is usually one polar

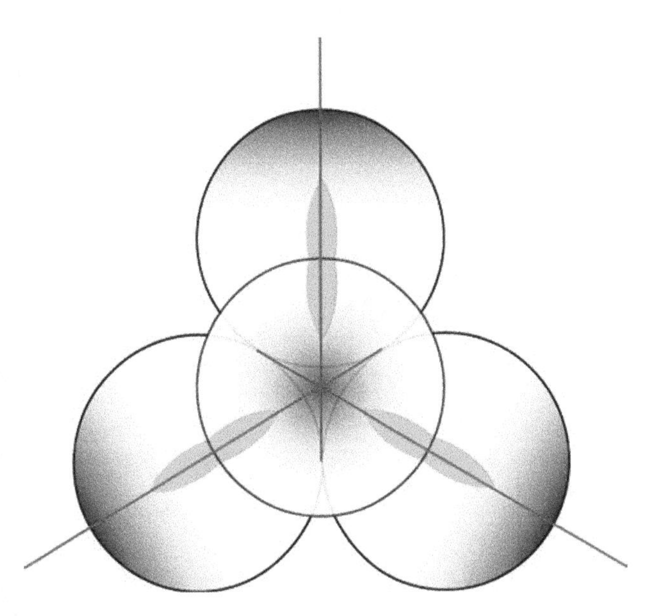

Fig. 1 Tetrad stage. Orientation of microspores/pollen in the tetrad; distal poles shaded green

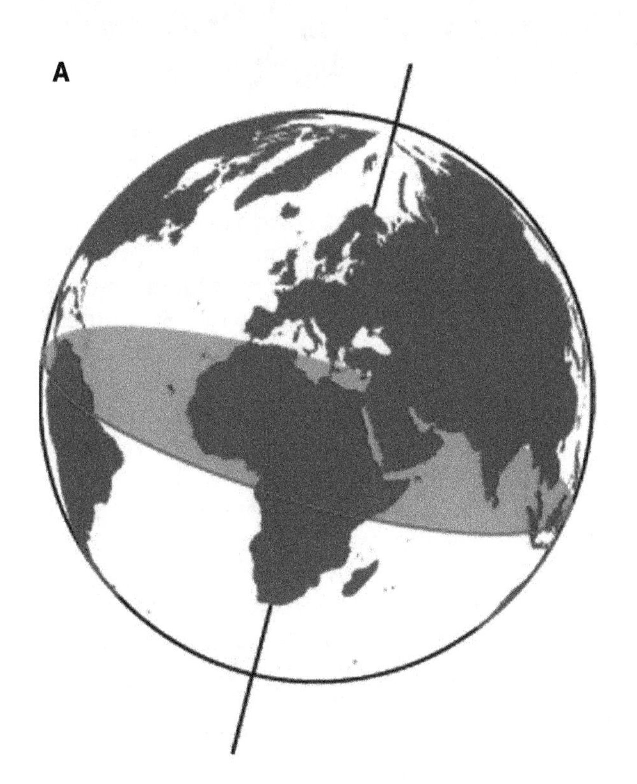

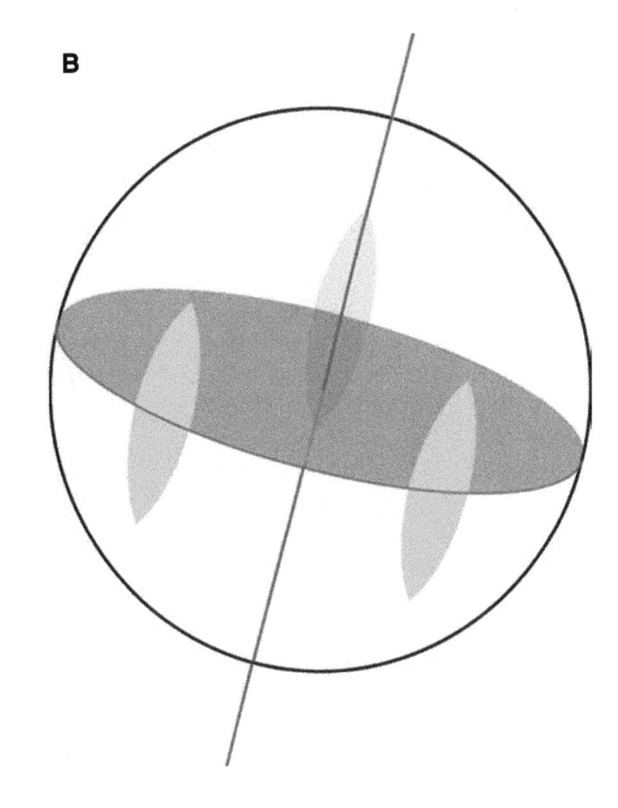

Fig. 2 Polar axis and equatorial plane. A-B. Polar axis and equatorial plane

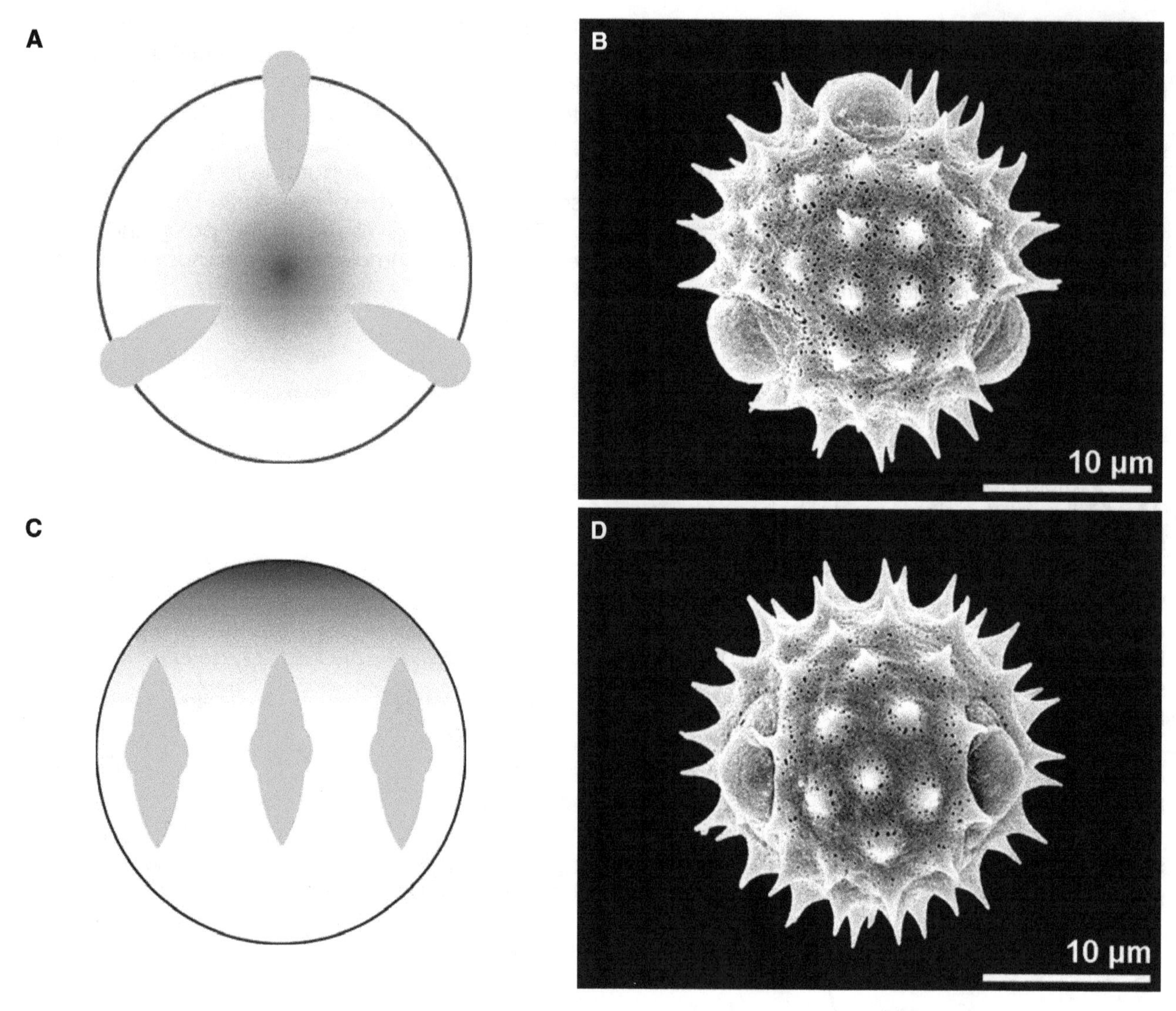

Fig. 3 Polarity of pollen in dicots. A-B. *Bellis perennis*, Asteraceae, polar view. **C-D.** *Bellis perennis*, Asteraceae, equatorial view

and one equatorial view (Fig. 3). In monocots, due to the mostly distal position of apertures, there are four views: distal polar, proximal polar, and two different equatorial views (Fig. 4).

Isopolar pollen has identical proximal and distal poles, thus the equatorial plane is a symmetry plane. In **heteropolar** pollen the proximal and distal halves differ (Fig. 5).

The various arrangements of the four microspores within **tetrads** depend on the simultaneous or successive type of cytokinesis and on the type of intersporal wall formation. The spatial arrangement of microspores after **simultaneous** cytokinesis is a **tetrahedral** (or rarely decussate) **tetrad** (Fig. 6A). This tetrad types may have systematic relevance, e.g., all species within the genus *Rhododendron* are characterized by tetrahedral tetrads. The spatial arrangement of microspores after **successive** cytokinesis leads to different morphological tetrad types, which can be differentiated into **planar** (tetragonal, linear, T-shaped) and/or **non-planar** (decussate or tetrahedral) tetrads (Fig. 6B). These morphotypes have no systematic relevance, as tetrads may vary within a genus/species, e.g., in *Typha* tetrads may be tetragonal, T-shaped and/or linear (Furness and Rudall 2001; Copenhaver 2005; see also "Illustrated Pollen Terms").

P/E ratio (Fig. 7) refers to the length of the polar axis (P) between the two poles compared to the equatorial diameter (E). In **isodiametric** pollen the polar axis is ± equal to the equatorial diameter. In **prolate** pollen the polar axis is longer than the equatorial diameter. In **oblate** pollen the polar axis is shorter than the equatorial diameter. **Pollen shape** refers to the 3-dimensional form of a pollen grain in relation to the P/E ratio. A pollen grain can, for

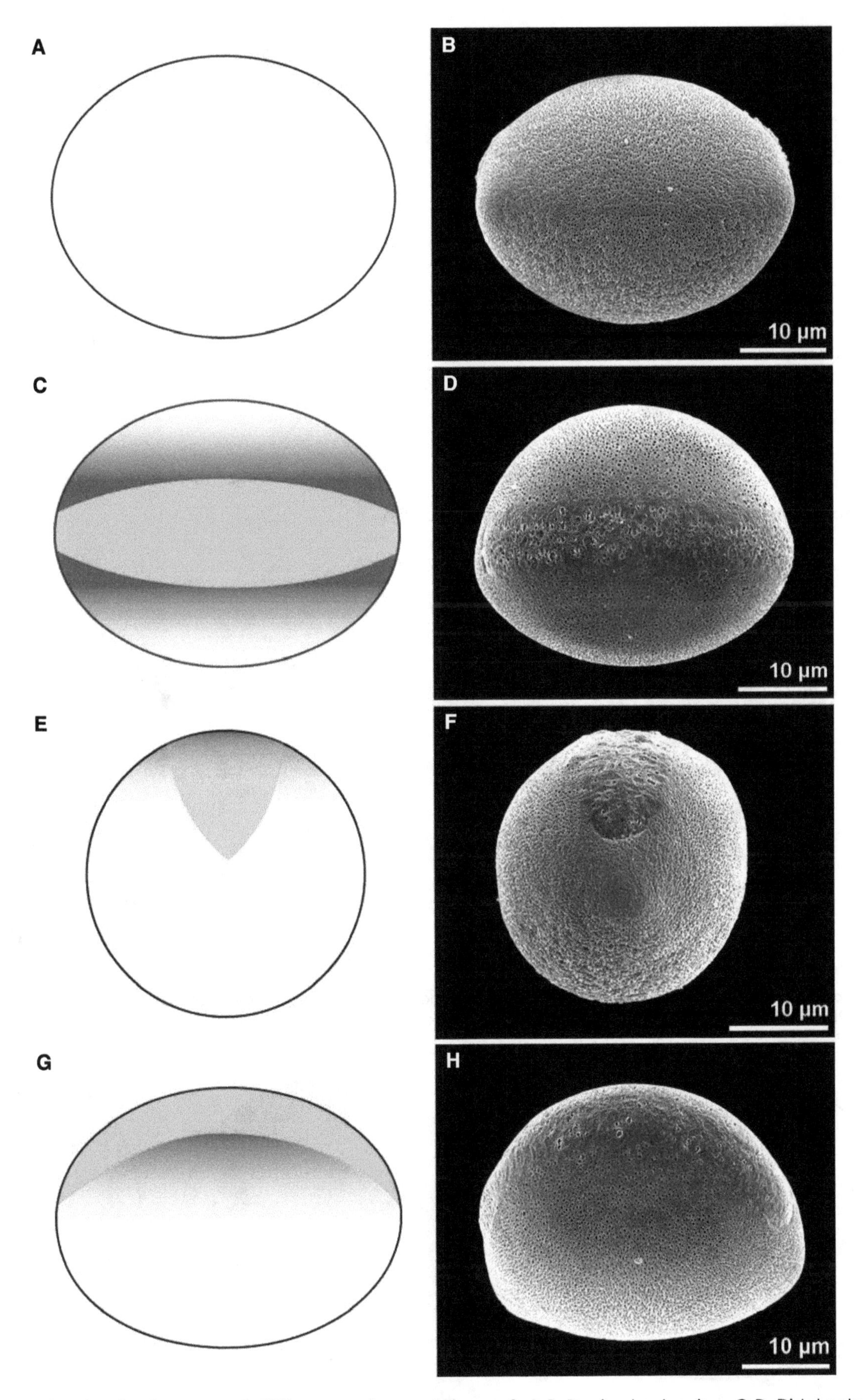

Fig. 4 Polarity of pollen in monocots (*Allium paradoxum*, Alliaceae). A-B. Proximal polar view. **C-D.** Distal polar view. **E-F.** Equatorial view (short axis). **G-H.** Equatorial view (long axis)

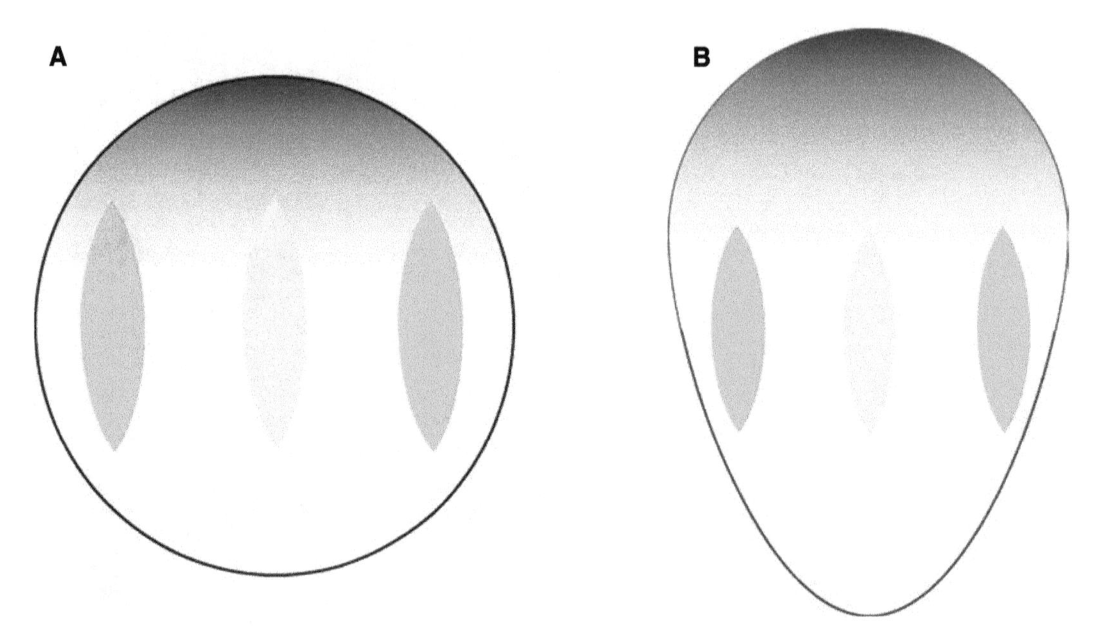

Fig. 5 Pollen symmetry. A. Isopolar pollen. **B.** Heteropolar pollen

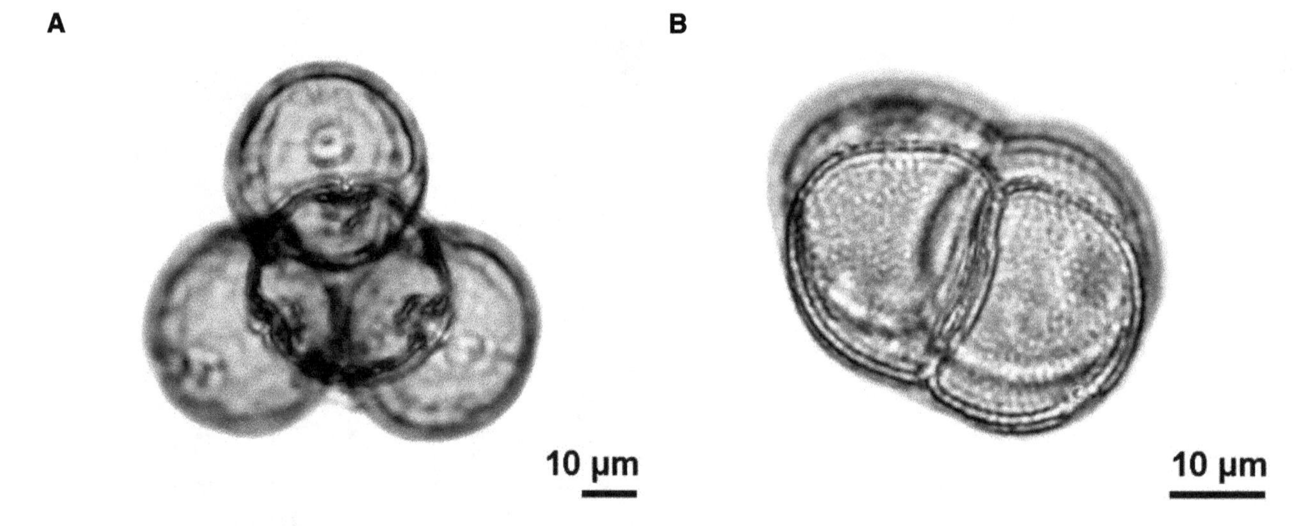

Fig. 6 Pollen arrangements in tetrads. A. Tetrahedral tetrad, *Fagus* sp., Fagaceae, fossil, Quaternary, Austria; apical view. **B.** Planar tetrad, *Typha latifolia*, Typhaceae, fossil, Quaternary, Austria

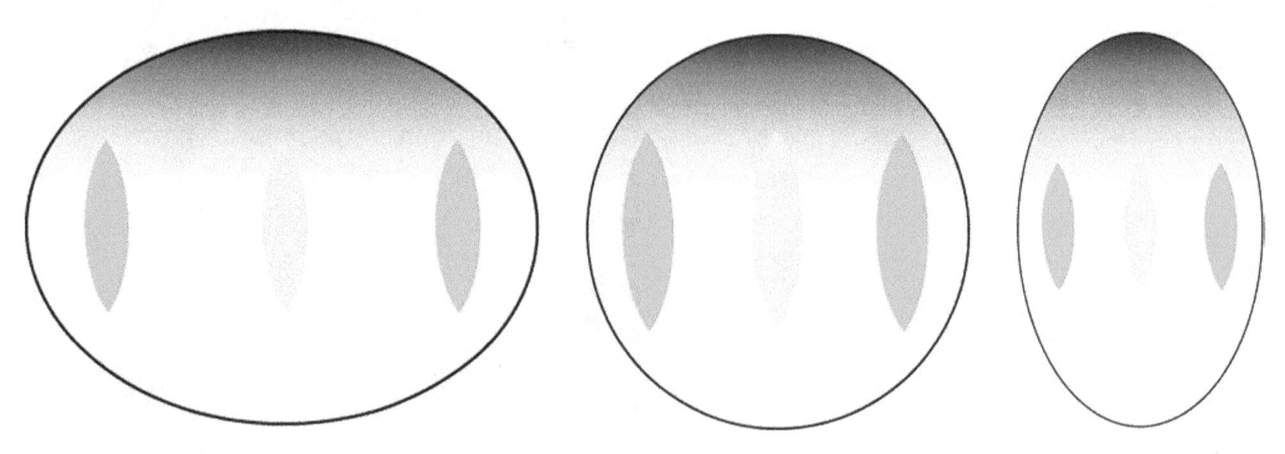

Fig. 7 P/E ratio of pollen. Schematic drawings of oblate (left), isodiametric (middle), and prolate (right) pollen

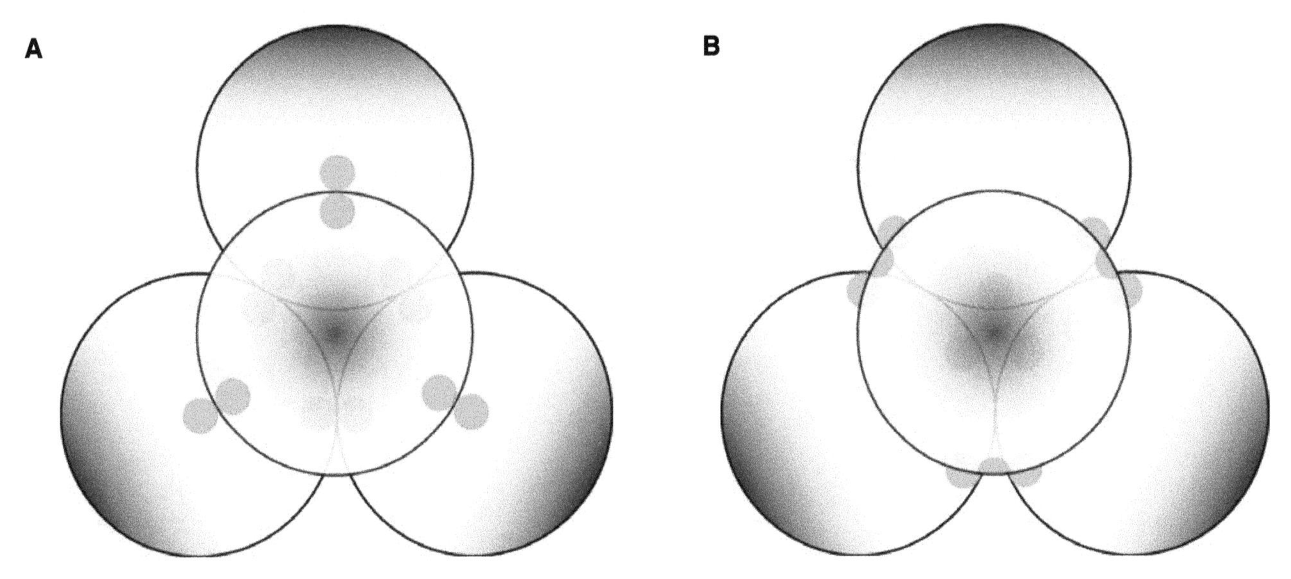

Fig. 8 Aperture arrangement. A. Fischer's law, apertures in pairs. **B.** Garside's law, apertures in a group of three

example, be spheroid-, cup-, boat-, cube-, tetrahedral-, triangular dipyramid-, hexafoil dipyramid-, triangular prism-, pentagonal prism-, or hexagonal prism shaped (see "Illustrated Pollen Terms").

In pollen grains with three apertures, two types of aperture arrangement occur after simultaneous cytokinesis (Fig. 8). **Fischer's law** refers to the most frequent arrangement where a pair of apertures occurs at six points in a tetrad (e.g., Ericaceae, permanent tetrads). **Garside's law** refers to the unusual arrangement of apertures where a group of three apertures occur at four points in the tetrad (probably restricted to Proteaceae, no permanent tetrads; Blackmore and Barnes 1995) (Fig. 8).

Apertures

An **aperture** is a region of the pollen wall that differs significantly from its surroundings in morphology and/or anatomy. The aperture is presumed to function as the site of germination and to play a role in harmomegathy. Pollen grains lacking apertures are called **inaperturate** (Furness 2007). The aperture definition fits both angiosperm and gymnosperm pollen, but in gymnosperms the type of aperture (e.g., leptoma; germination area) usually differs from that in angiosperms.

The polarity of the pollen grain determines the aperture terminology. A circular aperture is termed a **porus** if situated equatorially or globally; if situated distally, it is called an **ulcus.** An elongated aperture is termed a **colpus** if situated equatorially or globally; if situated distally, it is termed a **sulcus.** A combination of porus and colpus is termed a **colporus**; colpori are situated equatorially or globally. In **heteroaperturate** pollen two different types of apertures (single and/or combined) are present in a combination of colpi with colpori or pori. A circular or elliptic aperture with indistinct margins is termed a **poroid.** Additional rare combinations of ekto- and endoapertures, mostly observed in LM, include pororate and colporoidate (Fig. 9). Pollen grains that have compound apertures composed of circular ektopori and endopori are termed **pororate.** Compound apertures composed of a colpus (ektoaperture) with an indistinct endoaperture are termed **colporoidate.** When the colpus has a clear bulge in the equatorial region of a pollen grain it is termed **geniculum** (Fig. 9D).

The number of equatorial apertures (pori, colpi, colpori) is indicated by the prefixes di-, tri-, tetra-, penta- or hexa-. Writing numbers instead of prefixes is in common use, e.g., 4-porate or tetraporate, 6-colpate or hexacolpate. In this book we prefer the use of prefixes. For pollen grains with more than three apertures, positioned at the equator, the term **stephanoaperturate (stephanoporate, stephanocolpate, stephanocolporate)** is used together with the aperture number (e.g., stephano(4)porate or 4-porate, stephanoporate). Pollen grains with globally distributed apertures are termed **pantoaperturate.**

Apertures are normally covered by an exinous layer, the **aperture membrane.** The aperture membrane can be **ornamented**, e.g., covered with various exine elements, or it is **psilate** (smooth). The aperture can also be covered by an **operculum**, a distinctly delimited exine structure, covering the aperture like a lid (Halbritter and Hesse 1995; Furness and Rudall 2003).

Number, type, and position of apertures are genetically determined and usually the same within

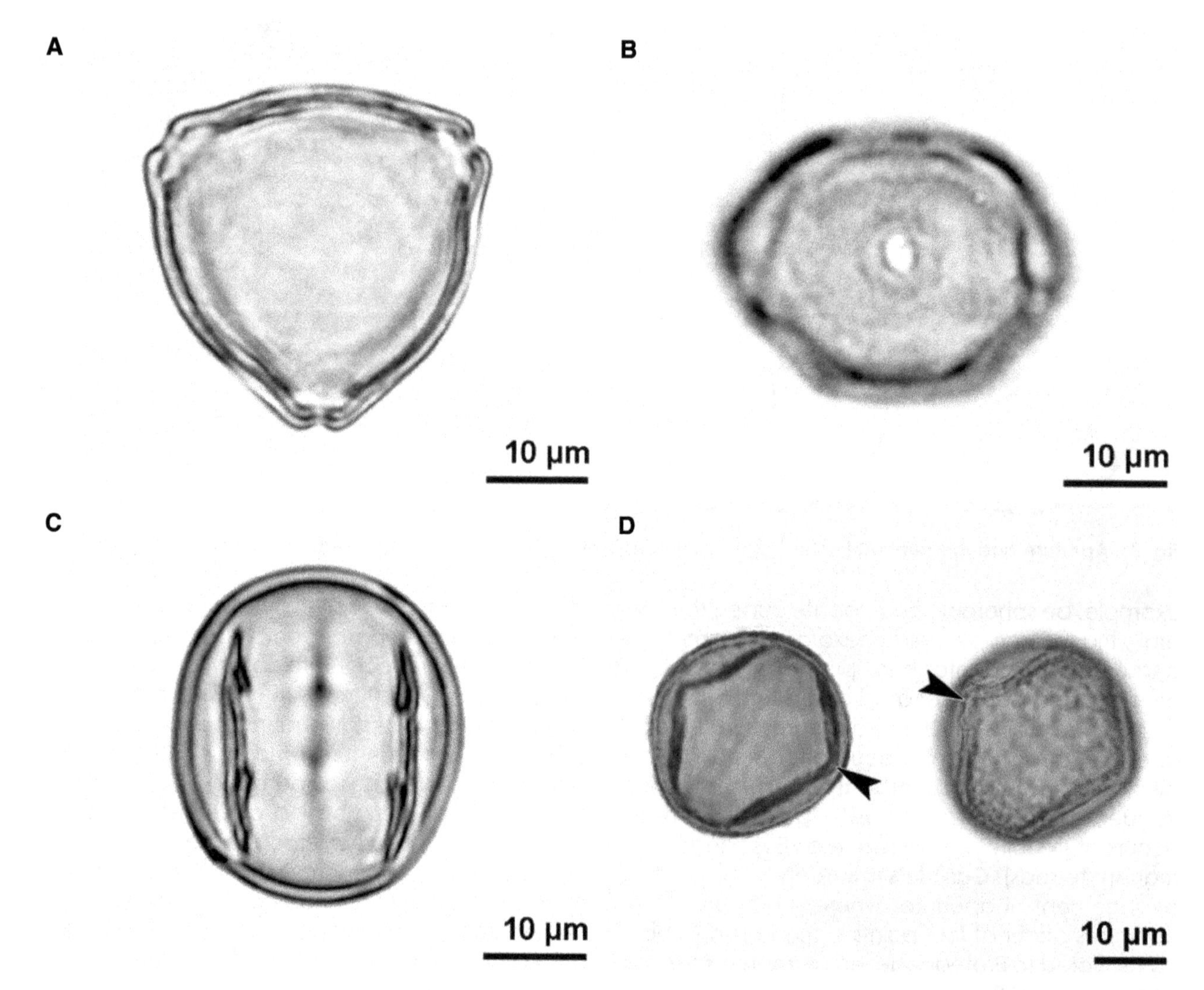

Fig. 9 Special aperture features observed in LM. A-B. *Corylus* sp., Betulaceae, fossil, Quaternary, Austria, pororate pollen in polar and equatorial view. **C.** *Eucommia* sp., Eucommiaceae, fossil, middle Miocene, Austria, colporoidate pollen in equatorial view. **D.** *Quercus petrea*, Fagaceae, pollen with geniculum (arrowhead), optical section (left) and upper focus (right)

a species, but may also vary (e.g., *Alnus* is usually 5-porate, but number of pori can vary from 3 to 6).

A **pseudocolpus** occurs in heteroaperturate pollen and is presumed to be non-functional. Pseudocolpi mostly alternate with colpori (e.g., in Boraginaceae, Lythraceae) or are flanking each colporus (in Acanthaceae). For examples, see "Illustrated Pollen Tems." Pseudocolpi are believed to play a role in **harmomegathy**, but their effect has been poorly studied.

Pre-(prae-)pollen (Fig. 10) is characterized by proximal and sometimes additional distal apertures, and by presumed proximal germination. Pre-pollen are microspores of certain extinct basal seed plants occurring from the Late Devonian until the Cretaceous. Proximal germination is typical for spores.

Spores germinate at the **tetrad mark** (Fig. 11), the so-called **laesura** (for an extensive overview, see

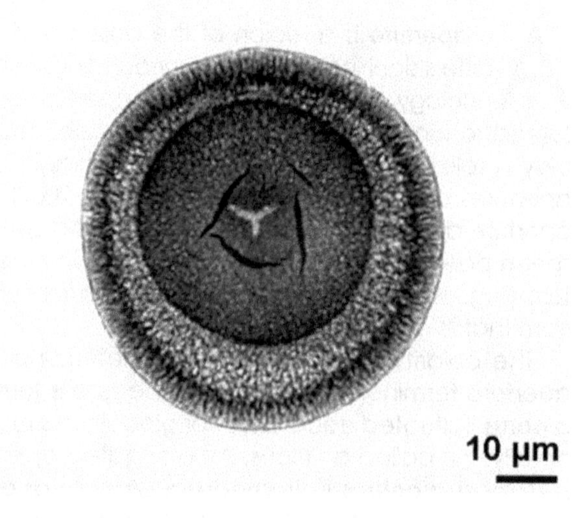

Fig. 10 Pre-pollen. *Nuskoissporites* sp., fossil, Permian, Austria, polar view

Tryon and Lugardon 1991). The tetrad mark is situated at the proximale pole (proximal germination).

Proximal germination is a rare exception in seed plants (Fig. 12), e.g., *Beschorneria yuccoides* (Agavaceae) and *Annona muricata* (Annonaceae). In the two cases, this proximally situated aperture (germination area) is functionally replacing the dysfunctional sulcus (Hesse et al. 2009). In *Beschorneria*, pollen grains forming the tetrads are loosely interconnected and separate frequently. In this special case, the sulcus (distal) is not functional, whereas the proximal face, with a highly reduced exine, functions as germination site. In *Annona*, the microspores rotate within the tetrad during development and the original distally placed sulcus becomes proximally positioned (Tsou and Fu 2002).

The aperture usually acts as the (exclusive) **germination** site. In inaperturate angiosperm pollen the pollen tube can protrude at any given site. In taxoid gymnosperm pollen the exine ruptures during hydration at a specialized region, the leptoma, and is subsequently shed (Fig. 13A-B). The protoplast (enclosed by the intine) is released and a pollen tube can be formed anywhere (resembling functionally an inaperturate pollen grain). Furthermore some angiosperm taxa shed the exine before pollen tube formation, e.g., in some Annonaceae, Araceae. Within the Araceae, a shed pollen wall has been observed in several taxa, e.g., *Amorphophallus*, *Taccarum* (Ulrich et al. 2017). The outer pollen wall (composed of polysaccharide) splits immediately in water and sheds soon afterwards. Subsequently, the naked protoplast is

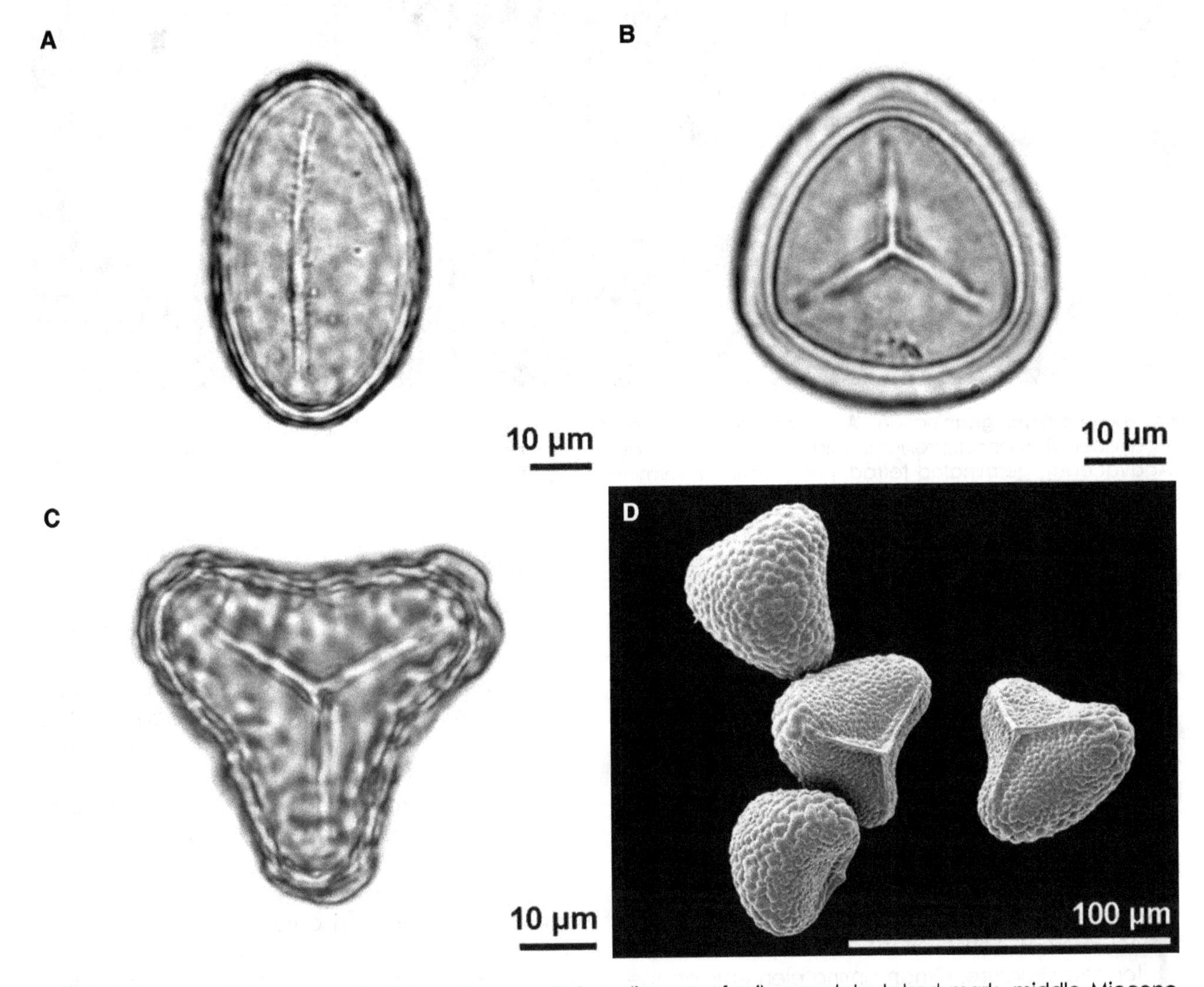

Fig. 11 Tetrad mark in spores. A. *Polypodium* sp., Polypodiaceae, fossil, monolete tetrad mark, middle Miocene, Austria, polar view. **B.** *Sphagnum* sp. Sphagnaceae, fossil, trilete tetrad mark, middle Miocene, Austria, polar view. **C.** Pteridaceae indet., fossil, middle Miocene, Austria, trilete tetrad mark, polar view. **D.** *Cryptogamma crispa*, Pteridaceae, trilete tetrad mark

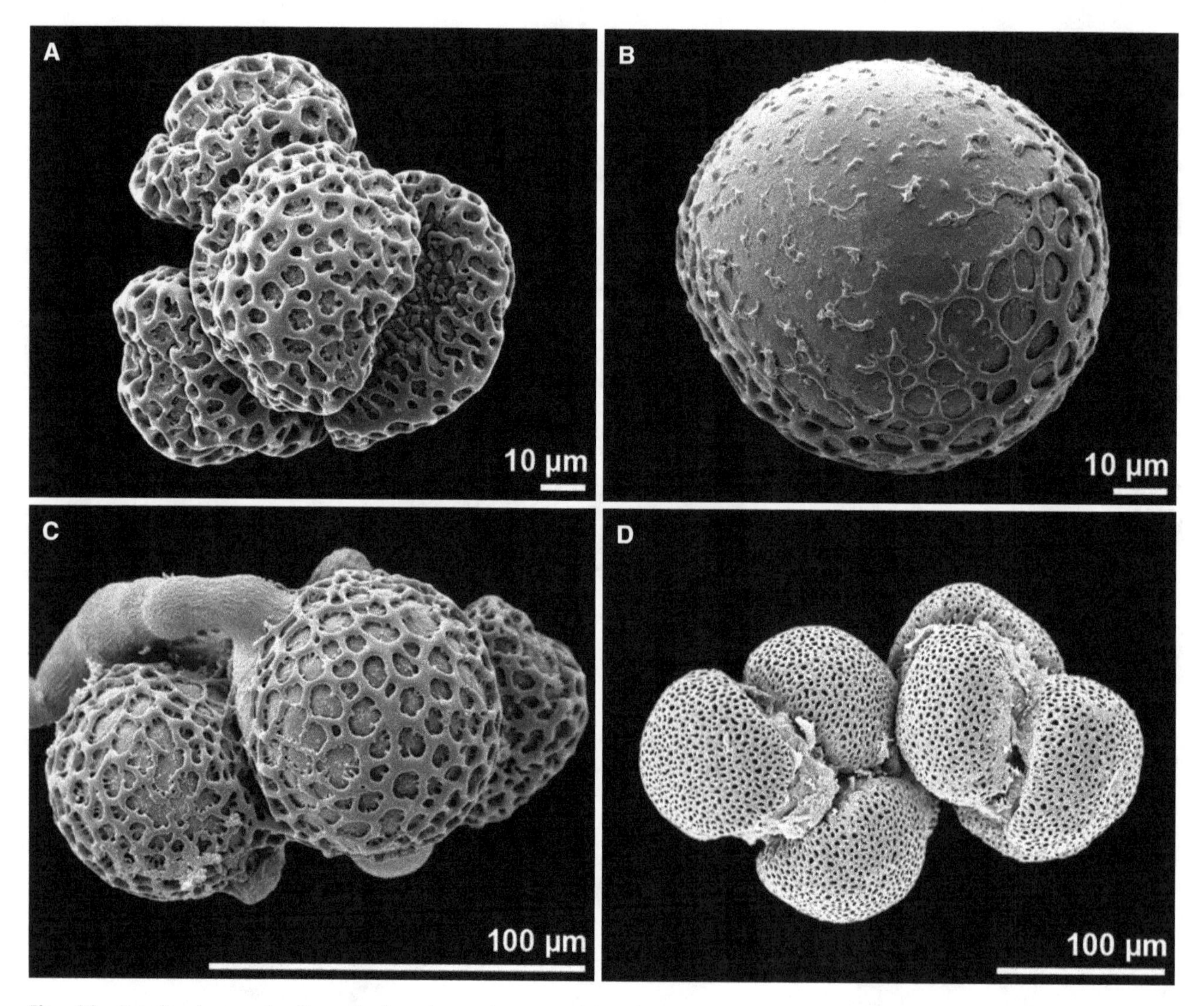

Fig. 12 Proximal germination. A. *Beschorneria yuccoides* (Agavaceae), dry tetrad. **B.** *Beschorneria yuccoides* (Agavaceae), monad, proximal polar view, proximal face functions as germination site. **C.** *Beschorneria yuccoides* (Agavaceae), germinated tetrad, note proximal germination. **D.** *Annona muricata* (Annonaceae), mature tetrads, sulcus hidden in proximal position

floating in water and germinates about 1 hour after shedding (Fig. 13C-D).

During germination, usually a single pollen tube is formed. In some cases, instant pollen **tube-like structures** are simultaneously developed at all apertures (Fig. 14). The formation of these pollen tube-like structures, in relation with moisture, is interpreted as a pre-germinative process that takes place during dehiscence (Blackmore and Cannon 1983).

Pollen Wall

The internal construction of the pollen wall is termed **structure.** Ornamenting elements on the pollen surface (ornamentation) are summarized under the term **sculpture** or sculpturing. However, it is not always possible to distinguish between structure and sculpture (e.g., free-standing columellae).

Structure

In general, the **pollen wall** (**sporoderm**) of seed plants is formed by two main layers: the outer **exine** and the inner **intine** (Fig. 15). The exine consists mainly of **sporopollenin**, which is an acetolysis- and decay-resistant biopolymer. The intine is mainly composed of cellulose and pectin. Commonly, the pollen wall in aperture regions is characterized by the reduction of exine structures or by a deviant exine, and a thick, often bilayered intine.

Two layers within the exine are distinguished: an inner endexine and an outer ektexine. In **tectate** pollen the ektexine usually consists of a basal **foot layer**, an **infratectum** (e.g., columellae) and a **tectum**, the **endexine** is a mainly unstructured layer (Fig. 16A–C). There are many deviations from this principal construction: layers may be thickened, variably structured or lacking. When the pollen

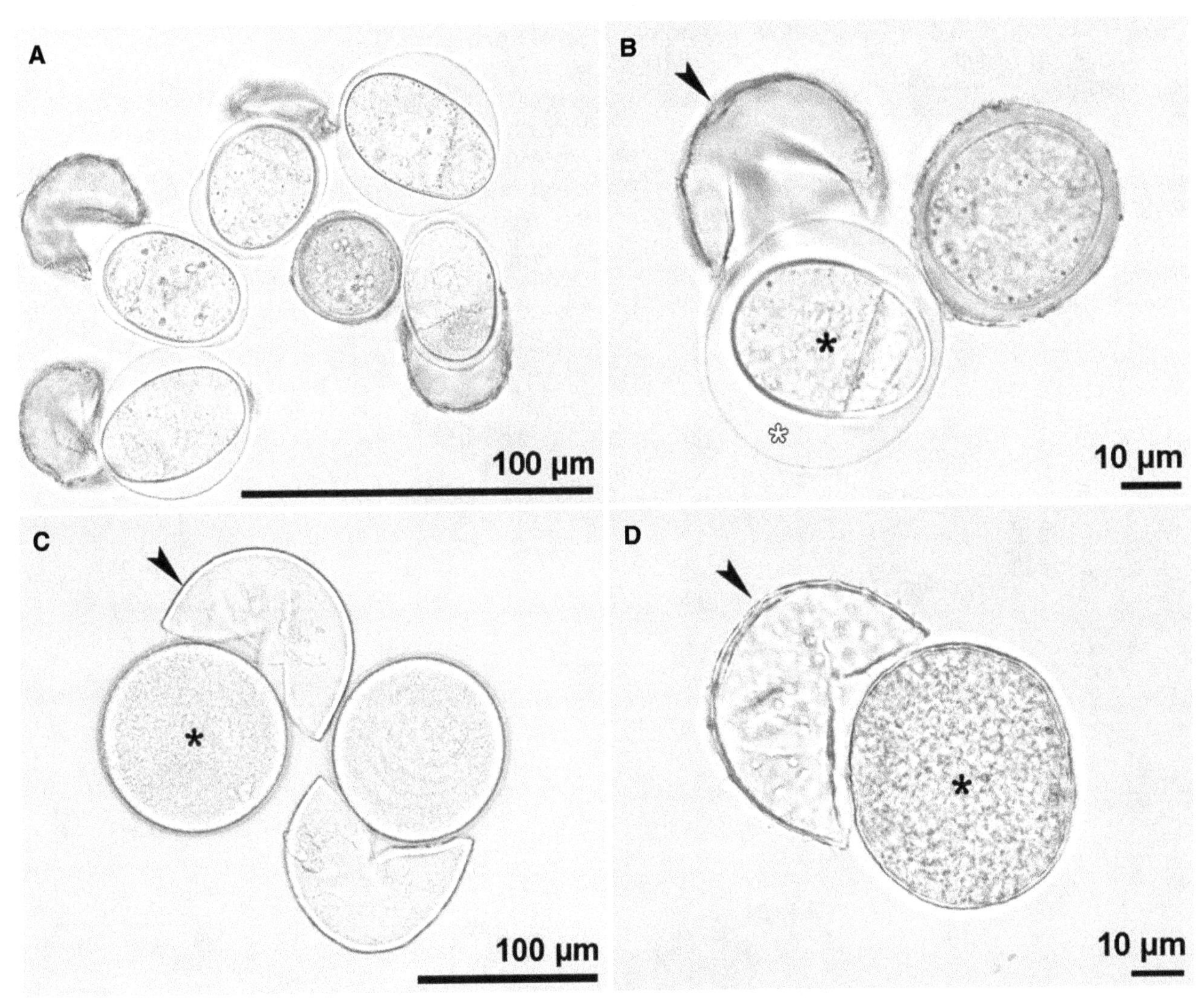

Fig. 13 Exine/pollen wall shedding. A. *Cephalotaxus* sp., Cephalotaxaceae, fresh pollen in water. **B.** exine (*arrowhead*) shedding prior to pollen tube formation, released protoplast (black asterisk) enclosed by a thick swelled intine (white asterisk). **C.** *Taccarum weddellianum*, Araceae, pollen wall shedding, released protoplast (black asterisk) and shed outer pollen wall (arrowhead). **D.** *Amorphophallus mangelsdorffii*, Araceae, pollen wall shedding, released protoplast (black asterisk) and shed outer pollen wall (arrowhead)

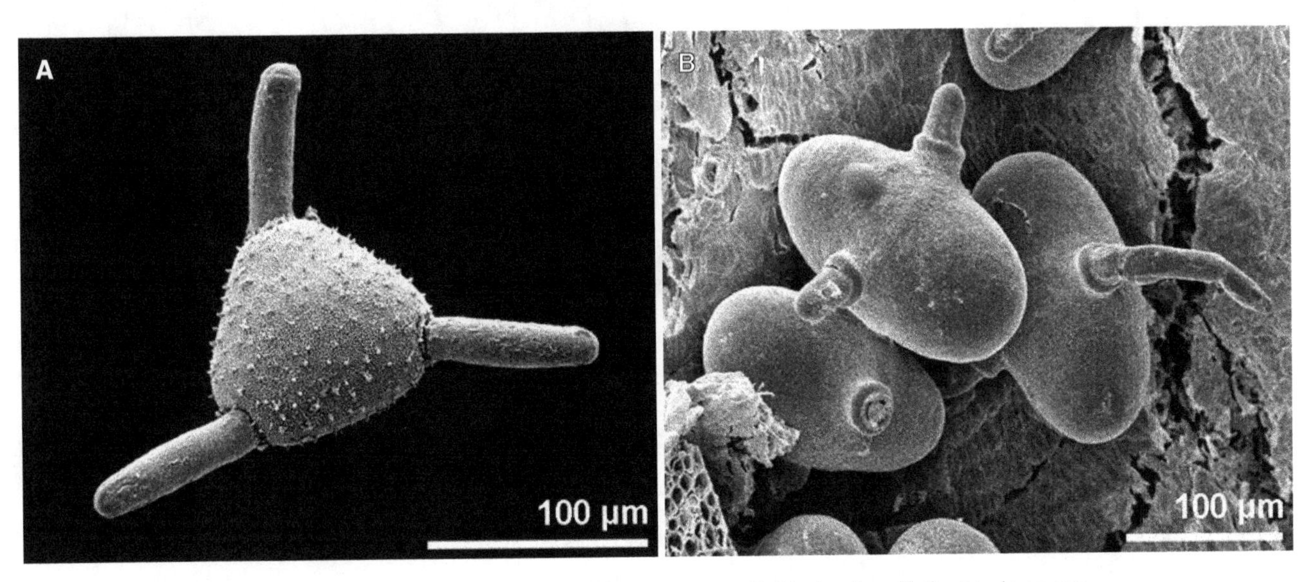

Fig. 14 Instant pollen tubes. A. *Scabiosa caucasica*, Dipsacaceae. **B.** *Morina longifolia*, Morinaceae

grain is lacking a tectum it is termed **atectate** (Fig. 16D–F). In apertural regions the pollen wall is generally characterized by a different exine construction.

The terms **sexine** for the outer, structured, and **nexine** for the inner, unstructured exine layer are widely used in light microscopy, but do not fully correspond to ekt- and endexine, respectively. When a cavity between the sexine and nexine is present in the interapertural area, this is termed **cavea** (Fig. 17).

Sporopollenin

John (1814) and Braconnot (1829) introduced the terms "pollenin" and "sporonin" for the resistant exine material of pollen and spores. Zetzsche et al. (1931) then combined the terms into "sporopollenin," that is the major component of the exine found in most pollen and spores, except in filiform seagrass pollen (e.g., Dobritsa et al. 2009; Jardine et al. 2015). Sporopollenin is a complex biopolymer and extremely resistant to

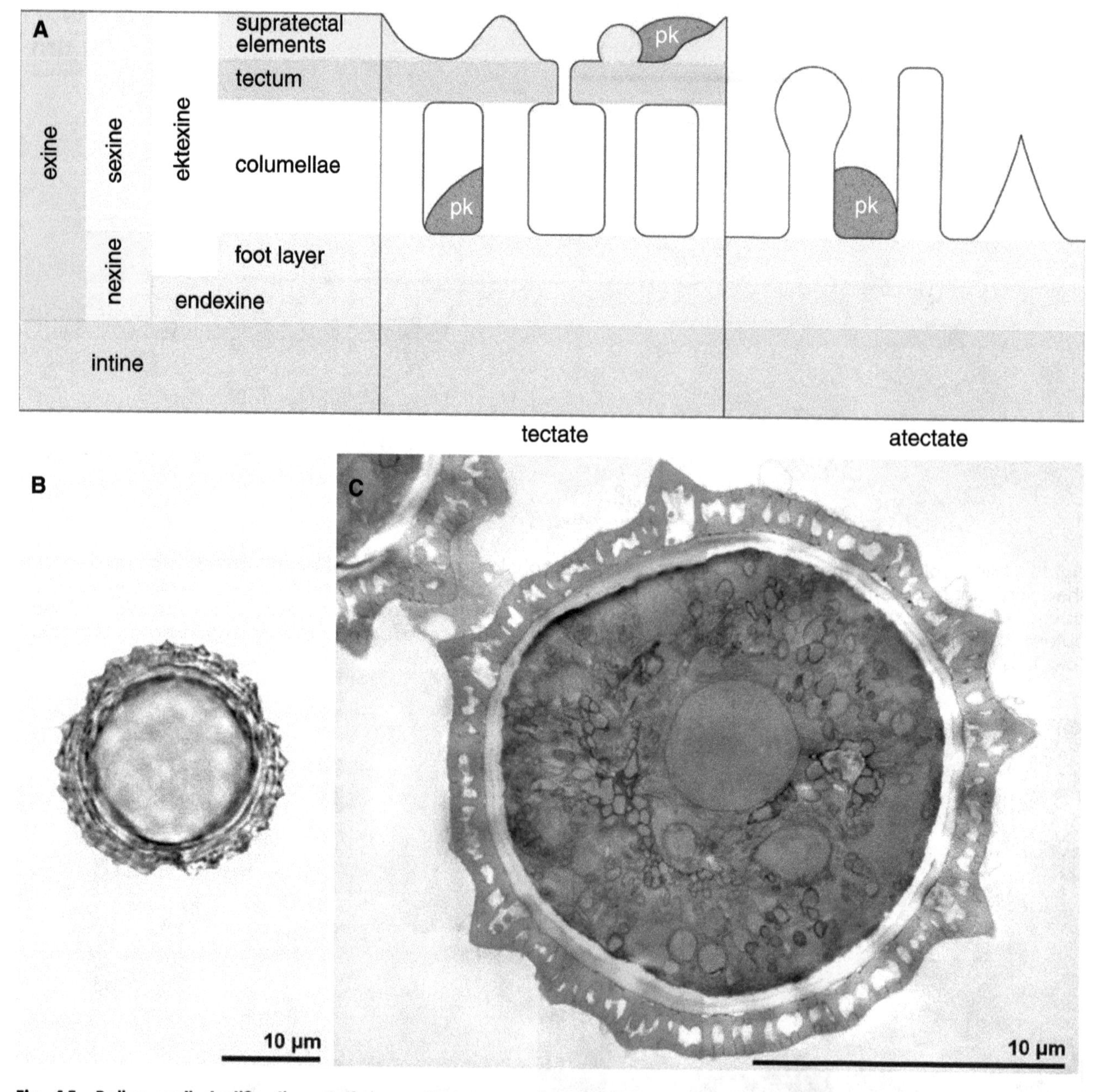

Fig. 15 Pollen wall stratification. A. Schematic cross section of pollen wall, pk: pollenkitt. **B.** *Ambrosia artemisiifolia*, Asteraceae, optical view showing both intine and exine; acetolyzed. **C.** *Ambrosia artemisiifolia*, Asteraceae, cross section showing both intine (yellow) and exine (blue); modified Thiery-test

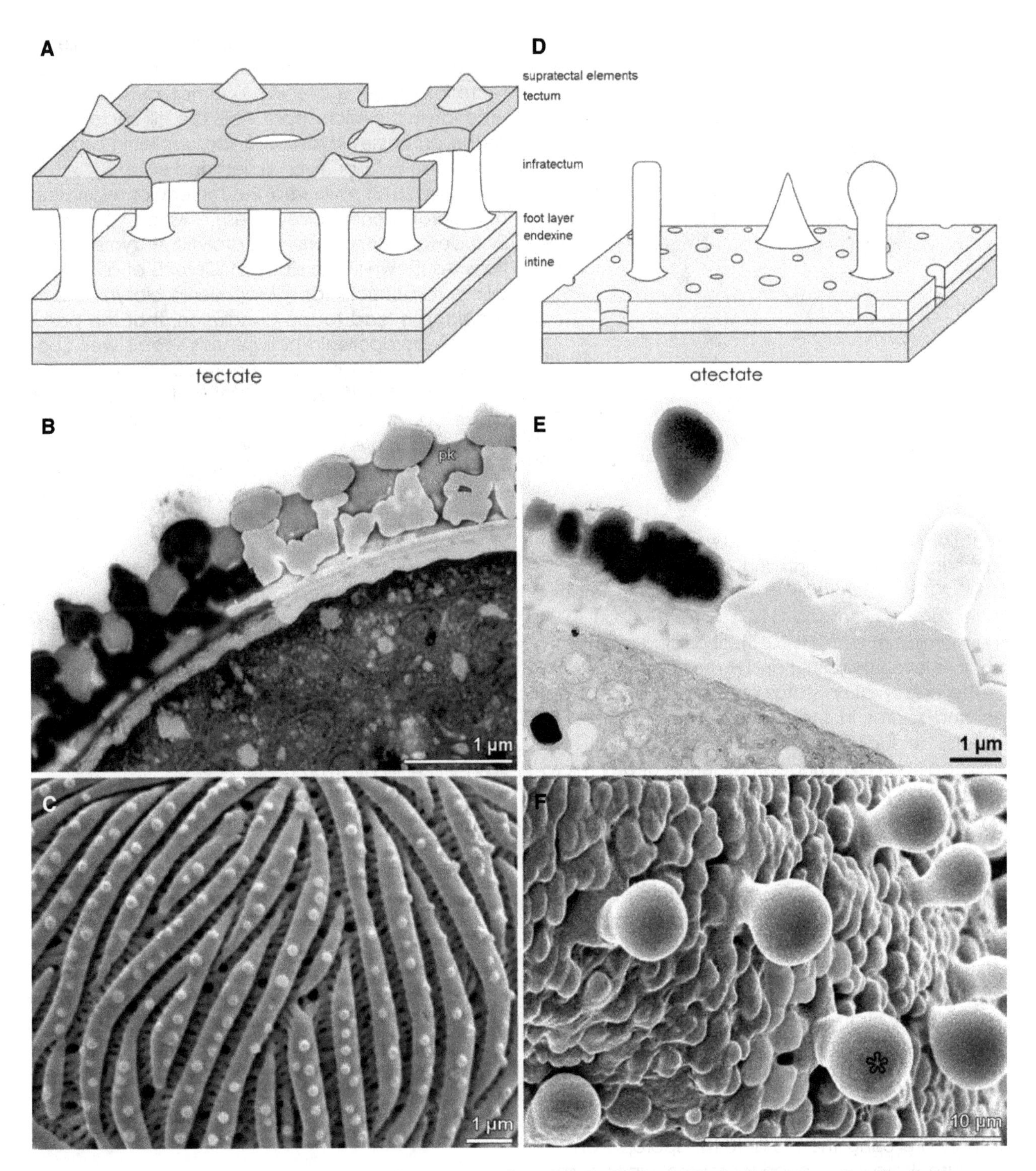

Fig. 16 Tectate vs atectate pollen wall. A-C. Tectate pollen wall. **A.** 3D-model. **B.** *Saxifraga scardica*, Saxifragaceae, cross section showing pollen wall stratification: tectum with internal tectum and supratectal elements, columellate infratectum, very thin footlayer, thin compact-continuous endexine, monolayered intine (colors refer to picture **A**), *pk* pollenkitt. **C.** *Saxifraga scardica*, Saxifragaceae, exine surface in SEM, sculpture striate with nanoechinate suprasculpture (colored). **D-F.** Atectate pollen wall. **D.** 3D-model. **E.** *Iris pumila*, Iridaceae, cross section showing pollen wall stratification: tectum and infratectum lacking, compact-continuous footlayer, monolayered intine (colors refer to picture **D**). **F.** *Iris pumila*, Iridaceae, exine surface (foot layer) in SEM, sculpture verrucate (colored) and clavate (asterisk)

decay as well as to chemical and mechanical damage (e.g., Steemans et al. 2010). However, in the environment, both biotic and abiotic factors are involved in pollen decomposition. Biotic factors are, for instance, the intrusion of bacteria and fungi (e.g., Elsik 1971; Havinga 1971, 1984; Skvarla et al. 1997; Phuphumirat et al. 2011). Abiotic factors include the pH-value of the substrate (e.g., Bryant and Hall 1993),

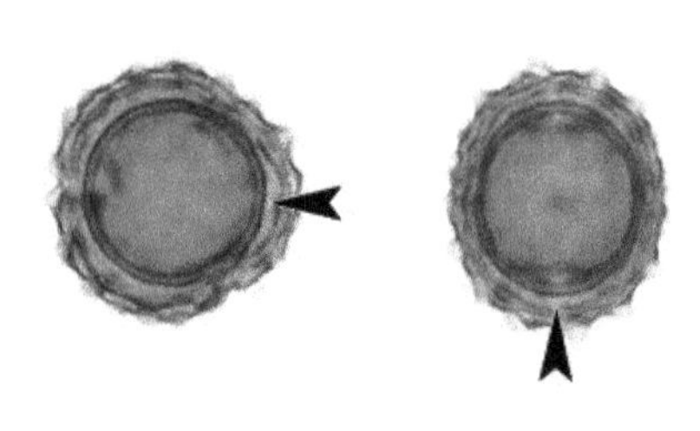

Fig. 17 Cavea. *Xanthium spinosum*, Asteraceae, acetolyzed pollen in polar (left) and equatorial (right) view showing exine cavity (cavea) between sexine and nexine (arrowheads)

oxidation/reduction (e.g., Twiddle and Bunting 2010), autoxidation by UV-light and oxygen (e.g., Jardine et al. 2015), destruction due to mechanic impact, water or fire (Cushing 1967; Bryant et al. 1994; Phuphumirat et al. 2011, 2015), and rapid changes in moisture levels (Halbritter and Hesse 2004).

The preservation status and the amount of pollen and spores in sediments depends on several factors, including rapid anaerobic burial and embedding in mud or peat, absence of any microbial destruction or sapropel, and the exclusion of oxygen (Klaus 1960, 1987; Playford and Dettmann 1996; Traverse 1988, 2007).

Recent studies on the **composition of sporopollenin** suggest that it may have two different types of chemical structures, oxygenated aromatic compounds and aliphatic compounds (e.g., Wiermann et al. 2001; Dobritsa et al. 2009; Gabarayeva and Grigorjeva 2010; Gabarayeva et al. 2010; Steemans et al. 2010; Colpitts et al. 2011). Although its exact structure remains unknown, sporopollenin is believed to compose oxidative polymers of carotenoids, polyunsaturated fatty acids, and conjugated phenols (Diego–Taboada et al. 2014). Some authors are using the plural form "sporopollenins," because there is evidence for several types of sporopollenin in ferns, gymnosperms, and angiosperms (Hemsley et al. 1993; de Leeuw et al. 2006). According to Diego–Taboada et al. (2014) sporopollenin in plants share a common aliphatic core, but depending on the taxon, contain different aromatic side chains. The **chemical constitutional formula** of sporopollenin is also unknown. The **empirical formula** of sporopollenin has highly variable amounts of H- and O-numbers. A generalized formula is $C_{90}H_{142}O_{36}$ (Traverse 1988; Riding and Kyffin–Hughes 2004).

The precise location of **synthesis** of sporopollenin precursors in tapetal cells and the mechanisms of secretion of sporopollenin monomers before polymerization in the microspore walls are still unclear, just as the processes involved in sporopollenin production at the cellular level (Lallemand et al. 2013). Liu and Fan (2013) reviewed the molecular regulation of sporopollenin biosynthesis, which probably includes a framework of catalytic enzyme reactions. As shown in the study by Colpitts et al. (2011), genes responsible for sporopollenin biosynthesis in *Arabidopsis* lead to the conclusion, that the pathway of sporopollenin biosynthesis seems well conserved in land plants since nearly 500 mya.

The question if sporopollenin is of sporophytic or gametophytic origin is still controversial. Probably both sources are involved. Most authors agree that sporopollenin is predominantly produced by the tapetum (Pacini and Franchi 1991; Blackmore et al. 2000; Wallace et al. 2015; Ariizumi and Toryama 2011; Quilichini et al. 2014).

The investigation of fossil pollen and spores revealed that **fossilized sporopollenin** appears chemically very different to sporopollenin found in modern plants (Fraser et al. 2011). During fossilization (coalification) and by diagenetic processes the chemical composition of sporopollenin is modified. Especially at high temperatures, above 200 °C, sporopollenin undergoes a series of chemical changes (Yule et al. 2000; Fraser et al. 2014).

Sporopollenin biochemistry appears to have remained relatively stable since at least the Middle Pennsylvanian (approx. 310 mya). Fraser et al. (2012, 2014) postulated that the structure of sporopollenin has remained constant since plants invaded land during the Middle Ordovician (470-458 mya). A recent comprehensive review on sporopollenin and other biopolymers (de Leeuw et al. 2006) suggests that there may have been multiple forms and configurations of sporopollenin over geological time.

The sporopollenin wall is regarded as a synapomorphy in land plants and allowed land dispersal during the Silurian, perhaps already during the Middle Ordovician (Rubinstein et al. 2010; Wellman 2010).

Chemically Related Biomacromolecules

Sporopollenin is not unique in pollen/spore walls. Cell walls of some algae and dinoflagellates may contain chemically related biomacromolecules, named **algaenan** and **dinosporin** (Versteegh et al. 2012; Bogus et al. 2012). Like sporopollenin these resistant biomacromolecules may also fossilize. They have been reported in, e.g., *Chlorella* (He et al. 2016), *Spirogyra* (Simons et al. 1983), and *Coleochaete* (Ueno 2009). Furthermore, "sporopollenin-like" biomacromolecules have been found in megaspores and "massulae" of water ferns (Salviniales) (van Bergen et al. 1993), as well as in fruiting bodies of cellular slime molds (Maeda 1984).

The Angiosperm Pollen Wall

In angiosperms the **ektexine** consists in general of **tectum, infratectum,** and **foot layer.** The outer layer, the more-or-less continuous tectum, can be covered by **supratectal elements.** The infratectum beneath is **columellate** or **granular** (a second layer of columellae may form an internal tectum). However, as, e.g., Doyle (2005) has pointed out intermediate conditions are common. Even the alveolate infratectum, that by definition is restricted to gymnosperms, can also be found in some angiosperms (see "Illustrated Pollen Terms"). The foot layer may be either continuous, discontinuous or absent. The **endexine** can be described as continuous or discontinuous, spongy or compact, overall present, in apertures only, or even completely absent. Some typical deviations of the wall thickness are termed: **arcus, annulus, tenuitas** (see "Illustrated Pollen Terms") and **costa** (a thickening of the nexine/ endexine bordering an endoaperture; Fig. 18).

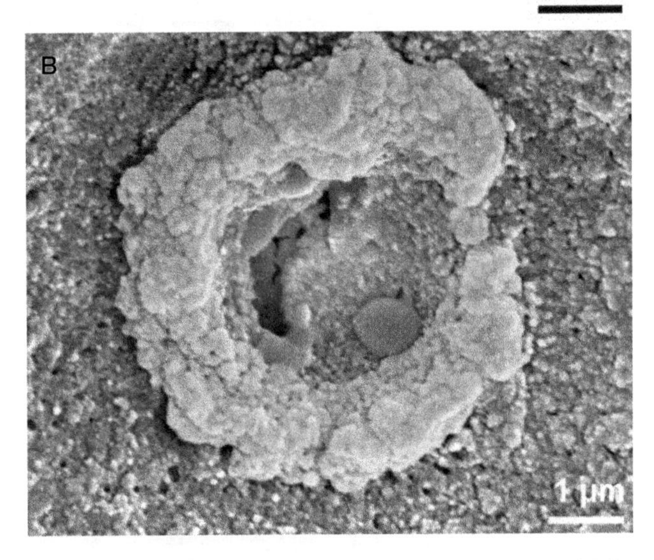

Fig. 18 Costa. A. *Nyssa* sp. Nyssaceae, fossil, middle Miocene, Austria, equatorial view (costa highlighted). **B.** *Austobuxus nitidus*, Picrodendraceae, view on the thickening around the endoaperture on the inner side of the wall

The Gymnosperm Pollen Wall

The gymnosperms comprise cycads, *Ginkgo*, conifers and Gnetales. The basic stratification (ektexine, endexine, and intine) of the gymnosperm pollen wall is identical to that of angiosperms. Still, the gymnosperm pollen wall differs from that of an angiosperm by having (1) a lamellate endexine in mature pollen, and (2) an infratectum that is never columellate (Van Campo and Lugardon 1973). The infratectum is either **alveolate** or **granular.**

A special terminology applies to saccate pollen, i.e. in Pinaceae and Podocarpaceae (Fig. 19). **Saccus** is an exinous expansion forming an air sac, with an alveolate infratectum. **Corpus** is the central body of a saccate pollen grain. **Cappa** is the thick walled proximal face of the corpus. **Leptoma** in conifer pollen refers to a thinning of the pollen wall on the distal face, presumed to function as germination area. Most frequently, two sacci are present (e.g., *Abies, Pinus, Picea;* Pinaceae), in some taxa even three (*Dacrycarpus, Microstrobus;* Podocarpaceae), or only a single one (*Tsuga;* Pinaceae).

The function and evolutionary significance of saccate pollen have been subject of much confusion. The sacci of Pinaceae and Podocarpaceae are reported to play an aerodynamic role, thus being of adaptive significance for wind pollination (Schwendemann et al. 2007; Grega et al. 2013). In fact, their functional role is to float in a liquid pollination droplet towards the ovule ("flotation hypothesis" by Leslie 2010). The flotation system is interpreted as ancestral in conifers. The absence of sacci in, e.g., Cupressaceae and Taxaceae might reflect the loss of "drop mechanism," correlated with the change of pollination mode (shift to upwards orientation of the ovules) (Doyle 2010).

In *Pinus,* pollen can be grouped into two morphotypes (Fig. 20) of systematic value (Grímsson and Zetter 2011). The *Pinus* subgenus *Strobus* **(haploxylon) type** is characterized by pollen grains with broadly attached half-spherical air sacs—in LM the leptoma shows dotted thickenings (seen as dark spots). The *Pinus* subgenus *Pinus* **(diploxylon) type** is characterized by pollen grains with narrowly attached, spherical air sacs often with nodula on nexine area—the leptoma does not show any thickenings.

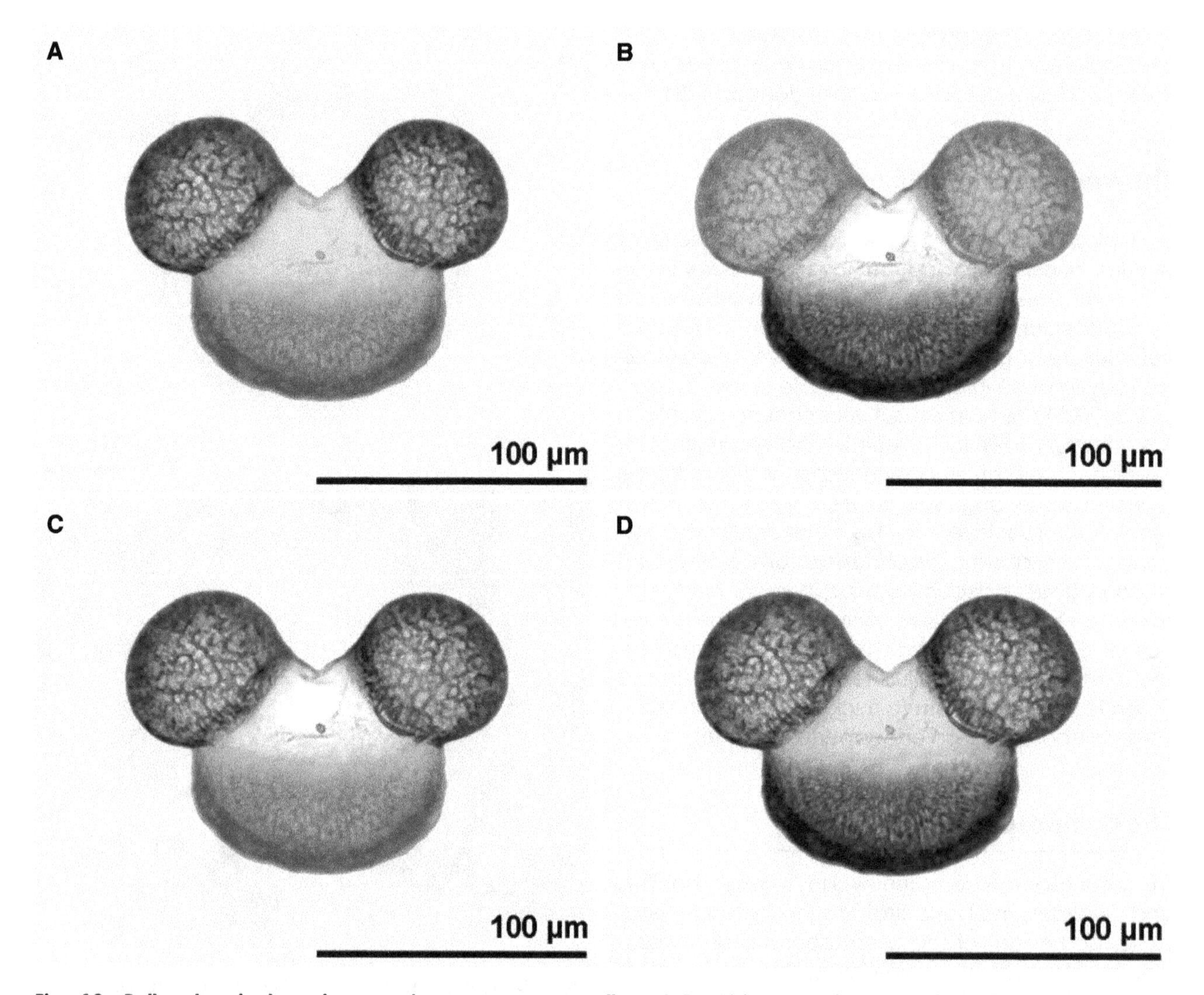

Fig. 19 Pollen terminology in saccate gymnosperm pollen. A-D. *Abies* sp., Pinaceae, bisaccate pollen, fossil, Quaternary, Austria, equatorial view. **A.** Corpus highlighted. **B.** Sacci highlighted. **C.** Cappa highlighted. **D.** Leptoma highlighted

Sculpture: Ornamentation

The terms ornamentation and sculpture applies to surface features of a pollen. The term sculpture is restricted by some authors to surface features in tectate pollen grains (e.g., Praglowski 1975; Punt et al. 2007). **Sculpture elements** (areola, clava, echinus, foveola, fossula, granulum, gemma, plicae, reticulum, rugulae, striae, verruca) can be extremely variable in both size and shape. Based on size many sculpture/ornamentation elements smaller than 1 µm can be described with the prefix micro- (1–0.5 µm) or nano- (0.5–0.1 µm). Also, the boundary between two ornamentation types can be diffuse. For example, "gemmae" and "clavae" are very variable and sometimes hard to differentiate. Combinations of different sculpture/ornamentation elements are common, such as the combination reticulate and foveolate, or echinate and perforate. With a combined sculpture, the pollen ornamentation should then be described in a defined order, with the most eye-catching feature mentioned first, followed by the others. For example, *Aristolochia* pollen is verrucate-perforate, as the verrucae are more prominent than the small perforations (Fig. 21). In the Caryophyllaceae, there are numerous, more-or-less regularly arranged microechini and perforations. In some taxa the microechini are more prominent (microechinate-perforate), in others the perforations (perforate-microechinate) (Fig. 22). In case none of the features are eye-catching, the dominant feature might be a subjective decision of the palynologist e.g., in taxa, where two features are on a par (microechinate and perforate). A more complex example is *Sanchezia nobilis* (Acanthaceae, Fig. 23): is it plicate and reticulate? Should the rod-like elements

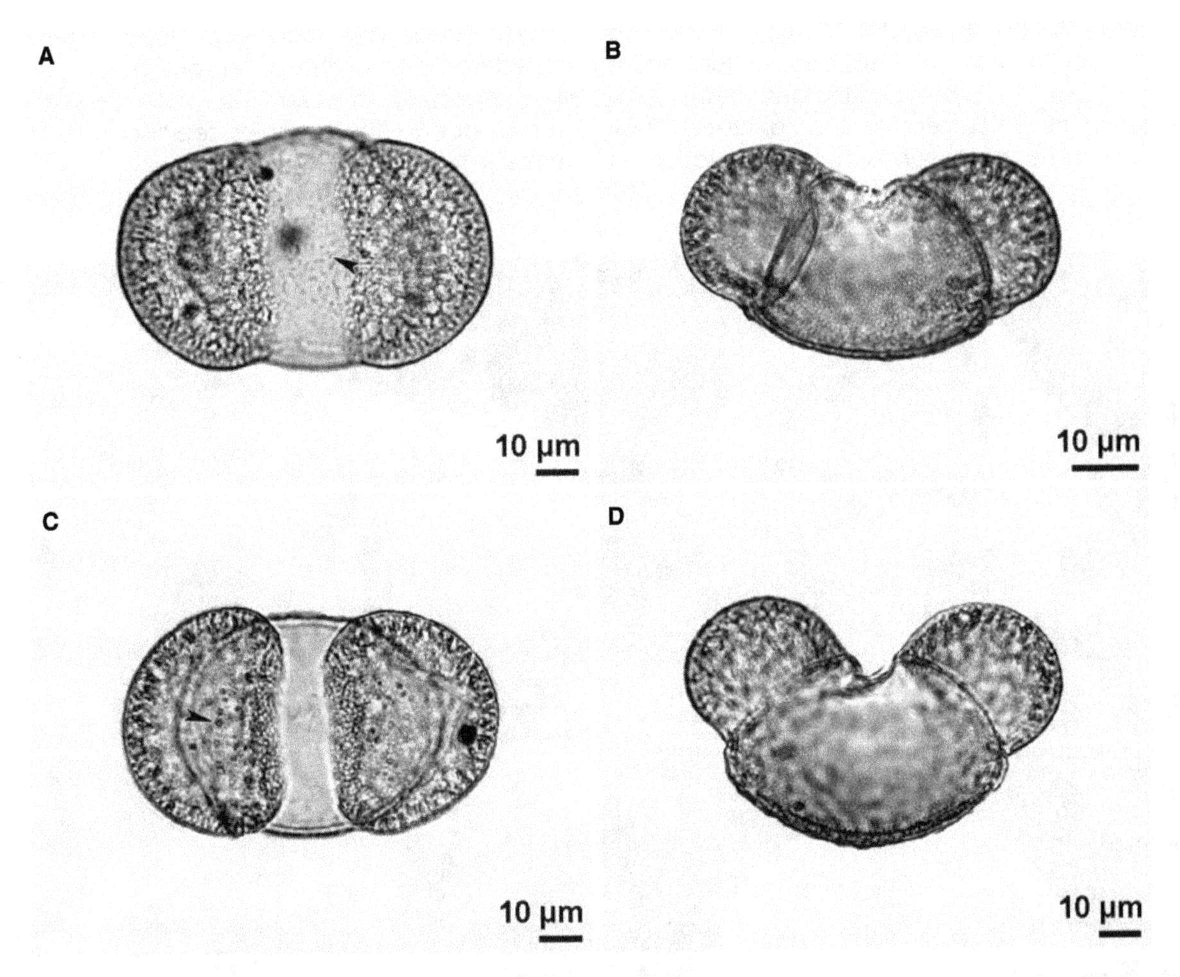

Fig. 20 Pollen types in saccate Pinus pollen (fossil, middle Miocene, Austria). **A.** *Pinus* subgenus *Strobus* (haploxylon), polar view, thickenings (arrowhead). **B.** *Pinus* subgenus *Strobus* (haploxylon), equatorial view. **C.** *Pinus* subgenus *Pinus* (diploxylon), polar view, nodula (arrowhead). **D.** *Pinus* subgenus *Pinus* (diploxylon), equatorial view

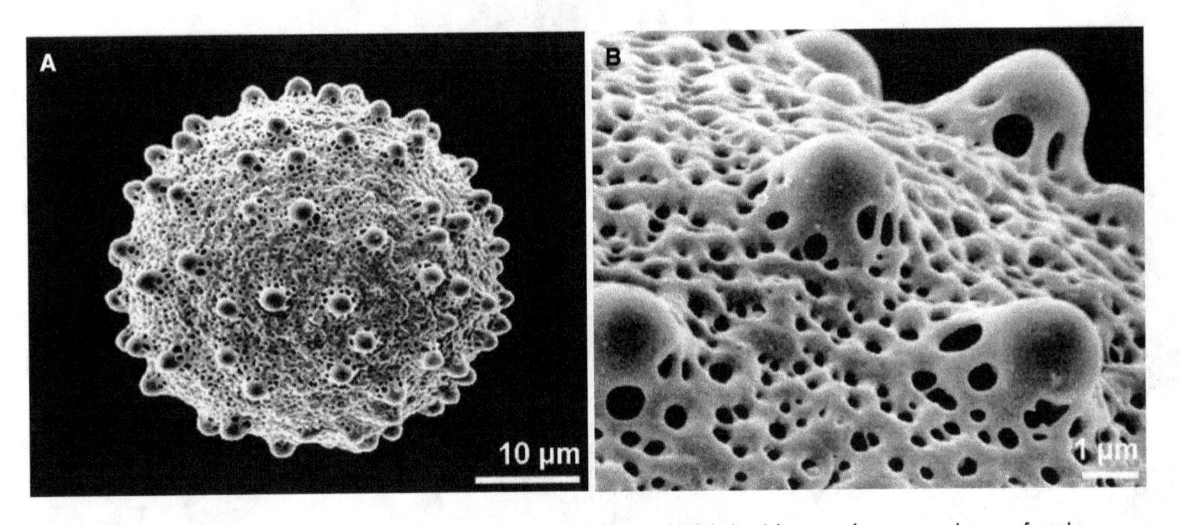

Fig. 21 Combined sculpture elements. **A-B.** *Aristolochia arborea* (Aristolochiaceae), verrucate, perforate

be termed clavae or free-standing columellae? Is the aperture a porus or a colporus? *PalDat* (www.paldat.org) might provide the answers?

Sculpture/ornamentation elements are often deviating and can be distributed regularly or irregularly over the pollen surface, restricted or absent from distinct areas (polar vs equatorial, interapertural vs aperture area; Fig. 24).

Ubisch bodies (orbicules) are sporopollenin elements produced by the tapetum. Ubisch bodies are usually found as isolated particles lining the mature locular wall, or between pollen grains (Huysmans et al. 1998; Halbritter and Hesse 2005; Vinckier et al. 2005; Verstraete et al. 2014). They often resemble the pollen wall ornamentation. In Cupressaceae and Taxaceae, Ubisch bodies are considered part of the pollen ornamentation and are especially frequent on the leptoma of Cupressaceae (for examples, see "Illustrated Pollen Terms").

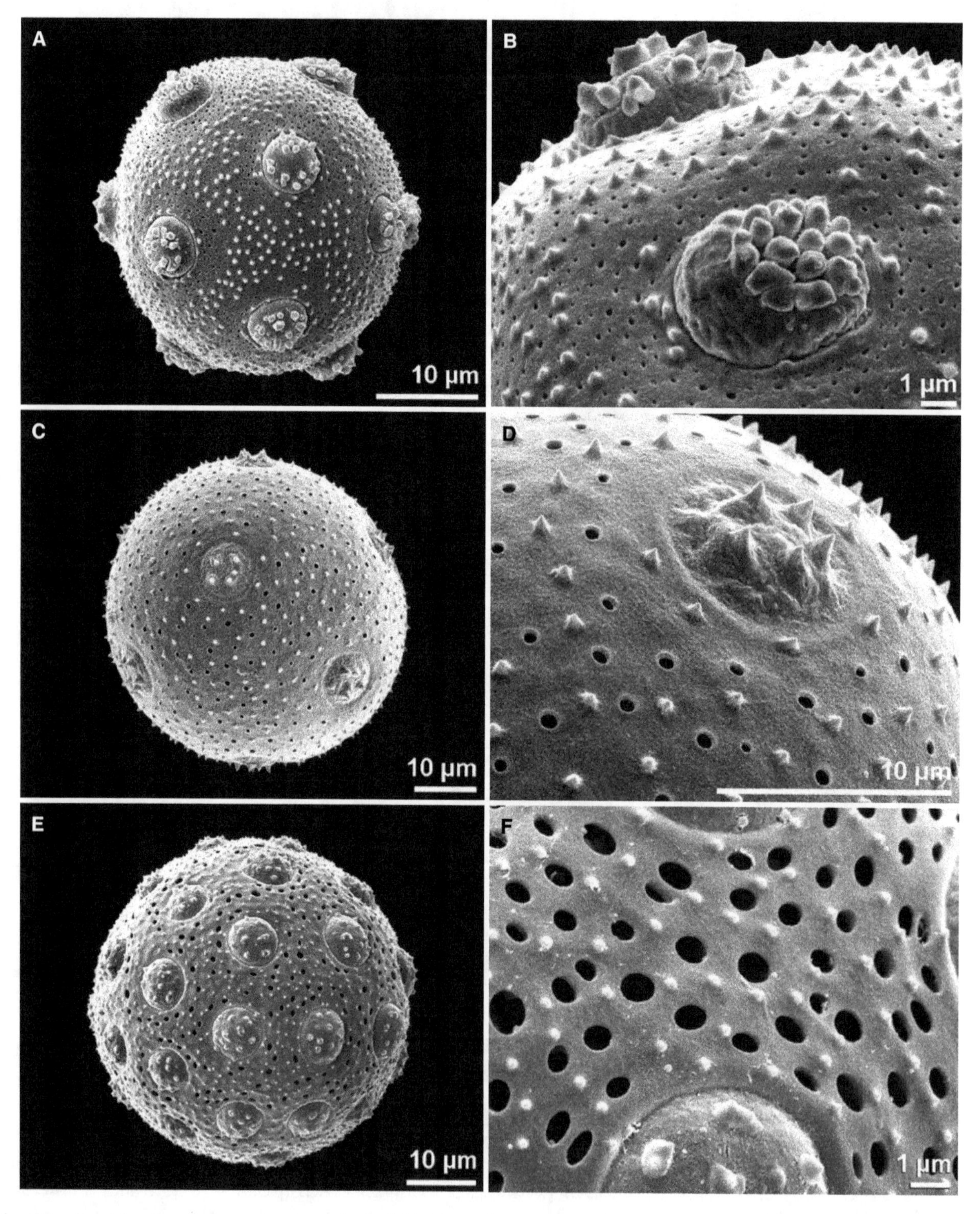

Fig. 22 Combined sculpture elements. A-B. *Stellaria media*, Caryophyllaceae, microechinate and perforate. **C-D.** *Saponaria officinalis*, Caryophyllaceae, microechinate and perforate. **E-F.** *Silene succulenta*, Caryophyllaceae, perforate and nanoechinate

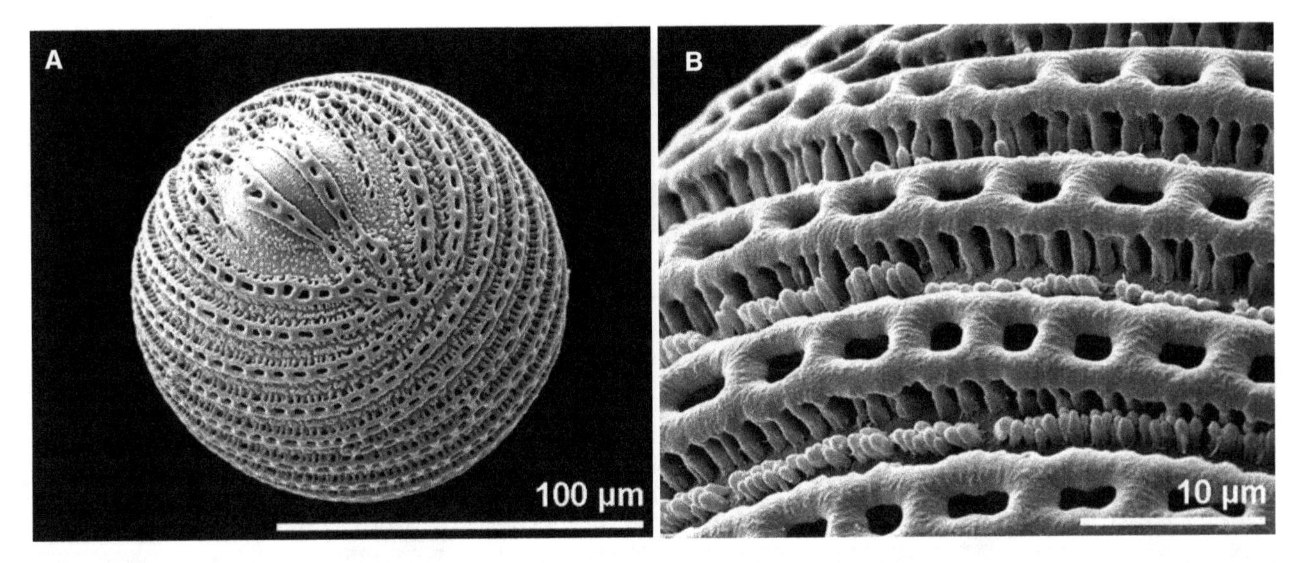

Fig. 23 **Interpretation of sculpture elements. A-B.** *Sanchezia nobilis*, Acanthaceae, oblique equatorial view and surface detail

Ornamentation in LM vs. SEM

An accurate description of pollen ornamentation depends on the optical magnification used and particularly on the point resolution. Even the SEM at low resolution may not be sufficient to distinguish pollen grains unequivocally (see "Methods in Palynology"). Depending on the type of microscope used for pollen analysis, some pollen features may remain hidden. For LM studies, the term **scabrate** is used, describing minute sculpture elements of undefined shape and size close to the resolution limit of the LM. For example, *Juglans* pollen is scabrate in LM as well as under low magnification SEM, but is nanoechinate at high resolution SEM (Fig. 25A-B).

The descriptive terms may differ whether LM or SEM is used and should be described for both. For example, *Ulmus* pollen seen in LM is described as **verrucate**. Using low SEM magnification the ornamentation is **rugulate to verrucate** (Fig. 25C-E). High SEM magnification shows additional **granula** (≤0.1 µm).

Another example for different interpretations in LM vs SEM is the term **psilate**. Many pollen grains that appear psilate in LM show a distinct ornamentation using high SEM magnification. For example, pollen of *Allium ursinum* is psilate in LM, but is striate and perforate in SEM (Fig. 25F-G).

Terms with nano- or micro- can only be observed in SEM (see "Methods in Palynology"). For example, the term **granulate** should only be used when describing pollen ornamentation under SEM. When minute sculptural elements are observed under high resolution SEM, it is possible to distinguish real "granula" (sculpture element of different/indefinable shape, ≤ than 0.1 µm) from other nano- and/or micro-sculpture elements. For example, the allegedly granulate ornamentation of many Poaceae is in fact nanoechinate, the pointed ends of the echini are seen best in profile and not from top view (see "Illustrated Pollen Terms").

Role of Pollen Ornamentation in Pollination

Depending on the pollination mode the outer pollen wall may be either highly ornamented, often with plenty of pollen coatings (mainly pollenkitt; Pacini and Hesse 2005), or with a more or less psilate pollen surface. The pollen wall of zoophilous plants, as well as autogamous plants, is usually highly ornamented and the thick exine consists of high amounts of sporopollenin (Fægri and Iversen 1989). Pollen of anemophilous plants are known to have less ornamentation and less sporopollenin (Friedman and Barrett 2009). Usually psilate pollen in temperate and boreal zones is indicative for anemophily (Fægri and Iversen 1989), whereas in the tropics it is also indicative for zoophily (Furness and Rudall 1999). For example, in Aroideae (e.g., *Montrichardia*, *Dieffenbachia*, *Philodendron*, *Gearum*) psilate pollen usually equipped with pollenkitt is adapted for entomophily (Weber and Halbritter 2007).

Functional Value of Exine Reduction

Layers of the basic pollen wall type may vary and be partly or totally reduced (for examples, see

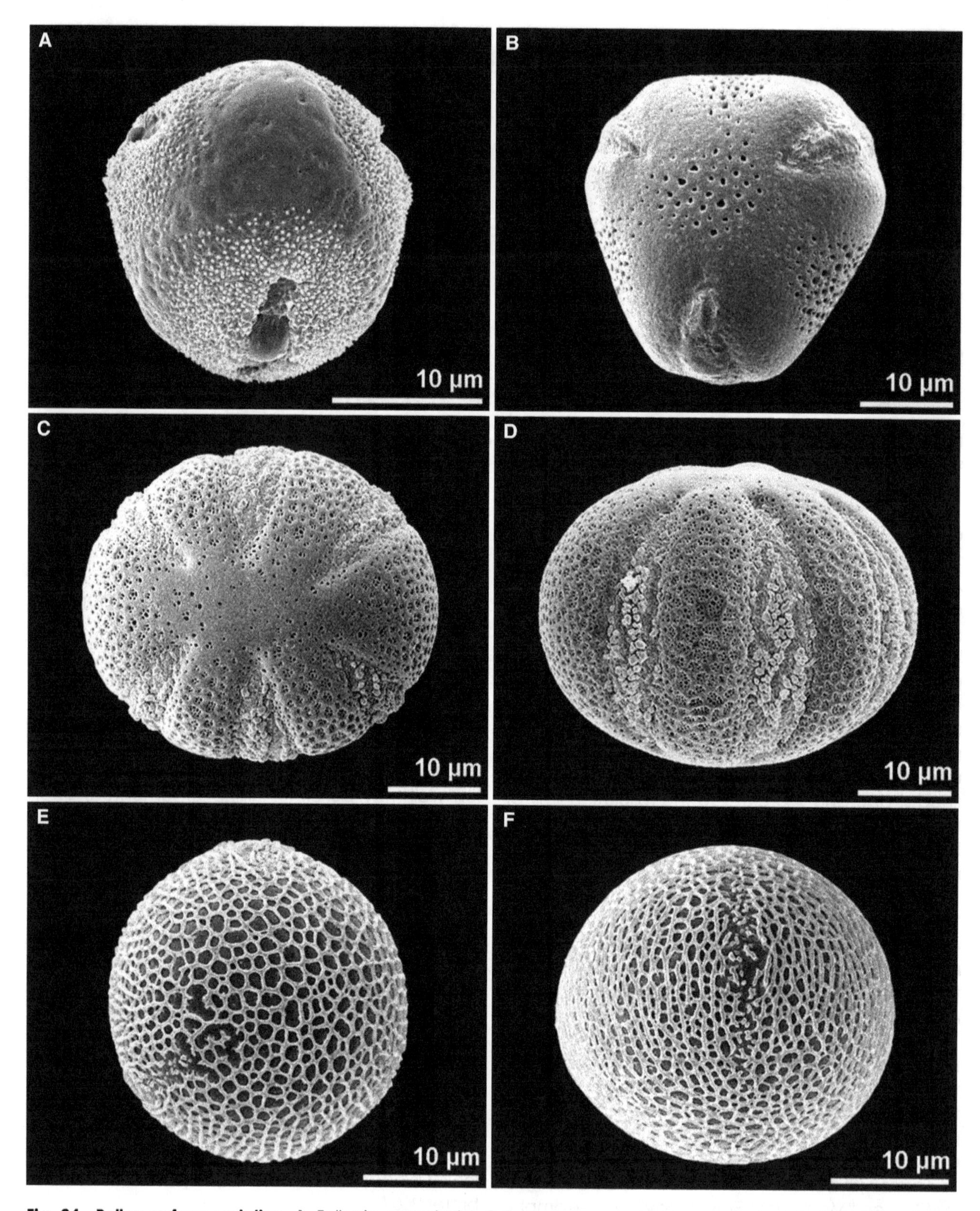

Fig. 24 Pollen surface variation. A. *Fallopia convolvulus*, Polygonaceae, polar view, polar area psilate to perforate and regions around apertures microechinate. **B.** *Sideritis montana*, Lamiaceae, polar view, polar and interapertural areas perforate to foveolate and regions around apertures psilate. **C-D.** *Salvia austriaca*, Lamiaceae, pollen bireticulate, except psilate polar areas (polar and equatorial view). **E-F.** *Solandra longiflora*, Solanaceae, polar area reticulate, equatorial region striato-reticulate (polar and equatorial view)

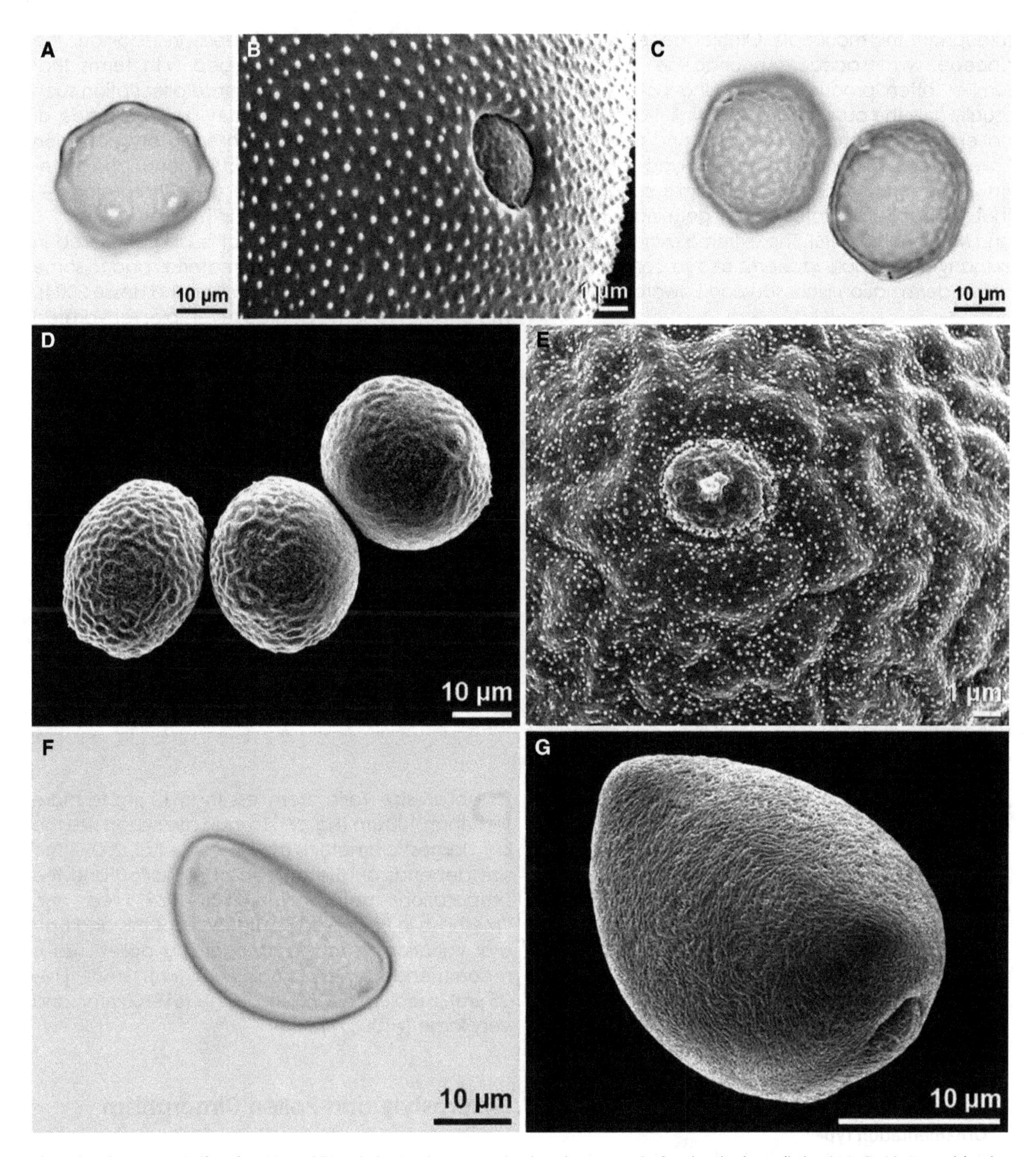

Fig. 25 Ornamentation in LM vs SEM. A-B. *Juglans* sp., Juglandaceae. **A.** Scabrate to psilate, LM. **B.** Nanoechinate, SEM. **C-E.** *Ulmus laevis*, Ulmaceae. **C.** Rugulate, LM. **D.** Rugulate to verrucate, low magnification, SEM. **E.** Verrucate, granulate, high magnification SEM. **F-G.** *Allium ursinum*, Amaryllidaceae. **F.** Psilate to scabrate, LM. **G.** Rugulate-perforate, low magnification, SEM

"Pollen Wall" in "Illustrated Pollen Terms"). The sporopollenin ektexine is lacking e.g., in some genera of Monimiaceae and Lauraceae (Walker 1976), in the aquatic Ceratophyllaceae (Takahashi 1995), in many genera of Aroideae, and in the inaperturate filiform pollen of seagrasses, *Posidonia*. An absent exine is an adaptation to hydrophily and correlated with, e.g., aquatic habits, anemophily, and pollinia (Furness 2007). Interestingly, exine reduction has evolved iteratively in angiosperms, especially throughout the monocots. Orchidaceae, Asclepiadaceae, Mimosaceae, Annonaceae, and other families often produce compound pollen, where usually only the outermost pollen wall show the typical ektexine structure with tectum and columellae. Pollen grains within calymmate polyads or tetrads have extremely reduced and fragile pollen walls, that probably facilitates pollen germination (Knox and McConchie 1986). The extreme exine reduction in many orchid pollinia seems also to correlate with pollen germination (Johnson and Edwards 2000).

Harmomegathy: The Harmomegathic Effect

Pollen grains are able to absorb and release water (+ various liquids); thus, each pollen grain exists in two morphologically different conditions, **dry** and **hydrated** (Fig. 26). Harmomegathic mechanisms, e.g., infolding of the pollen wall (Rowley and Skvarla 2000), accommodate the change of the osmotic pressure in the cytoplasm during hydration or dehydration. These mechanisms are denoted as harmomegathic effect, also known as Wodehouse effect. The main purpose of the harmomegathic effect is to protect the male gametophyte against desiccation during pollen presentation and dispersal, and is often related to pollination biology.

In mature anthers, pollen is turgescent before shedding. After anther dehiscence and during pollen presentation, water loss takes place and the pollen grain becomes typically infolded. Various pollen wall features are involved in the harmomegathic effect:

- **Position, number, and type of apertures**: the most important features
- **Thinned or thickened regions within the pollen wall**: in particular, internal belts or endoapertures. If the ektexine is considerably reduced, its role is taken over by other wall strata, namely, by a thick endexine or intine. On the other hand, if the exine is extremely rigid, then the harmomegathic effect is only marginal
- **Ornamentation type**
- **Pollen size**: small, thin-walled pollen grains which are usually less infolded
- **Pollen coatings**: if abundant, pollen coatings have an insulating influence that reduces the harmomegathic effect

The combination of these features is influencing the mode of infolding. Terms used for common morphotypes of dry pollen include: apertures sunken, boat-shaped, cup-shaped, interapertural area infolded, irregularly infolded, not infolded. In addition, the pollen shape can be described with terms that might be helpful for an adequate description such as barrel-like, disk-like, or kidney-like. The mode of infolding and/or shape of pollen in dry condition may be typical for a family and/or genus and therefore of systematic relevance (see "Palynology — History and Systematic Aspects").

The harmomegathic effect is also observed in pollen taken from herbarium material, and to some degree in fossil material (Halbritter and Hesse 2004). This effect is to some degree reversible: rehydrated pollen at the stigma, or under laboratory conditions (various liquids), is again turgescent and largely recalls the shape before shedding. A second dehydration does not necessarily result in the typical dry shape but, if pollen walls are sufficiently stable, the harmomegathic effect can be induced several times in the same way. In pollen with thin walls, the susceptible internal structure may become damaged, and the harmomegathic effect may result in different and randomly shaped pollen. Infoldings of the pollen wall after acetolysis treatment are mostly not comparable with those observed in dry condition.

Size

Pollen **size** varies from less than 10 µm to more than 100 µm (Fig. 27). To indicate pollen size the largest diameter is used (Hesse et al. 2009). The size depends on the degree of hydration and the preparation method (Reitsma 1969, see also "Methods in Palynology"). Because of this and natural variation, a range categorizing pollen size is recommended: very small (<10 µm), small (10–25 µm), medium (26–50 µm), large (51–100 µm), and very large (>100 µm).

Heterostyly and Pollen Dimorphism

In **heterostylous** (long-styled and short-styled) species two different pollen types occur, where pollen size and number of apertures or the ornamentation may differ. In *Linum flavum* (Linaceae) pollen of the short-styled morph is baculate, and the long-styled morph clavate (Fig. 28). In *Primula veris* (Primulaceae) the pollen of the short-styled morph is larger and has more apertures than pollen of the long-styled morph (Fig. 29A). In the tristylous species *Lythrum salicaria*

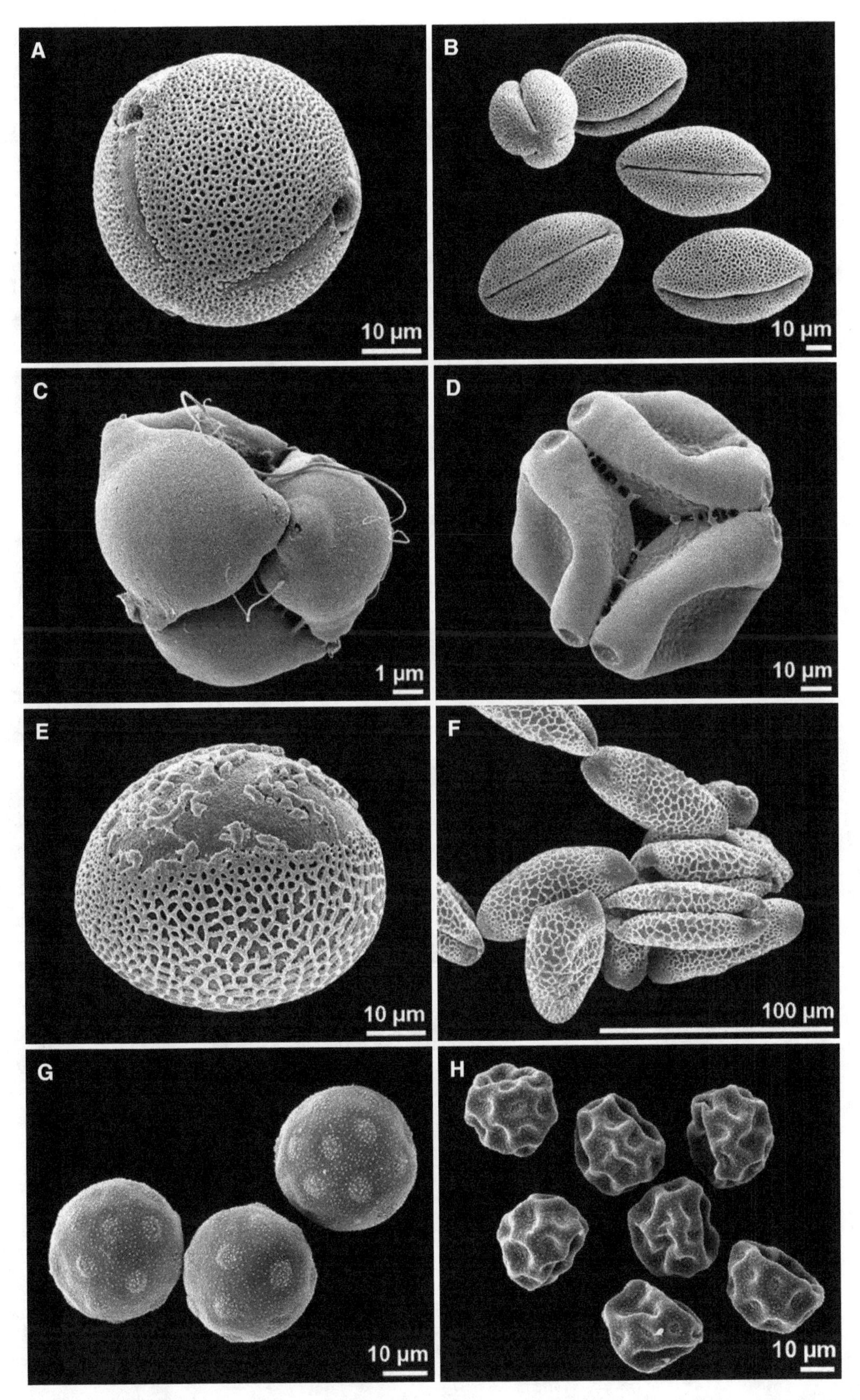

Fig. 26 Harmomegathic effect—hydrated vs dry pollen. A-B. *Cistus creticus*, Cistaceae. **A.** Spheroidal, outline circular. **B.** Prolate, outline lobate, apertures infolded. **C-D.** *Epilobium palustre*, Onagraceae, tetrad. **C.** Oblate, outline triangular. **D.** Interapertural area sunken. **E-F.** *Vriesea pabstii*, Bromeliaceae. **E.** Oblate, outline elliptic. **F.** Boat-shaped. **G-H.** *Alisma lanceolatum*, Alismataceae. **G.** Spheroidal, outline circular. **H.** Irregularly infolded

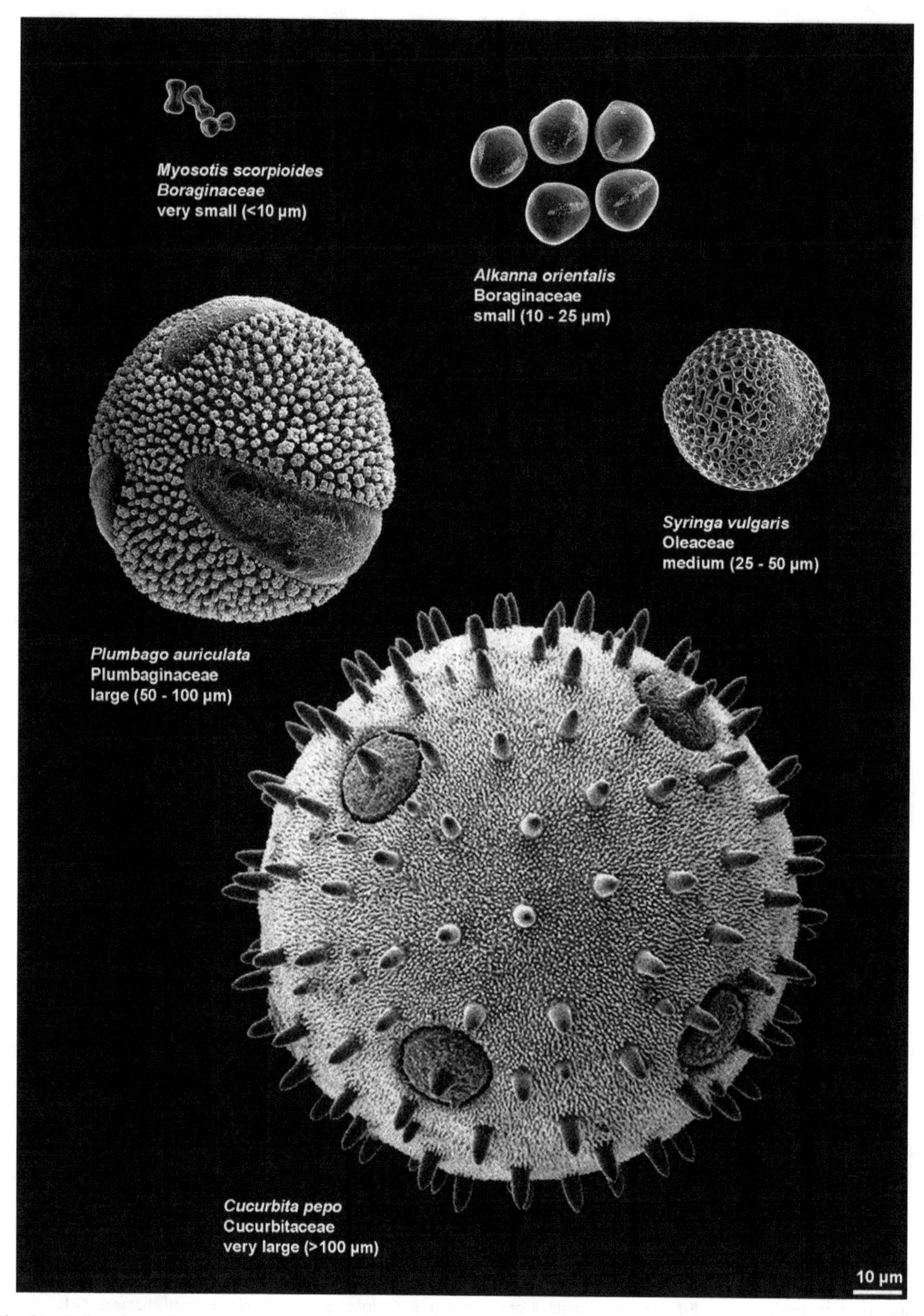

Fig. 27 Pollen size categories.

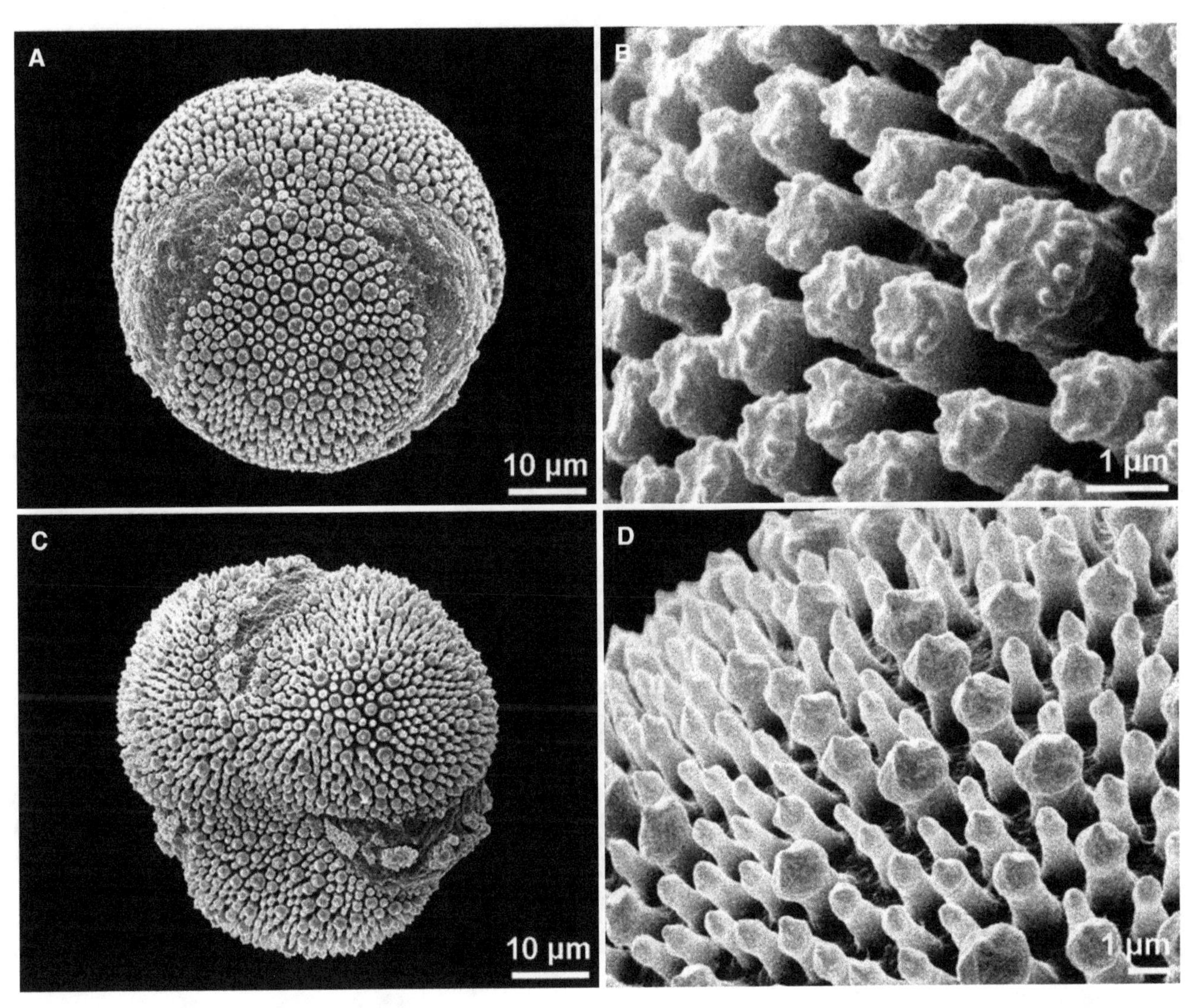

Fig. 28 Pollen dimorphism — different ornamentation. A-D. *Linum flavum*, Linaceae, **(A-B)** short-styled morph, baculate, **(B-C)** long-styled morph, clavate

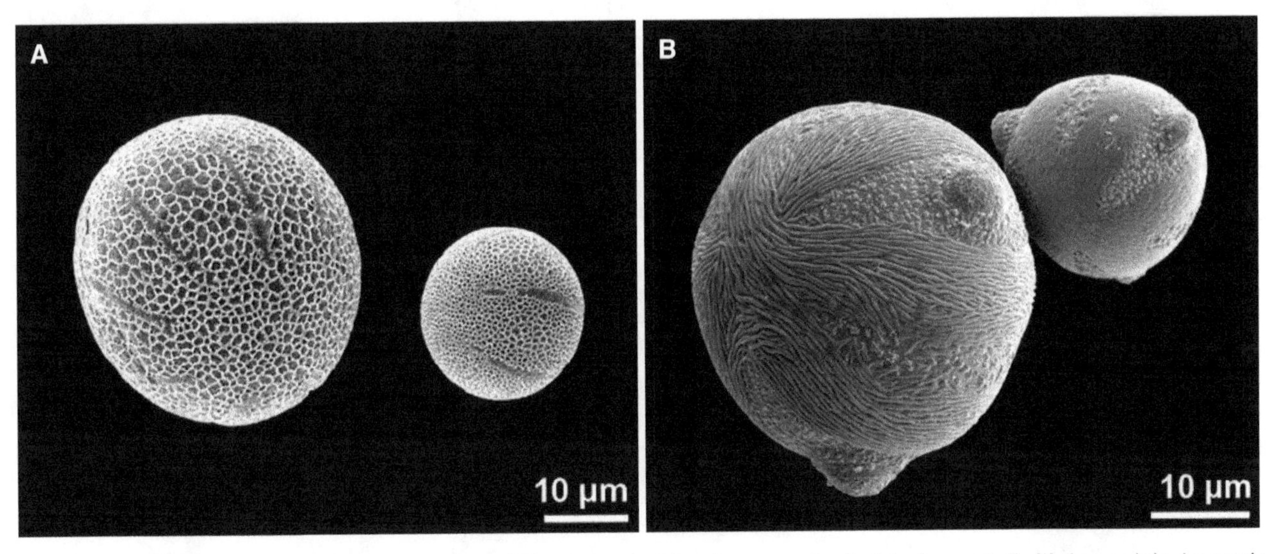

Fig. 29 Pollen dimorphism — different size. A. *Primula veris*, Primulaceae, short-styled morph (left), long-styled morph (right). **B.** *Lythrum salicaria* (Lythraceae), medium-styled morph, dimorphic pollen

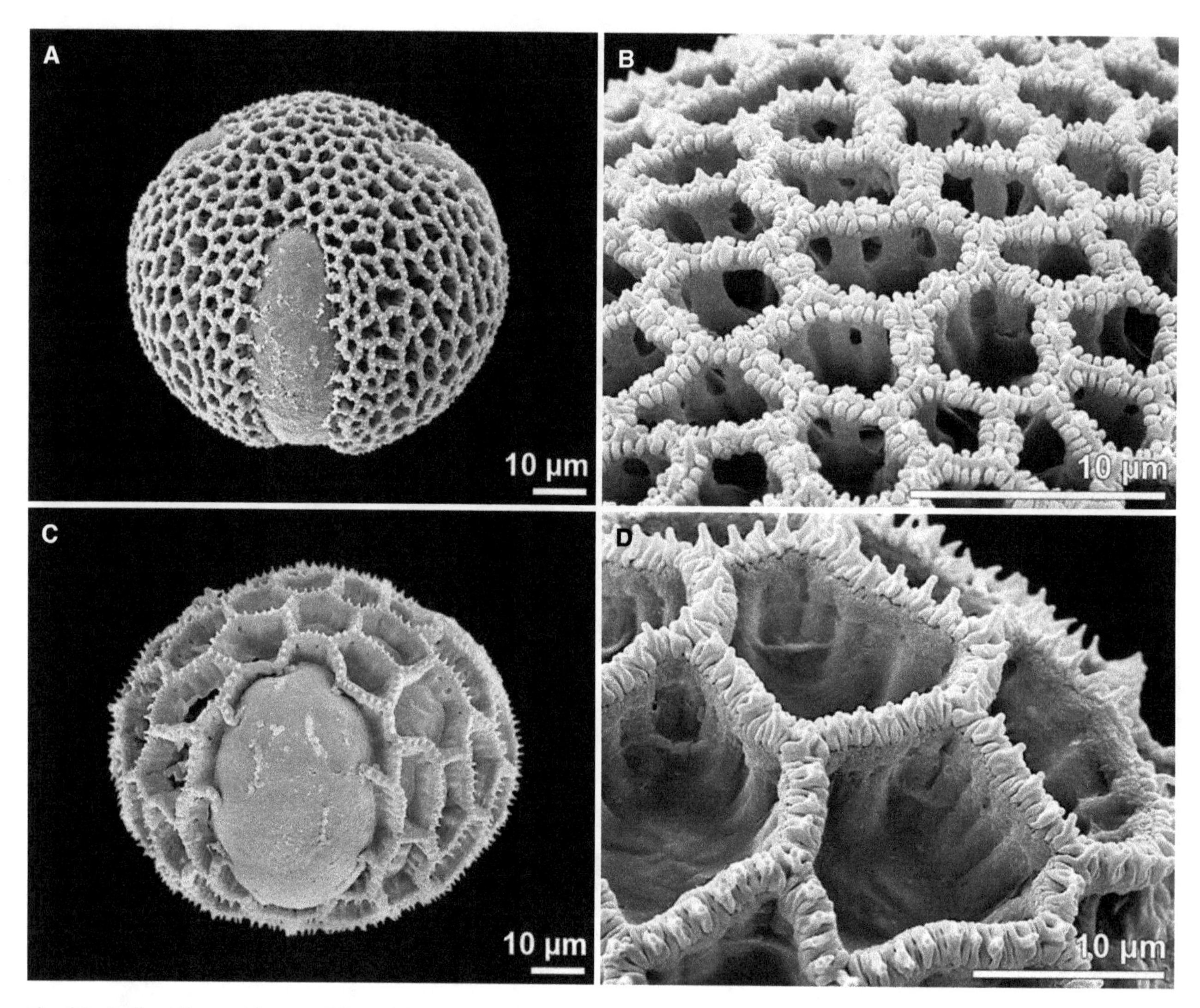

Fig. 30 Pollen dimorphism — different ornamentation. A-D. *Armeria alpina*, Plumbaginaceae. **A-B.** Morph 1, reticulate. **C-D.** Morph 2, reticulate

(Lythraceae) pollen is dimorphic, with different size, ornamentation, and even color of the pollen grains (blue and yellow). Pollen of the long-styled morph is small sized (about 20 µm), short styled morph is medium sized (about 35 µm) and medium-styled morph is small to medium sized (within a single anther) (Fig. 29B). In some Plumbaginaceae, for example in distylous species of *Armeria*, pollen dimorphism (reticulate, different size of lumina and suprasculpture elements) is correlated with dimorphic stigmatic papillae, but style and stamen lengths are monomorphic (Ganders 1979; Fig. 30).

Aberrant Pollen Grains

Aberrant pollen grains are often ignored but occur regularly in small percentages in nearly all anthers and may vary from one individual to another (Pozhidaev 2000a, b; Banks et al. 2007). These aberrant pollen grains can differ from the typical pollen type of the species in shape and dimension, in number and arrangement of apertures, and in ornamentation type (Fig. 31). Reasons for the production of deviating pollen forms are genetically (polyploidy),

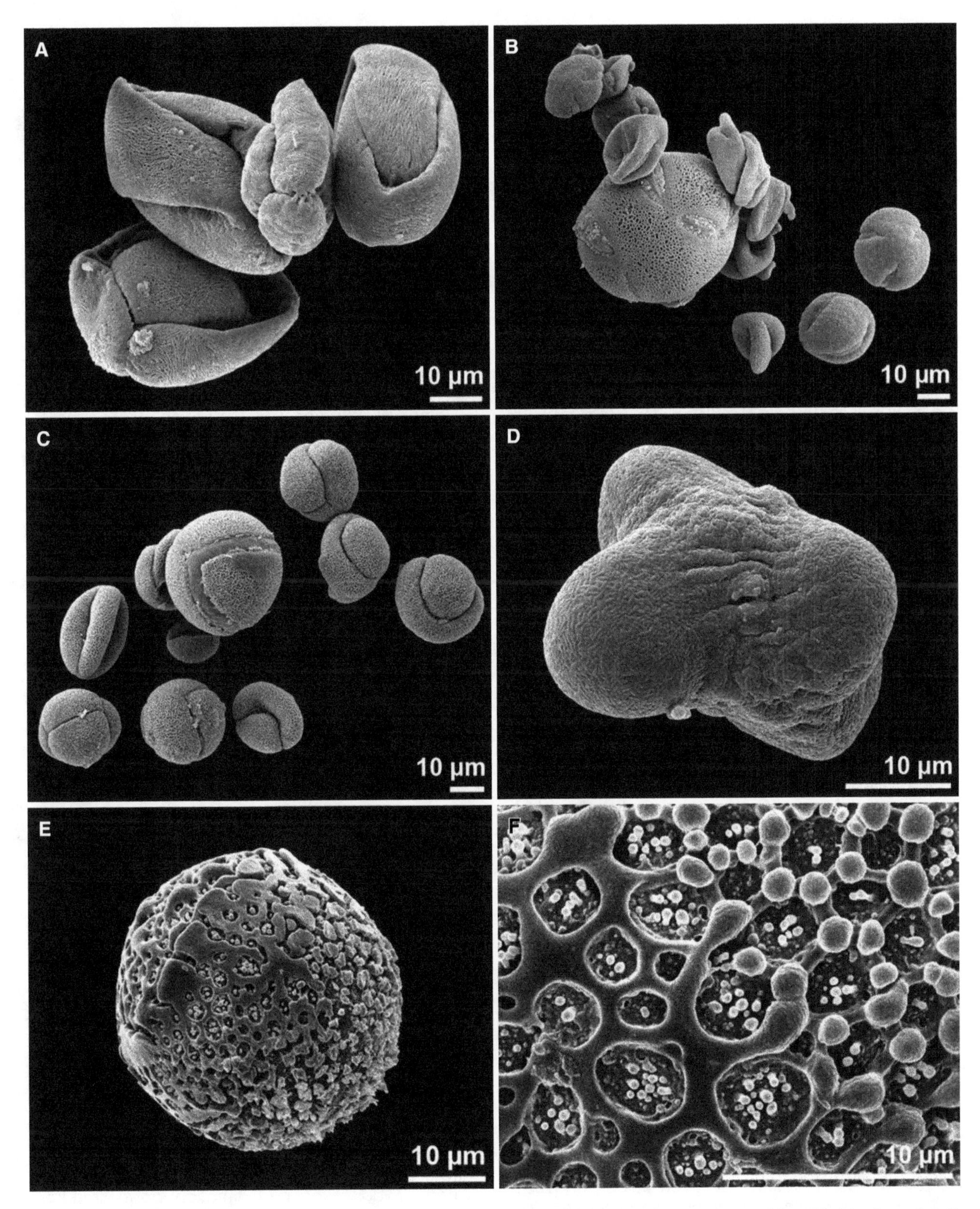

Fig. 31 Aberrant pollen grains. A. *Malus sieboldii*, Rosaceae, irregular aperture arrangement (usually tricolporate). **B.** *Oxalis* sp., Oxalidaceae, many aborted pollen grains, giant pollen (usually tricolpate). **C.** *Scaevola* sp., Goodeniaceae, pollen varies in size and aperture arrangement. **D.** *Scandix pecten-veneris*, Apiaceae, "double" pollen grain (usually tricolporate). **E.** *Codiaeum*-hybrid, Euphorbiaceae, ornamentation intermediate between parent plant species e.g., croton pattern and reticulate with free-standing columellae. **F.** *Codiaeum*-hybrid, Euphorbiaceae, surface detail

chemically, or environmentally induced. Such deviating, malformed pollen is frequently found in cultivated plants, ornamental plants, agricultural crops, annual plants, plants with asexual reproduction (autogamic plants, apomicts), and hybrids. Some species of apomicts, agricultural crops or cultivated plants (e.g., *Malus sieboldii*) produce only malformed pollen.

References

Ariizumi T, Toryama K (2011) Genetic regulation of sporopollenin synthesis and pollen exine development. Annu Rev Plant Biol 62: 437–460

Banks H, Stafford P, Crane PR (2007) Aperture variation in the pollen of *Nelumbo* (Nelumbonaceae). Grana 46: 157–163

Blackmore S, Barnes SH (1995) Garside's rule and the microspore tetrads of *Grevillea rosmarinifolia* A. Cunningham and *Dryandra polycephala* Bentham (Proteaceae). Rev Palaeobot Palynol 85: 111–121

Blackmore S, Cannon SM (1983) Palynology and systematics of Morinaceae. Rev Palaeobot Palynol 40: 207–226

Blackmore S, Takahashi M, Uehara K (2000) A preliminary phylogenetic analysis of sporogenesis in pteridophytes. In: Harley MM, Morton CM, Blackmore S (eds) Pollen and spores: morphology and biology. Royal Botanic Gardens, Kew, p. 109–124

Bogus K, Harding IC, King A, Charles AJ, Zonneveld KAF, Versteegh GJM (2012) The composition and diversity of dinosporin in species of the *Apectodinium complex* (Dinoflagellata). Rev Palaeobot Palynol 183: 21–31

Braconnot H (1829) Recherches chimiques sur le pollen du *Typha latifolia*, Lin., famille de typhacées. Ann Chim Phys 42: 91–105

Bryant VM, Hall SA (1993) Archaeological palynology in the United States: A critique. Am Antiquity 58: 277–286

Bryant VM, Holloway RG, Jones JG, Carlson DL (1994) Pollen preservation in alkaline soils of the American southwest. In: Traverse A (ed) Sedimentation of organic particles. Cambridge University Press, Cambridge, New York, Melbourne, Madrid, Cape Town, Singapore, Sao Paolo, p. 47–58

Colpitts CC, Kim SS, Posehn SE, Jepson C, Kim SY, Wiedemann G, Reski R, Wee AGH, Douglas CJ, Suh D-Y (2011) PpASCL, a moss ortholog of anther-specific chalcone synthase-like enzymes, is a hydroxyalkylpyrone synthase involved in an evolutionarily conserved sporopollenin biosynthesis pathway. New Phytol 192: 855–868

Copenhaver GP (2005) A compendium of plant species producing pollen tetrads. J North Carolina Acad Sci 12: 17–35

Cushing EJ (1967) Evidence for differential pollen preservation in late Quaternary sediments in Minnesota. Rev Palaeobot Palynol 4: 87–101

De Leeuw JW, Versteegh GJM, Van Bergen PF (2006) Biomacromolecules of algae and plants and their fossil analogues. Plant Ecology 182: 209–233

Diego-Taboada A, Beckett ST, Atkin SL, Mackenzie G (2014) Hollow pollen shells to enhance drug delivery. Pharmaceutics 6: 80–96

Dobritsa AA, Shrestha J, Morant M, Pinot F, Matsuno M, Swanson R, Lindberg Møller B, Preuss D (2009) CYP704B1 is a Long-Chain Fatty Acid v-Hydroxylase essential for sporopollenin synthesis in pollen of Arabidopsis. Plant Physiol 151: 574–589

Doyle J (2005) Early evolution of angiosperm pollen as inferred from molecular and morphological phylogenetic analyses. Grana 44: 227–251

Doyle JA (2010) Function and evolution of saccate pollen. New Phytol 188: 6–9

Elsik WC (1971) Microbial degradation of sporopollenin. In: Brooks J, Grant PR, Muir MD, Van Gijzel P, Shaw G (eds) Sporopollenin. Academic Press, London New York, p. 480–511

Fægri K, Iversen J (1989) Textbook of Pollen analysis. 4th edition, John Wiley & Sons, Chichester

Fraser WT, Scott AC, Forbes AES, Glasspool IJ, Plotnick RE, Kenig F, Lomax BH (2012) Evolutionary stasis of sporopollenin biochemistry revealed by unaltered Pennsylvanian spores. New Phytol 196: 397–401

Fraser WT, Sephton MA, Watson JS, Self S, Lomax BH, James DI, Wellman CH, Callaghan TV, Beerling DJ (2011) UV-B absorbing pigments in spores: biochemical responses to shade in a high-latitude birch forest and implications for sporopollenin-based proxies of past environmental change. Polar Res 30, 8312, https://doi.org/10.3402/polar.v30o0.8312

Fraser WT, Watson JS, Sephton MA, Lomax BH, Harrington G, Gosling WD, Self S (2014) Changes in spore chemistry and appearance with increasing maturity. Rev Palaeobot Palynol 201: 41–46

Friedman J, Barrett SCH (2009) Wind of change: new insights on the ecology and evolution of pollination and mating in wind-pollinated plants. Ann Bot 103: 1515–1527

Furness CA (2007) Why does some pollen lack apertures? A review of inaperturate pollen in eudicots. Bot J Linn Soc 155: 29–48

Furness CA, Rudall PJ (1999) Microsporogenesis in Monocotyledons. Ann Bot 84: 475–499

Furness CA, Rudall PJ (2001) Pollen and anther characters in monocot systematics. Grana 40: 17–25

Furness CA, Rudall PJ (2003) Apertures with lids: distribution and significance of operculate pollen in Monocotyledons. Int J Plant Sci 164: 835–854

Gabarayeva NI, Grigorjeva VV (2010) Sporoderm ontogeny in *Chamaedorea microspadix* (Arecaceae): self-assembly as the underlying cause of development. Grana 49: 91–114

Gabarayeva NI, Grigorjeva VV, Rowley JR (2010) A new look at sporoderm ontogeny in *Persea americana* and the hidden side of development. Ann Bot 105: 939–955

Ganders FR (1979) The biology of heterostyly. NZ J Bot 17(4): 607–635

Grega L, Anderson S, Cheetham M, Clemente M, Colletti A, Moy W, Talarico D, Thatcher SL, Osborn JM (2013) Aerodynamic characteristics of saccate pollen grains. Int J Plant Sci 174: 499–510

Grímsson F, Zetter R (2011) Combined LM and SEM study of the Middle Miocene (Sarmatian) palynoflora from the Lavanttal Basin, Austria: Part II. Pinophyta (Cupressaceae, Pinaceae and Sciadopityaceae). Grana 50: 262–310

Halbritter H, Hesse M (1995) The convergent evolution of exine shields in Angiosperm pollen. Grana 34: 108–119

Halbritter H, Hesse M (2004) Principal modes of infoldings in tricolp(or)ate Angiosperm pollen. Grana 43: 1–14

Halbritter H, Hesse M (2005) Specific ornamentation of orbicular walls and pollen grains, as exemplified by Acanthaceae. Grana 44: 308–313
Havinga AJ (1971) An experimental investigation into the decay of pollen and spores in various soil types. In: Brooks J, Grant PR, Muir MD, Van Gijzel P (eds) Sporopollenin. Academic Press, London, New York, p. 446–479
Havinga AJ (1984) A 20–year experimental investigation into the differential corrosion susceptibility of pollen and spores in various soil types. Pollen Spores 26: 541–558
He X, Dai J, Wu Q (2016) Identification of Sporopollenin as the Outer Layer of Cell Wall in Microalga *Chlorella protothecoides*. Front Microbiol 7: 1047
Hesse M, Halbritter H, Zetter R, Weber M, Buchner R, Frosch–Radivo A, Ulrich S (2009) Pollen Terminology. An illustrated Handbook. Springer, Vienna
Hemsley AR, Barrie PJ, Chaloner WG, Scott AC (1993) The composition of sporopollenin and its use in living and fossil plant systematics. Grana 32, Suppl 1: 2–11
Huysmans S, El–Ghazaly G, Smets E (1998) Orbicules in angiosperms: morphology, function, distribution, and relation with tapetum types. Bot Rev 64: 240–272
Jardine PE, Fraser WT, Lomax BH, Gosling WD (2015) The impact of oxidation on spore and pollen chemistry. J Micropalaeontol 24: 139–149
John JF (1814) Ueber den Befruchtungsstaub, nebst einer Analyse des Tulpenpollens. J Chem Phys 12: 244–252
Johnson ST, Edwards TJ (2000) The structure and function of orchid pollinaria. Plant Syst Evol 222: 243–269
Klaus W (1960) Sporen der karnischen Stufe der ostalpinen Trias. In: Oberhauser R, Kristan–Tollmann E, Kollmann K, Klaus W (eds) Beiträge zur Mikropaläontologie der alpinen Trias. Jahrb Geol Bundesanstalt, Sonderband 5: 107–184
Klaus W (1987) Einführung in die Paläobotanik. Fossile Pflanzenwelt und Rohstoffbildung, Band I. Grundlagen - Kohlebildung - Arbeitsmethoden/ Palynologie. Deuticke, Wien
Knox RB, McConchie CA (1986) Structure and function of compound pollen. In: Blackmore S, Ferguson IK (eds) Pollen and Spores, Form and Function. Linnean Society of London, London, p. 265–282
Lallemand B, Erhardt M, Heitz T, Legrand M (2013) Sporopollenin biosynthetic enzymes interact and constitute a metabolon localized to the endoplasmic reticulum of tapetum cells. Plant Physiol 162: 616–625
Leslie AB (2010) Flotation preferentially selects saccate pollen during conifer pollination. New Phytol 188: 273–279
Liu L, Fan X (2013) Tapetum: regulation and role in sporopollenin biosynthesis in *Arabidopsis*. Plant Mol Biol 83: 165–175
Maeda Y (1984) The presence and location of sporopollenin in fruiting bodies of the cellular slime moulds. J Cell Sci 66: 297–308
Pacini E, Franchi GG (1991) Role of the tapetum in pollen and spore dispersal. Plant Syst Evol, Suppl. 7: 1–11
Pacini E, Hesse M (2005) Pollenkitt – its composition, forms and functions. Flora 200: 399–415
PalDat – a palynological database (2000 onwards, www.paldat.org)
Phuphumirat W, Gleason FH, Phongpaichit S, Mildenhall DC (2011) The infection of pollen by zoosporic fungi in tropical soils and its impact on pollen preservation: a preliminary study. Nova Hedwigia 92: 233–244
Phuphumirat W, Zetter R, Hofmann C–C, Ferguson DK (2015) Pollen degradation in mangrove sediments: A short–term experiment. Rev Palaeobot Palynol 221: 106–116
Playford G, Dettmann ME (1996) Spores. In: Jansonius J, McGregor DC (eds) Palynology: principles and applications. American Association of Stratigraphic Palynologists Foundation, vol. 1, AASP Foundation, Dallas, p. 227–260
Pozhidaev AE (2000a) Pollen variety and aperture patterning. In: Harley MM, Morton CM, Blackmore S (eds) Pollen and Spores: Morphology and Biology. Royal Botanic Gardens, Kew, p. 205–225
Pozhidaev AE (2000b) Hypothetical way of pollen aperture patterning. 2: Formation of polycolpate patterns and pseudoaperture geometry. Rev Palaeobot Palynol 109: 235–254
Praglowski J (1975) Importance de la mise au point des terms "structure" de l'exine. Bull Soc Bot France, Coll Palynologie 122: 75–78
Punt W, Hoen PP, Blackmore S, Nilsson S, Le Thomas A (2007) Glossary of pollen and spore terminology. Rev Palaeobot Palynol 143: 1–81
Quilichini TD, Douglas CJ, Samuels AL (2014) New views of tapetum ultrastructure and pollen exine development in *Arabidopsis thaliana*. Ann Bot 114: 1189–120
Reitsma TJ (1969) Size modification of recent pollen grains under different treatments. Rev Palaeobot Palynol 9: 175–202
Riding JB, Kyffin–Hughes JE (2004) A review of the laboratory preparation of palynomorphs with description of an effective non–acid technique. Rev Bras Paleontolog 7: 13–44
Rowley JR, Skvarla JJ (2000) The elasticity of the exine. Grana 37: 1–7
Rubinstein CV, Gerrienne P, de la Puente GS, Astini RA, Steemans P (2010) Early Middle Ordovician evidence for land plants in Argentina (eastern Gondwana). New Phytol 188: 365–369
Schwendemann AB, Wang G, Mertz ML, McWilliams RT, Thatcher SL, Osborn JM (2007) Aerodynamics of saccate pollen and its implications for wind pollination. Am J Bot 94: 1371–1381
Simons J, Van Beem AP, De Vries PJR (1983) Structure and chemical composition of the spore wall in *Spirogyra* (Zygnemataceae, Chlorophyceae). Acta Bot Neerl 31: 359–370
Skvarla JJ, Rowley JR, Chissoe WF (1997) Exine resistance to fungal infestations in Strelitziaceae. Taiwania 42: 17–27
Steemans P, Lepot K, Marshall CP, Le Herisseé A, Javaux EJ (2010) FTIP characterisation of the chemical composition of Silurian miospores (cryptospores and trilete spores) from Gotland, Sweden. Rev Palaeobot Palynol 162: 577–590
Takahashi M (1995) Development of structure–less pollen wall in *Ceratophyllum demersum* L. (Ceratophyllaceae). J Plant Res 108: 205–208
Traverse A (1988) Paleopalynology. Unwin Hyman, Boston
Traverse A (2007) Paleopalynology. 2nd ed, Springer, Dordrecht
Tryon AF, Lugardon B (1991) Spores of the Pteridophyta: Surface, wall structure and diversity based on electron microscopy studies. Springer, New York
Tsou C–H, Fu Y–L (2002) Tetrad pollen formation in *Annona* (Annonaceae): Proexine formation and binding mechanism. Am J Bot 89: 734–747

Twiddle CL, Bunting MJ (2010) Experimental investigations into the preservation of pollen grains: A pilot study of four pollen types. Rev Palaeobot Palynol 162: 621–630

Ueno R (2009) Visualization of sporopollenin-containing pathogenic green micro-alga Prototheca wickerhamii by fluorescent in situ hybridization (FISH). Can J Micro 55: 465–472

Ulrich S, Hesse M, Weber M, Halbritter H (2017) Amorphophallus: New insights into pollen morphology and the chemical nature of the pollen wall. Grana 56: 1–36

Van Bergen PF, Collinson ME, de Leeuw JW (1993) Chemical composition and ultrastructure of fossil and extant salvinialean microspore massulae and megaspores. Grana 32, Suppl 1: 18–30

Van Campo M, Lugardon B (1973) Structure grenue infratectal de l'ectexine des pollens de quelques Gymnospermes et Angiospermes. Pollen Spores 15: 171–189

Versteegh GJM, Blokker P, Bogus KA, Harding IC, Lewis J, Oltmanns S, Rochon A, Zonneveld KAF (2012) Infra red spectroscopy, flash pyrolysis, thermally assisted hydrolysis and methylation (THM) in the presence of tetramethylammonium hydroxide (TMAH) of cultured and sediment-derived *Lingulodinium polyedrum* (Dinoflagellata) cyst walls. Org Geochem 43: 92–102

Verstraete B, Moon H-K, Smets E, Huysmans S (2014) Orbicules in flowering plants: A phylogenetic perspective on their form and function. Bot Rev 80: 107–134

Vinckier S, Cadot P, Smets E (2005) The manifold characters of orbicules: structural diversity, systematic significance, and vectors for allergens. Grana 44: 300–307

Walker JW (1976) Evolutionary significance of the exine in the pollen of primitive angiosperms. In: Ferguson IK, Muller J (eds) The evolutionary significance of the exine. Academic Press, London, p. 251–308

Wallace S, Chater CC, Kamisugi Y, Cuming AC, Wellman CH, Beerling DJ, Fleming AJ (2015) Conservation of Male Sterility 2 function during spore and pollen wall development supports an evolutionarily early recruitment of a core component in the sporopollenin biosynthetic pathway. New Phytol 205: 390–401

Weber M, Halbritter H (2007) Exploding pollen in *Montrichardia arborescens* (Araceae). Plant Syst Evol 263: 51–57

Wellman CH (2010) The invasion of the land by plants: when and where? New Phytol 188: 306–309

Wiermann R, Ahlers F, Schmitz-Thom I (2001) Sporopollenin. In: Hofrichter M, Steinbüchel A (eds) Biopolymers 1: Lignin, Humic Substances and Coal, Wiley-VCH Weinheim, p. 209–227

Yule BL, Roberts S, Marshall JEA (2000) The thermal evolution of sporopollenin. Org Geochem 31: 859–870

Zetzsche F, Kalt P, Leichti J, Ziegler E (1931) Zur Konstitution des Lycopodiumsporonins, des Tasmanins und des Lange-Sporonins. J Prakt Chem 148: 67–84

4

Microscopic Observations and their Misinterpretations

The description of pollen ornamentation depends on three major parameters (1) the interpretations of the palynologist (which are subjective), (2) the pollen terminology applied, and (3) the magnification, resolution, and methods used.

The application of different preparation and staining methods and a combined analysis with light microscopy, scanning- and transmission electron microscopy are essential for the interpretation of pollen characters. Investigation of recent and fossil pollen material often reveals interesting features that in some cases may be misinterpreted. To demonstrate the wide range of possible misinterpretations, the following examples are given:

Example 1: Tripartite Feature in Gymnosperms — Impression Mark

Mature pollen of conifers, such as *Abies*, *Larix*, and *Pseudotsuga*, often shows proximally a Y-shaped bulge on the proximal polar side, comparable to a tetrad mark, which is called an **impression mark** (Fig. 1; Harley 1999). The mark results from the close proximity of the four pollen grains at the post-meiotic tetrad phase and is retained afterwards and is not a germination feature. Impression marks are also found in palm pollen. Note: the term tetrad mark is restricted to spores, where it is a germination feature.

Example 2: Tripartite Feature in Angiosperms — Triangular Tenuitas

Superficially similar features in angiosperms are not comparable to those observed in gymnosperms. In recent and fossil Sapindaceae a three-armed feature (more precisely a triangle) is found. *Cardiospermum* has a narrow **triangular tenuitas** (thinning) at the proximal pole, whereas other recent and subfossil Sapindaceae show such a feature at both poles (Fig. 2).

Example 3: Tripartite Feature in Angiosperms — Synaperture

Triangular pollen as found in Myrtaceae, some Primulaceae (*Primula farinosa* or *P. denticulata*) and Loranthaceae is characterized by a tripartite feature in both polar areas (Fig. 3). These are in fact

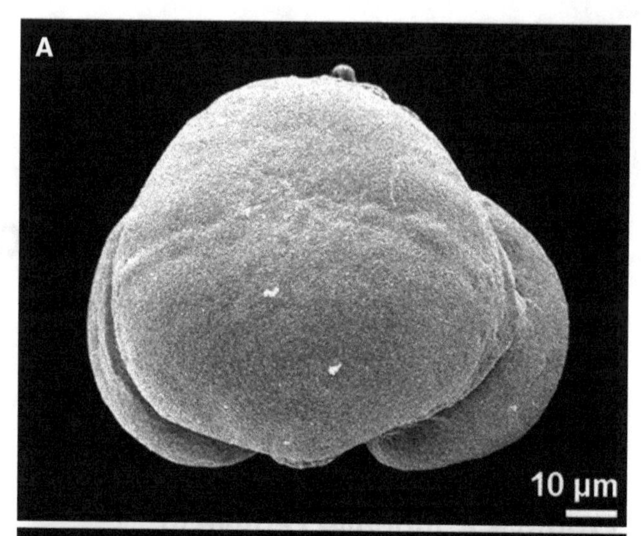

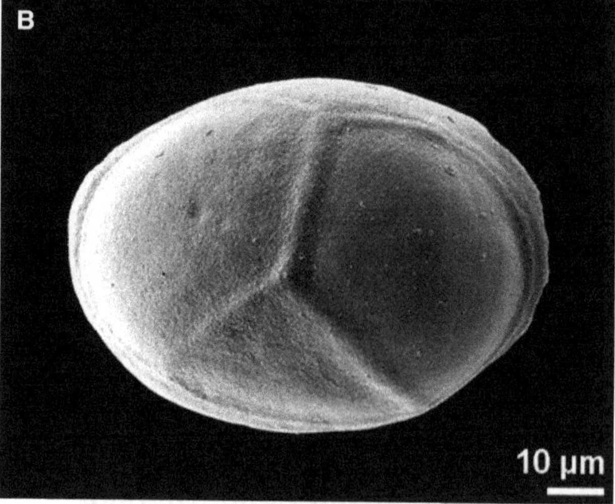

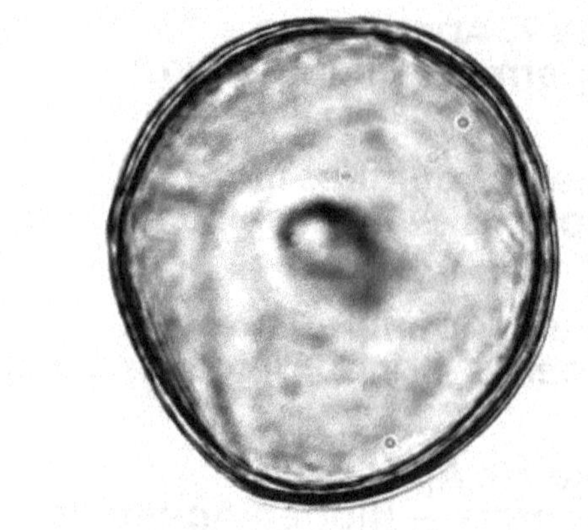

Fig. 1 Impression mark. A. *Abies cephalonica*, Pinaceae, proximal polar view, indistinct impression mark. **B-C.** *Larix* sp., Pinaceae, fossil, middle Miocene, Austria, proximal polar view, Y-shaped impression mark in SEM (**B**) and LM (**C**)

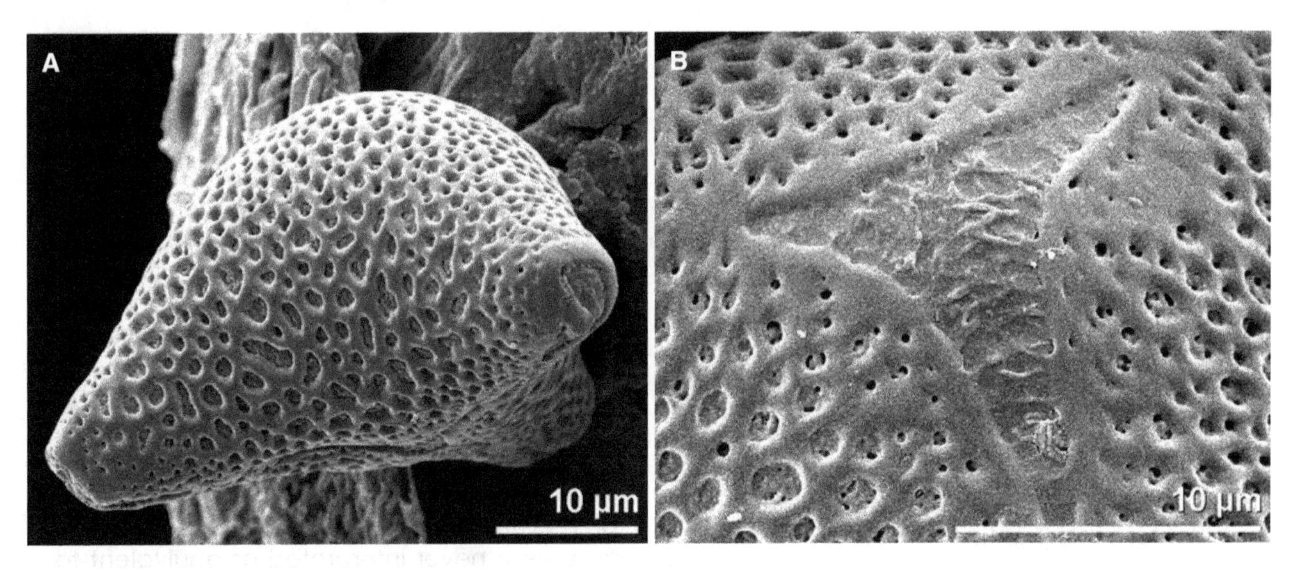

Fig. 2 Triangular tenuitas. A-B. *Cardiospermum corindum*, Sapindaceae, tricolporate, equatorial view (**A**), proximal pole with triangular thinning area (**B**)

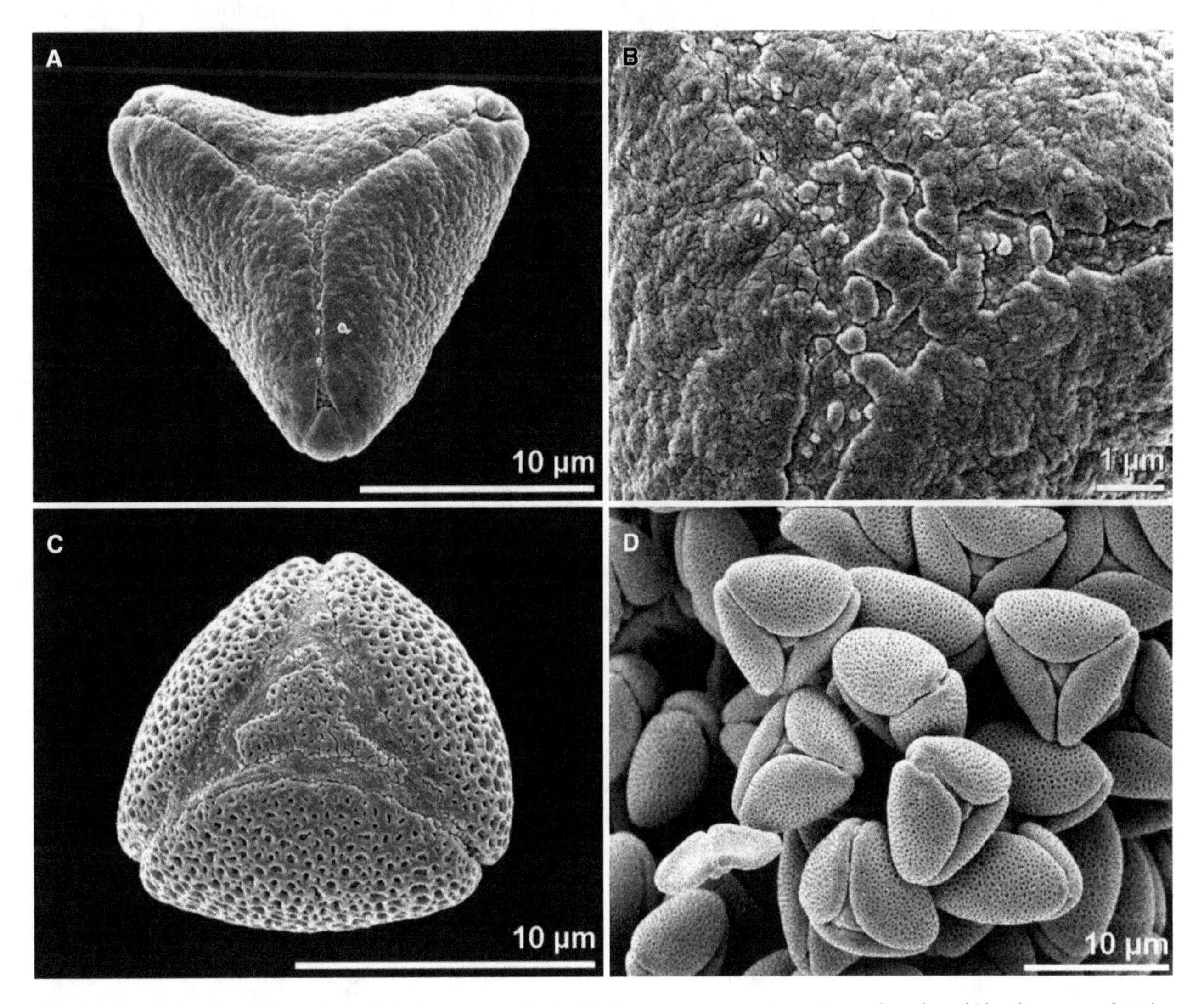

Fig. 3 Synaperturate pollen. A-B. *Melaleuca armillaris*, Myrthaceae, syncolporate, polar view (**A**), close-up of polar area (**B**). **C.** *Primula denticulata*, Primulaceae, syncolpate, polar view. **D.** *Primula farinosa*, Primulaceae, syncolpate, dry pollen

three colpi, extending towards and merging at the poles. The pollen is therefore synaperturate (syncolpate, syncolporate). In for example, *Primula* the colpi dissect in the polar area, leaving a triangular field at both poles.

Example 4: Tripartite Feature in Angiosperms — Trichotomosulcus

Another tripartite feature is the **trichotomosulcus** (Harley 2004), a three-armed sulcus occurring exclusively distally, as, e.g., in *Dianella*. Trichotomosulcate pollen has been discussed in relation to the evolution of the tricolpate dicot condition, but so far without success (Fig. 4).

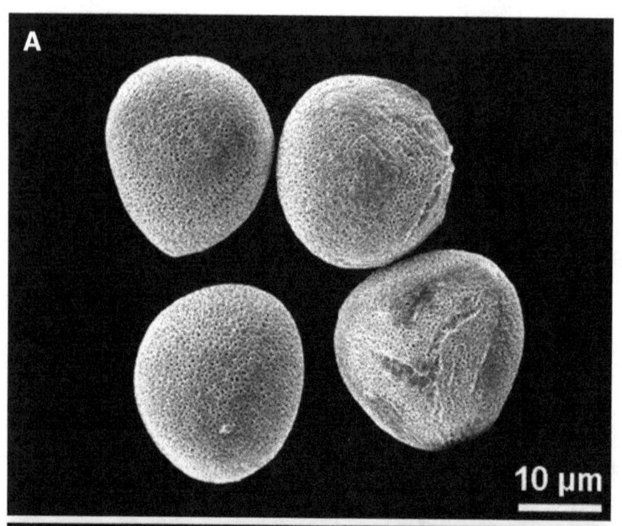

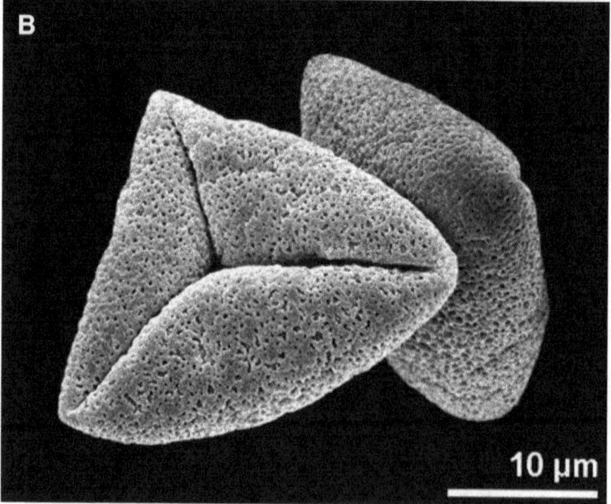

Fig. 4 Trichotomosulcus. A-B. *Dianella tasmanica*, Phormiaceae, trichotomosulcus (**A**), dry pollen, aperture infolded (**B**)

Example 5: Tripartite Feature in Angiosperms — Sulci vs. Colpi vs. Tenuitas

The angiosperm-like pollen of the fossil genus *Eucommiidites* is "trisulcate": a broad distal sulcus and two narrower additional "sulci" (at angles of c. 120° seen from the main sulcus; Fig. 5). This feature was erroneously interpreted as tricolpate pollen (with colpi equatorially situated).

A similar arrangement of a distal sulcus and two small additional sulci on the proximal face was described, for example, in some species of *Tulipa* (Liliaceae) and *Tinantia* (formerly *Commelinantia*, Commelinaceae), but these cases were never interpreted as equivalent to a tricolpate condition (Harley 2004) (Fig. 6). The two small additional sulci may also be interpreted as tenuitates. In some cases the three "sulci" are of similar size. The aperture condition is very similar to a tricolpate one.

Example 6: Tripartite Feature in Angiosperms — Triradiate Aperture

Another three-armed feature is the triradiate aperture in *Thesium alpinum* (Santalaceae) pollen. The heteropolar pollen is 3-aperturate, with apertures placed in the three tapered edges of a tetrahedron (Feuer 1977). Each aperture has a very inconspicuous triradiate outline, which is situated equatorially. Two of the arms point towards the neighboring tetrahedron edge and are rather short; the third, elongated arm is directed towards the rounded pole (Fig. 7).

Example 7: Apertures in Angiosperms — Planaperturate

Sometimes apertures are inconspicuous and not discernible at first sight. In pollen of *Pachira aquatica* (Malvaceae) three large, more-or-less hemispherical areas are seen equatorially, which may at first sight be interpreted as pores. However, a detailed observation reveals **planaperturate** pollen grains with three short colpi (Fig. 8).

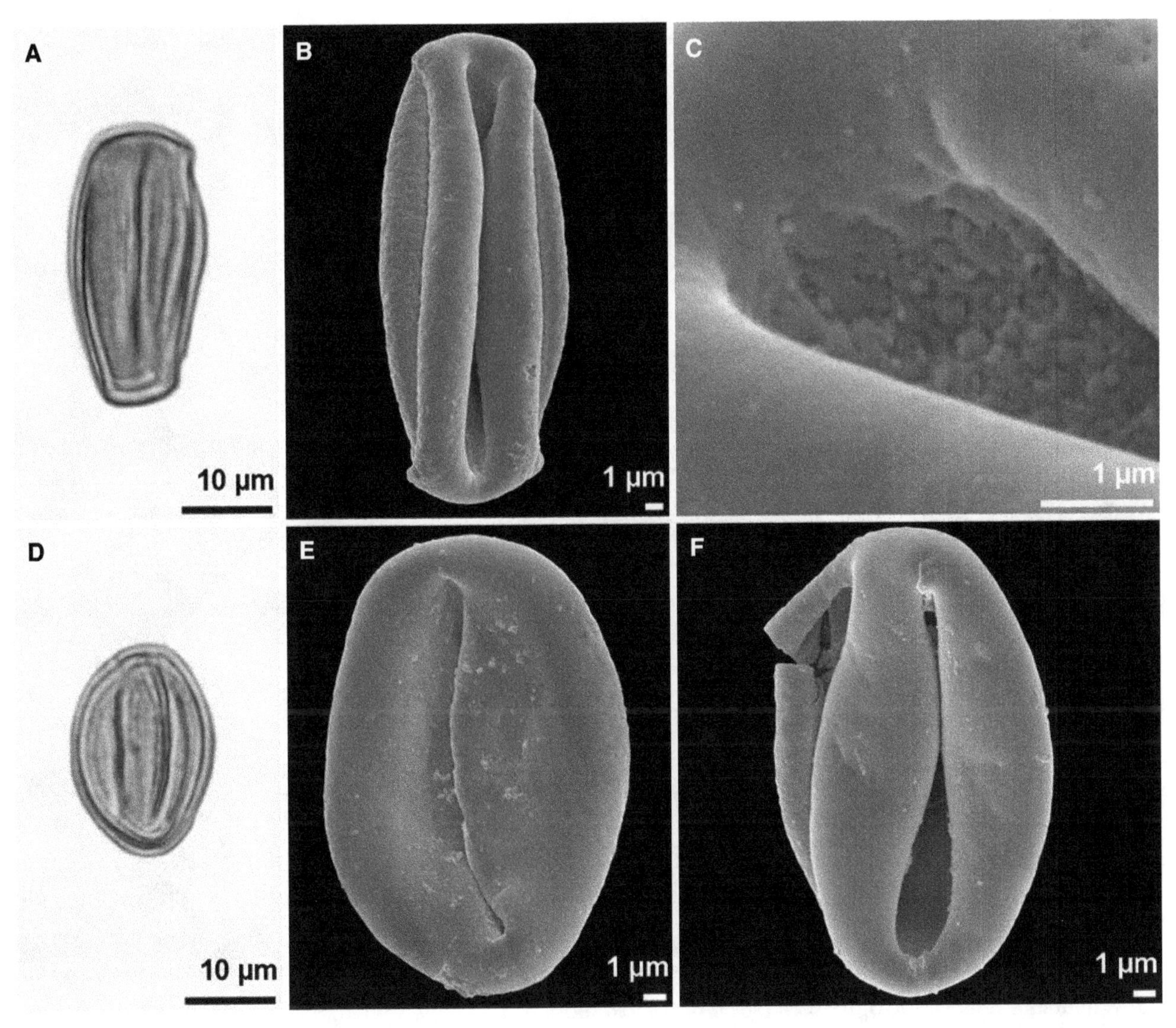

Fig. 5 Trisulcate pollen. A-C. *Eucommiidites* sp., fossil pollen, Lower Cretaceous of U.S.A., main sulcus with membrane seen in center of pollen grain, flanked by additional narrow sulci on each side (at angles of c. 120°, **A-B**), close-up showing sulcus membrane of main sulcus (**C**). **D-F.** *Eucommiidites* sp., fossil pollen, Lower Cretaceous of U.S.A., Narrow lateral sulcus (**E**), same grain turned showing the main broad sulcus and one narrow lateral sulcus (**F**)

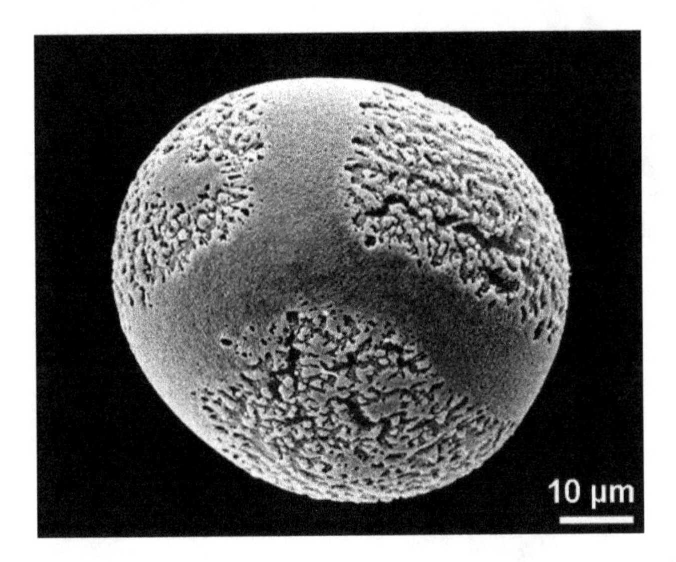

Fig. 6 Trisulcate pollen. *Tulipa kaufmanniana*, Liliaceae, trisulcate or sulcate with two tenuitates, equatorial view

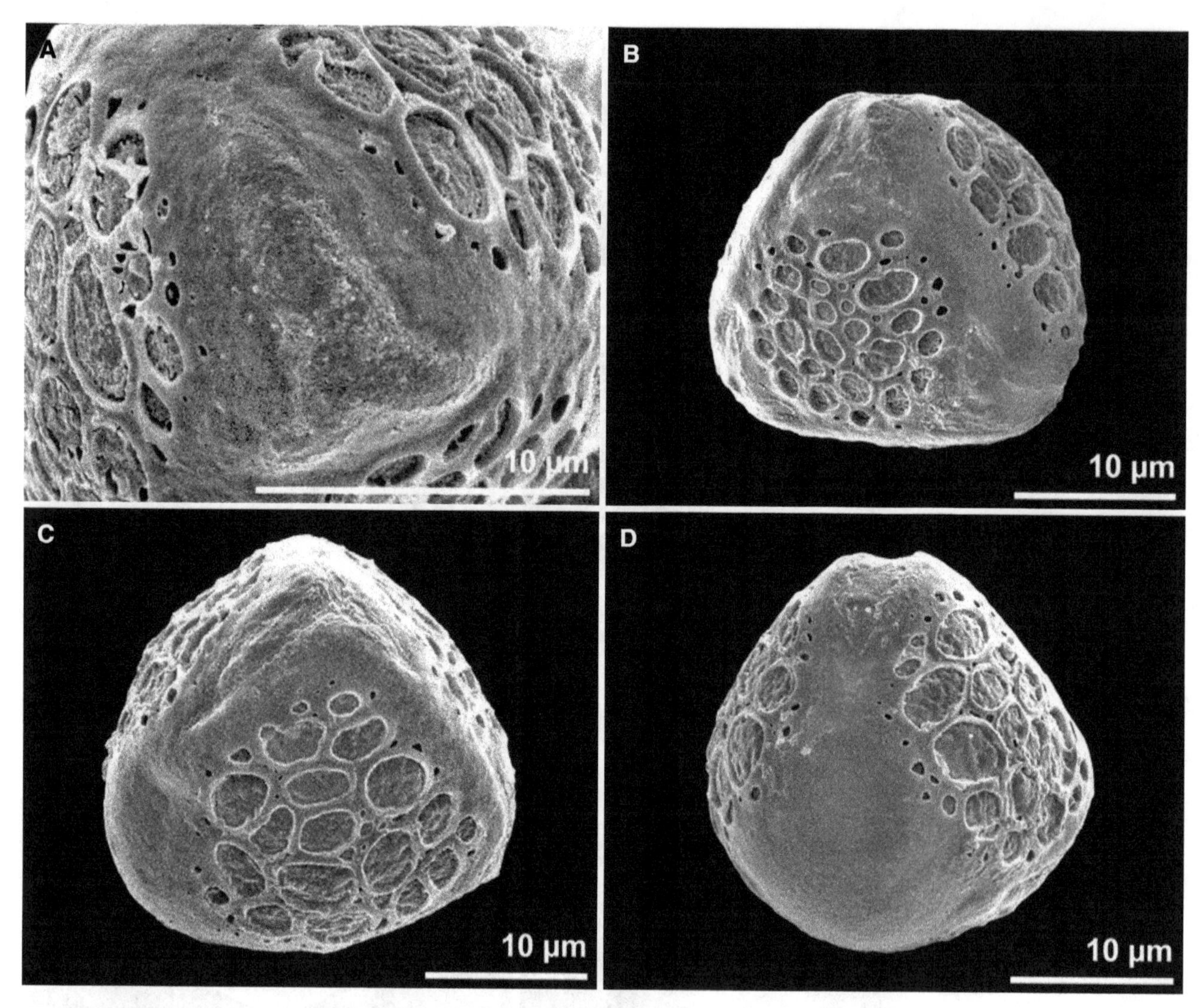

Fig. 7 Triradiate aperture. A-D. *Thesium alpinum*, Santalaceae. **A.** Tricolpate, heteropolar, triradiate colpus. **B.** Polar view (flattened pole). **C.** Equatorial view. **D.** Polar view (rounded pole)

Example 8: Apertures in Angiosperms — Inconspicuous Pori

In *Calliandra emarginata* (Mimosaceae) the monads forming a polyad are separated by narrow groove-like depressions. At low magnification the presence and localization of the apertures remain indistinct; high SEM magnification reveals that the apertures are very inconspicuous pores, situated equatorially, usually at the conjunction of three or four monads (Fig. 9 A, B).

Also, the aperture condition may be overlooked due to other eye-catching features. The clypeate pollen of *Phyllanthus* x *elongatus* (Euphorbiaceae) seems to be inaperturate. Only close-ups reveal the inconspicuous few pores between the exine shields (Fig. 9 C, D).

Example 9: Apertures in Angiosperms — Inconspicuous Colpi

The disc-like pollen of *Oryctanthus* sp. (Loranthaceae) shows at both poles conspicuous circular depressions that are not apertures (Feuer and Kuijt 1985; Grímsson et al. 2018). The pollen is according to Grímsson et al. (2018) demi(3)colpate, with

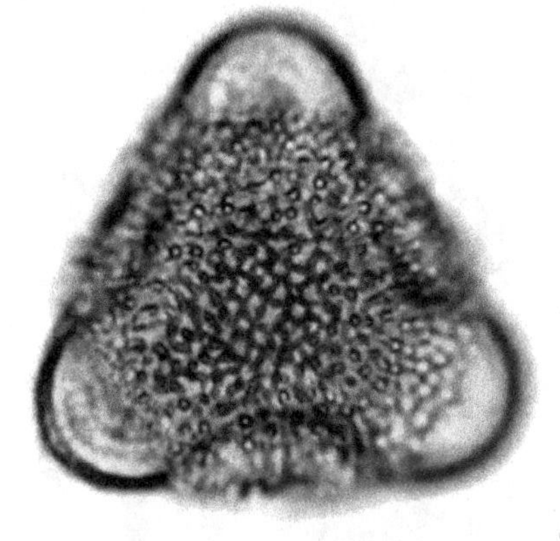

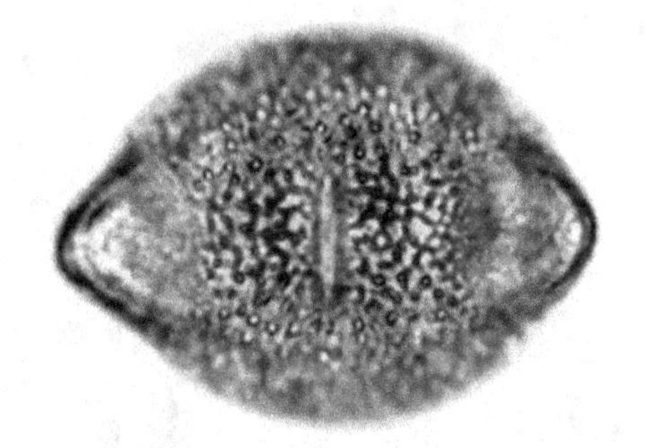

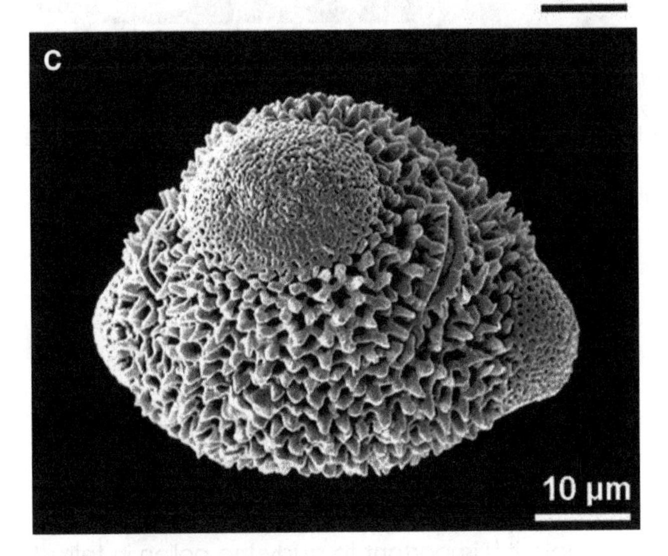

Fig. 8 Planaperturate pollen. A-C. *Pachira aquatica*, Malvaceae, polar view (**A**), equatorial view (**B**), oblique equatorial view (**C**)

inconspicuous slit-like colpi positioned between the polar depressions (Fig. 10). Another example are some Asteraceae pollen, where the colpi are often inconspicuous or not visible in SEM, but obvious in LM.

Example 10: Apertures in Angiosperms — Hidden Apertures

Recent and fossil triaperturate (colpate or porate) pollen of *Trapa* (Trapaceae) is distinguished by unique meridional exine ridges (crests) covering the apertures (Zetter and Ferguson 2001) (Fig. 11).

Example 11: Apertures in Angiosperms — Ring-like Apertures vs. Colpate-Operculate

The apertures in *Passiflora* cf. *incarnata* may be interpreted as three ring-like apertures or may be interpreted as pori (or colpi) each with an operculum. In other species of *Passiflora* e.g., *P. citrina* and *P. suberosa*, the apertures are both narrower and stephanocolpate (Fig. 12).

Example 12: Apertures in Angiosperms — Tenuitas vs. Poroid

Tenuitas is a general term for a pollen wall thinning (Kremp 1968; Harley 2004; Punt et al. 2007). It is normally found additional to apertures, e.g., in *Myosotis* (Fig. 13). A circular tenuitas can be mistaken for a **poroid**, which is a circular or elliptic aperture with an indistinct margin (see also "Illustrated Pollen Terms").

Example 13: Apertures in Angiosperms — Infoldings vs. Apertures

When pollen is infolded it can be hard to distinguish the apertures. Pollen of *Sparganium erectum* (Sparganiaceae) is in dry stage infolded, boat-shaped, and would be considered as sulcate. In fact, *Sparganium* pollen is ulcerate, the ulcus is seen clearly in the hydrated, spherical pollen stage (Fig. 14).

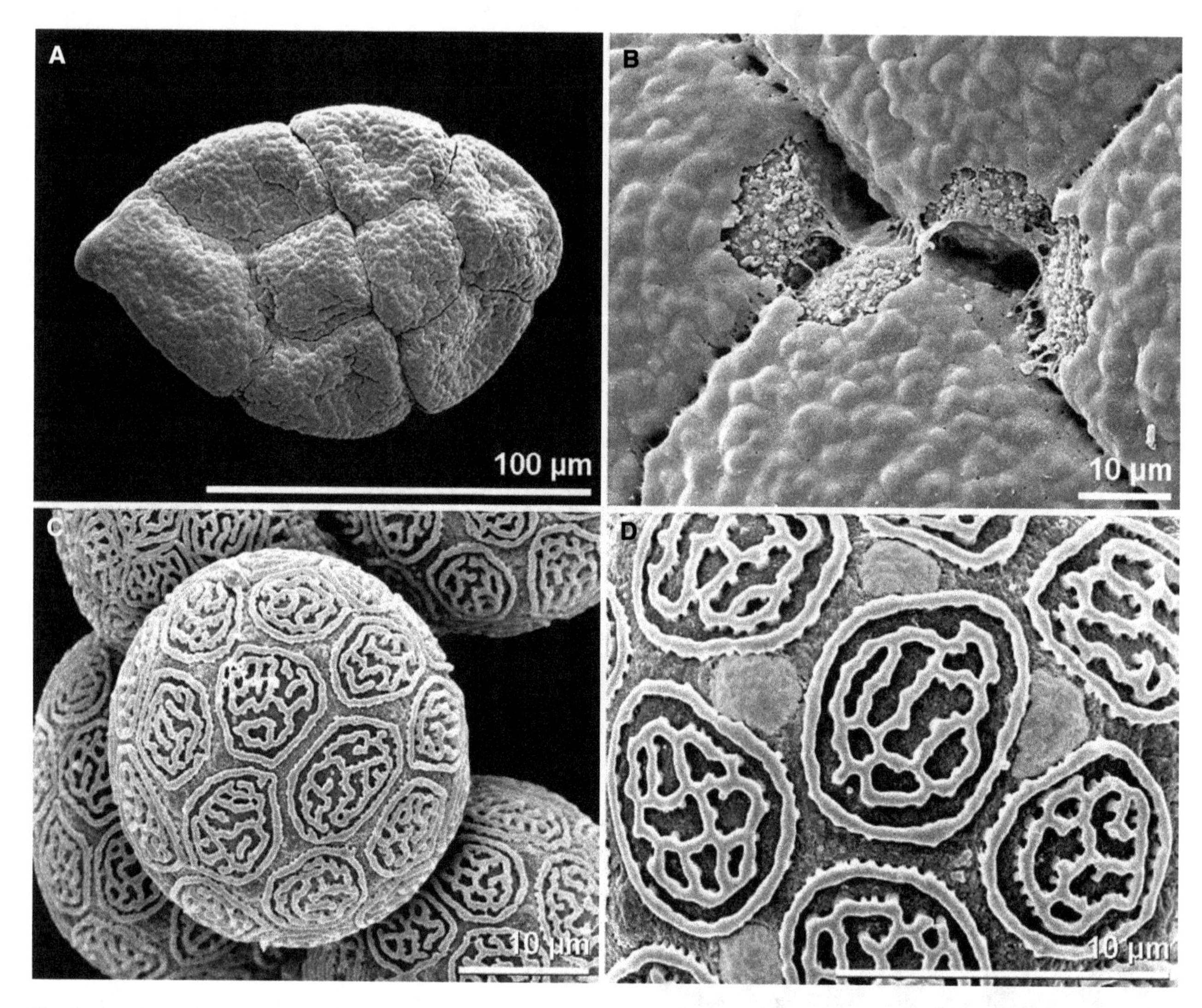

Fig. 9 Apertures in *Calliandra* and *Phyllanthus*. A-B. *Calliandra tergemina*, Fabaceae, polyad, dry state (**A**). Apertures (pori) at the junction of four monads (**B**). **C-D.** *Phyllanthus x elongatus*, Euphorbiaceae, clypeate, seemingly inaperturate (**C**), Inconspicuous pores (colored) between the exine shields (**D**)

Example 14: Apertures in Angiosperms — Ulcerate-Operculate vs. Ring-like Aperture

Nymphaea alba (Nymphaeaceae) pollen has asymmetrical halves divided by a ring-like aperture (Fig. 15). The features of the smaller distal half may be misinterpreted as a large ulcus with a conspicuous operculum. Ultrastructural studies and germination experiments support the interpretation of a ring-like aperture (Gabarayeva and Rowley 1994; Hesse and Zetter 2005).

Example 15: Apertures in Angiosperms — Disulcate vs. Dicolpate

The term disulcate defines two elongated apertures situated usually distally (but not directly at the distal pole), running parallel to or even in the equator (Fig. 16). If the apertures are running meridionally, pollen would be dicolpate (Halbritter and Hesse 1993). To distinguish if the pollen is disulcate or dicolpate it is important to study the pollen in tetrad arrangement to clarify the polarity and position of apertures (see Fig. 3 in "Methods in Palynology").

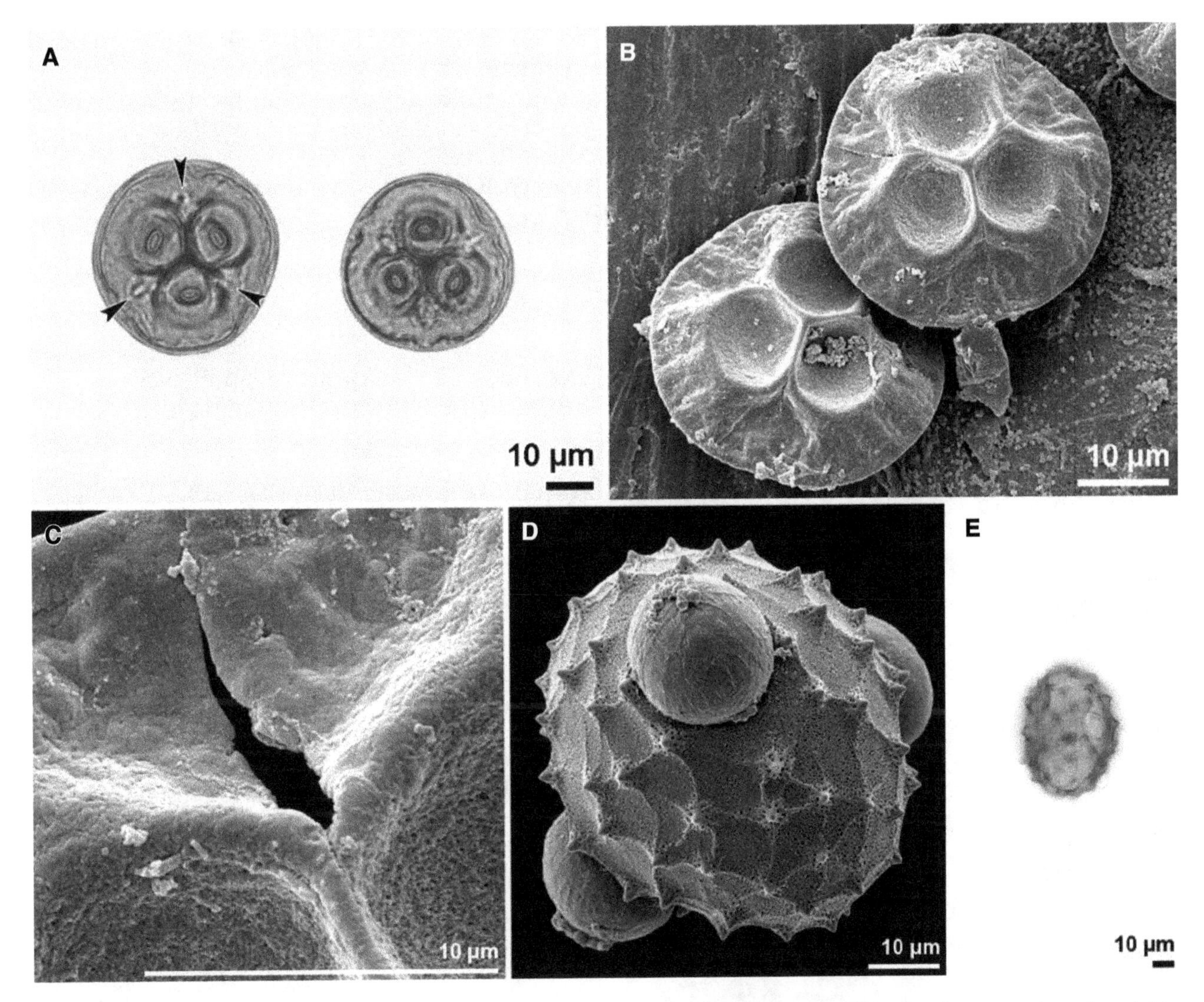

Fig. 10 Apertures in *Oryctanthus*. A-C. *Oryctanthus alveolatus*, Loranthaceae, acetolyzed pollen, arrowheads point to colpi, LM (**A**). two grains in polar view, SEM (**B**). close-up showing colpus (**C**). **D-E.** *Carthamus lanatus*, Asteraceae, hydrated pollen, pollen in SEM seem porate (**D**). Acetolyzed pollen, colporus (highlighted) only visible in LM (**E**)

Examples for taxa with disulcate pollen are the monocots *Tofieldia calyculata* with one sulcus distally, the other proximally, *Uvularia grandiflora*, *Eichhornia crassipes* (Hesse et al. 2009), some *Dioscorea* species (Schols et al. 2005), *Pontederia cordata* (Halbritter 2016), *Calla palustris* (Ulrich et al. 2013), and the magnoliid *Calycanthus floridus* (Huynh 1976).

Example 16: Apertures in Angiosperms — Zon-, Zono-, Zoni-, Zona- vs. Ring-like Aperture and Stephanoaperturate Pollen

Terms combining the basic prefix zon- together with its linguistic derivatives are a source of endless confusion, misunderstanding and superfluous inflation of terms. The prefix include **zon-** (in zonorate, for a ring-like endoaperture, the os, at the equator), the outdated, rarely used **zoni-** (however, with two quite different terminological applications), but especially **zona-** (indicating exclusively a ring-like feature situated anywhere) and **zono-** (indicating any feature located strictly equatorially).

Terms for ring-like (aperture) features include zona-aperturate, zona-sulculus (addressing the polarity by anazona-sulculus and catazona-sulculus), zona-sulcus, zonate, zono-aperturate, and also related names (e.g., "fully zonate condition" sensu Grayum 1992). Even the misleading and contradictory **zon<u>o</u>**-sulcus (a sulcus cannot be situated equatorially) is used instead of the correct, but phonetically confusable, **zon<u>a</u>**-sulcus. Even the

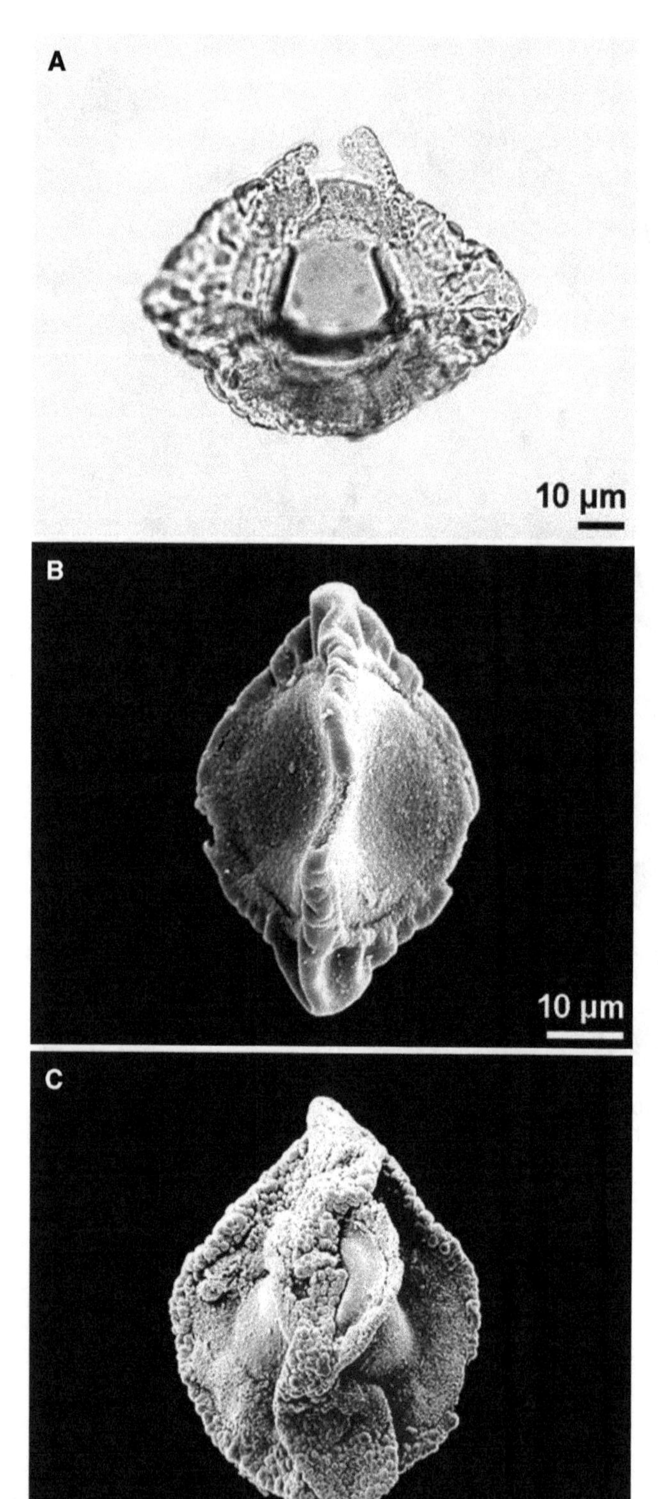

Fig. 11 Apertures in *Trapa*. A-C. *Trapa* sp., Trapaceae, fossil, late Miocene, Austria, equatorial view, crest broken, LM (**A**). Equatorial view, crest partly broken, colpus visible in SEM (**B**). Equatorial view, SEM (**C**)

trained palynologist may become confused. Therefore, all these terms should be avoided and we recommend the following two terms: **ring-like aperture and stephanoaperturate** (see "Illustrated Pollen Terms"). Any encircling aperture ("zona-aperturate"), irrespective of meridional or equatorial location, is simply called a **ring-like aperture**. Any case with more than three apertures at the equator ("zono-aperturate") is called **stephanoaperturate**.

Example 17: Magnification Effect — Retipilate vs. Reticulum Cristatum

The term retipilate (reticuloid) describes a reticulum formed by pila instead of muri (Erdtman 1952). Combined investigations based on LM and SEM have revealed that the given examples *Callitriche* (Punt et al. 2007) and *Cuscuta lupuliformis* (Erdtman 1952) do not fit the definition of retipilate. In fact, the reticulum consists of muri with prominent suprasculpture elements and are without isolated pilae. Such ornamentation is termed reticulum cristatum (a special type of reticulum; muri with prominent suprasculpture elements; Fig. 17, see also "Illustrated Pollen Terms"). So far no example for retipilate sensu Erdtman (1952) is currently known.

Example 18: Dispersal Units — Massula vs. Polyad

For a pollen dispersal unit of more than four pollen grains two terms are in use, **massula** and **polyad** (Fig. 18). The application of both terms is confusing and inconsistent in the literature. Often, the various authors employ the terms more or less interchangeably and do not provide a sharp delimitation (Walker 1971; Wagenitz 2003; Punt et al. 2007; Traverse 2007). These terms, however, are not exchangeable for historical and practical reasons (see extensive review by Teppner 2007).

The term massula was coined by Richard (1817) for parts of a pollinium in some Orchidaceae and should be used for the subunits of orchid sectile pollinia/pollinaria. Massulae within one and the same pollinium are variable and different in shape, size, and numbers of pollen grains. Unfortunately, the term massula has also been used to designate compound pollen in various other families, e.g.

Fig. 12 Apertures in *Passiflora*. A-B. *Passiflora* cf. *incarnata*, Passifloraceae; colpate, operculate aperture, polar view (**A**), equatorial view (**B**). **C.** *Passiflora citrina*, Passifloraceae, stephanocolpate, operculate, polar view. **D.** *Passiflora suberosa*, Passifloraceae, stephanocolpate, operculate, dry pollen

Fabaceae-Mimosoideae, producing dispersal units of more than four pollen grains (e.g., Wettstein 1907; Wagenitz 2003; Punt et al. 2007). For these the term polyad — coined by Iversen and Troels-Smith (1950) — should be used, denoting a symmetric dispersal unit of more than four regularly arranged and permanently united pollen grains. Polyads, currently known to occur in Fabaceae (Mimosoideae), Gentianaceae, Hippocrateaceae, Celastraceae and Annonaceae, contain a specific number of pollen grains (a multiple of four: 8, 12, 16, 24, 32, 48, 64) and show a species-specific shape.

Example 19: Preparation Effect — Psilate vs. Ornamented

Ornamentation sometimes depends on the **preparation method.** A striking example is pollen of many Aroideae (Araceae), that are ornamented (e.g., echinate, striate, verrucate) in fresh or dry condition, but become psilate following acetolysis (Fig. 19). The outer pollen wall layer and ornamentation elements are composed of polysaccharide (lack sporopollenin) and are therefore destroyed during acetolysis (Weber et al. 1999; Ulrich et al. 2017).

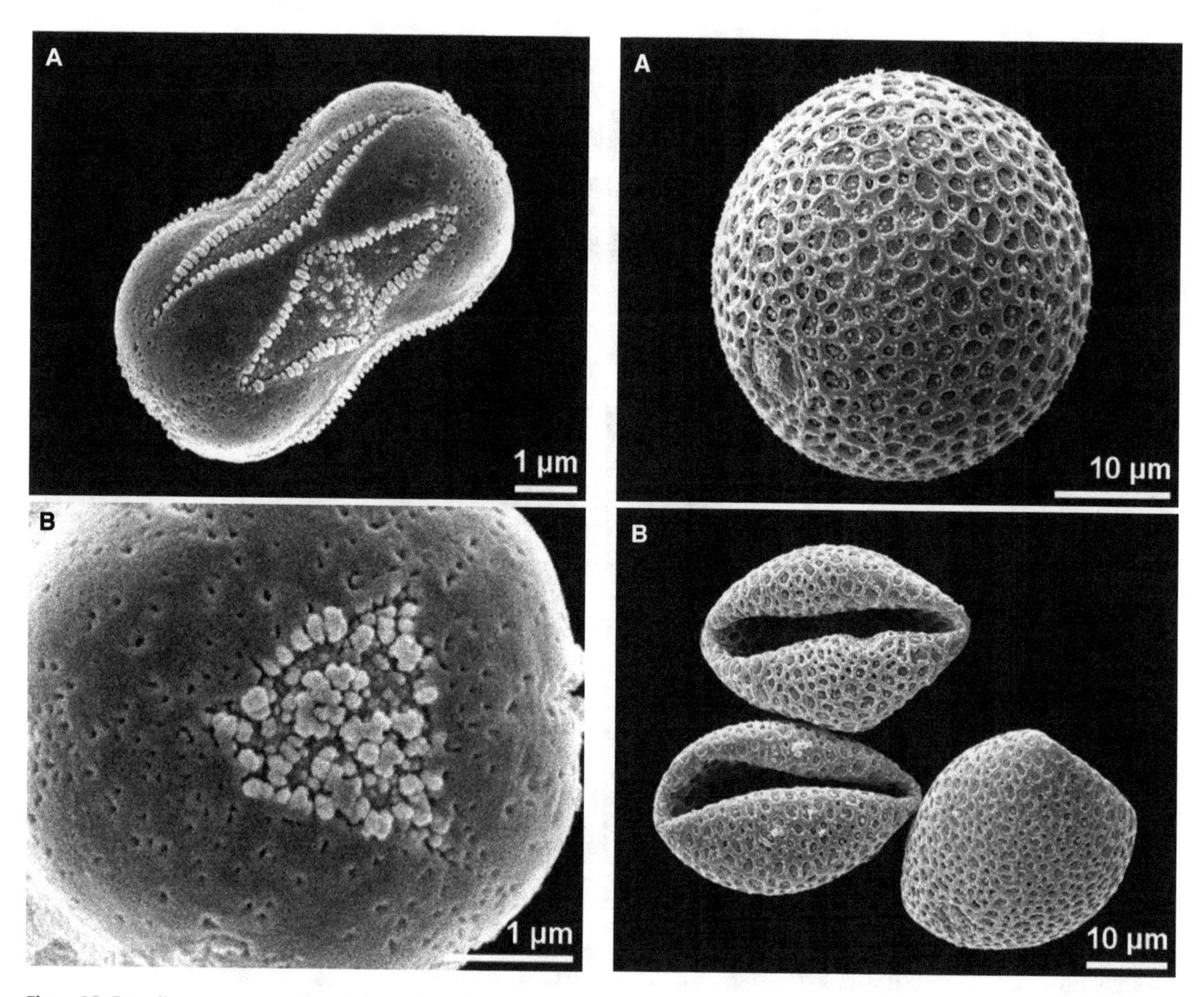

Fig. 13 Tenuitas vs. poroid. A-B. *Myosotis palustris*, Boraginaceae, equatorial view, heteroaperturate, alternating colpori and pseudocolpi (**A**), polar view, polar area with triangular tenuitas (**B**)

Fig. 14 Apertures in *Sparganium*. A-B. *Sparganium erectum*, Sparganiaceae, ulcerate, equatorial view hydrated pollen (**A**), boat-shaped, dry pollen (**B**)

Example 20: Preparation Effect — Areolate-Fossulate vs. Verrucate

The dehydration process with 2,2-dimethoxy-propane (DMP) and critical point drying (CPD) for SEM investigations can affect the ornamentation. An example for different interpretations in relation to a varying degree of hydration is *Trichosanthes anguina* (Cucurbitaceae), where the ornamentation can reflect different degrees of hydration. The ornamentation can be described as areolate and fossulate in partially hydrated condition or verrucate and perforate in fully hydrated condition (Fig. 20).

Example 21: Preparation Effect — Striate vs. Striato-reticulate

The ornamentation of *Amorphophallus longituberosus* pollen in dry condition or hydrated in water is striate, but after critical point drying it becomes striate to reticulate. The striate to reticulate ornamentation of *Amorphophallus longituberosus* is a result of an expanding thin surface layer (Fig. 21 D). During rehydration, the expansion of the thin layer itself forms a reticulum (Fig. 21 C), which finally ruptures partly or completely (Ulrich et al. 2017).

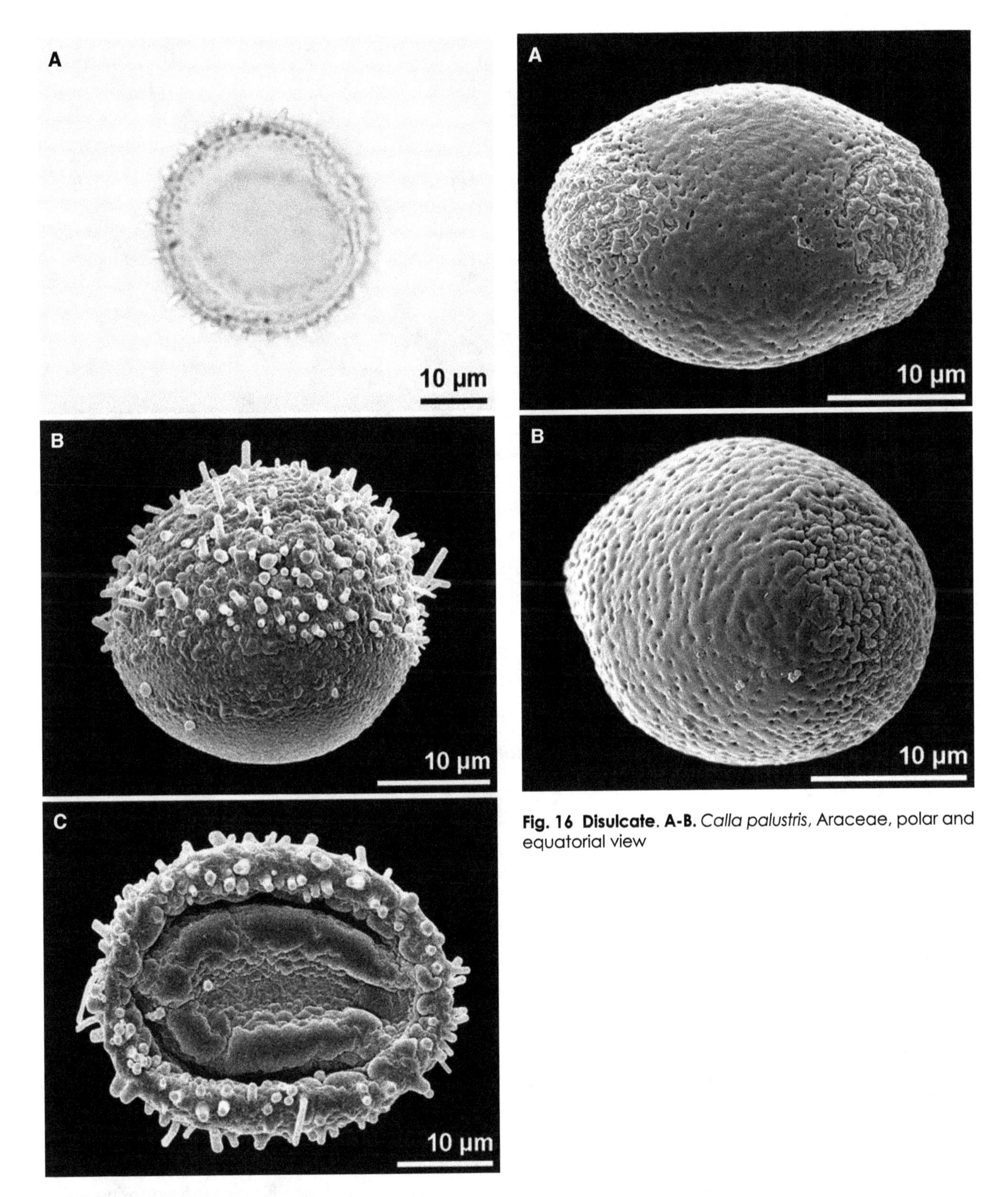

Fig. 15 Apertures in Nymphaea. A-C. *Nymphaea* sp., Nymphaeaceae; ring-like aperture, polar view (**A**), Ring-like aperture, equatorial view (**B**), dry pollen, cup-shaped (**C**)

Fig. 16 Disulcate. A-B. *Calla palustris*, Araceae, polar and equatorial view

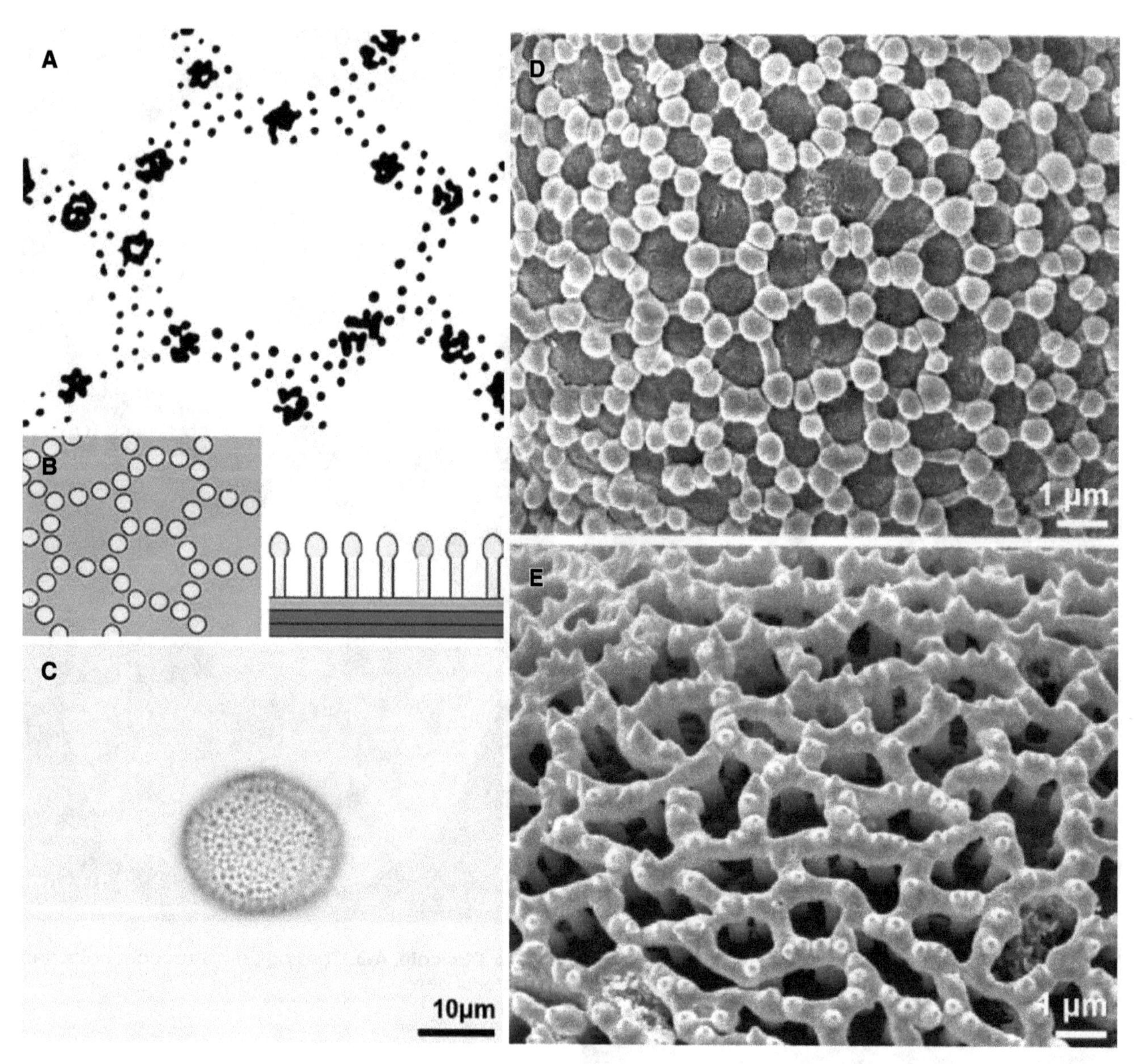

Fig. 17 Retipilate vs. reticulum cristatum. A. Drawing from Erdtman (1952). **B.** Drawings from Punt et al. (2007). **C.** *Callitriche palustris*, Plantaginaceae, acetolyzed pollen in LM. **D.** *Callitriche polymorpha*, Plantaginaceae, reticulum cristatum with small gemmae (suprasculpture) on thin muri. **E.** *Cuscuta lupuliformis*, Convolvulaceae, reticulum cristatum with nanoechini (suprasculpture)

Example 22: Staining Methods — Absence or Presence of Endexine

The staining behavior of the endexine is very heterogeneous, even within the same plant family or the same genus (Weber and Ulrich 2010). Therefore, the endexine is often reported as absent even though the layer is actually present. In most studies on pollen ultrastructure, sections are stained with uranyl acetate and lead citrate only. To truly distinguish the presence of endexine one should/must apply potassium permanganate which stains the endexine electron dense (Fig. 22, see also "Methods in Palynology").

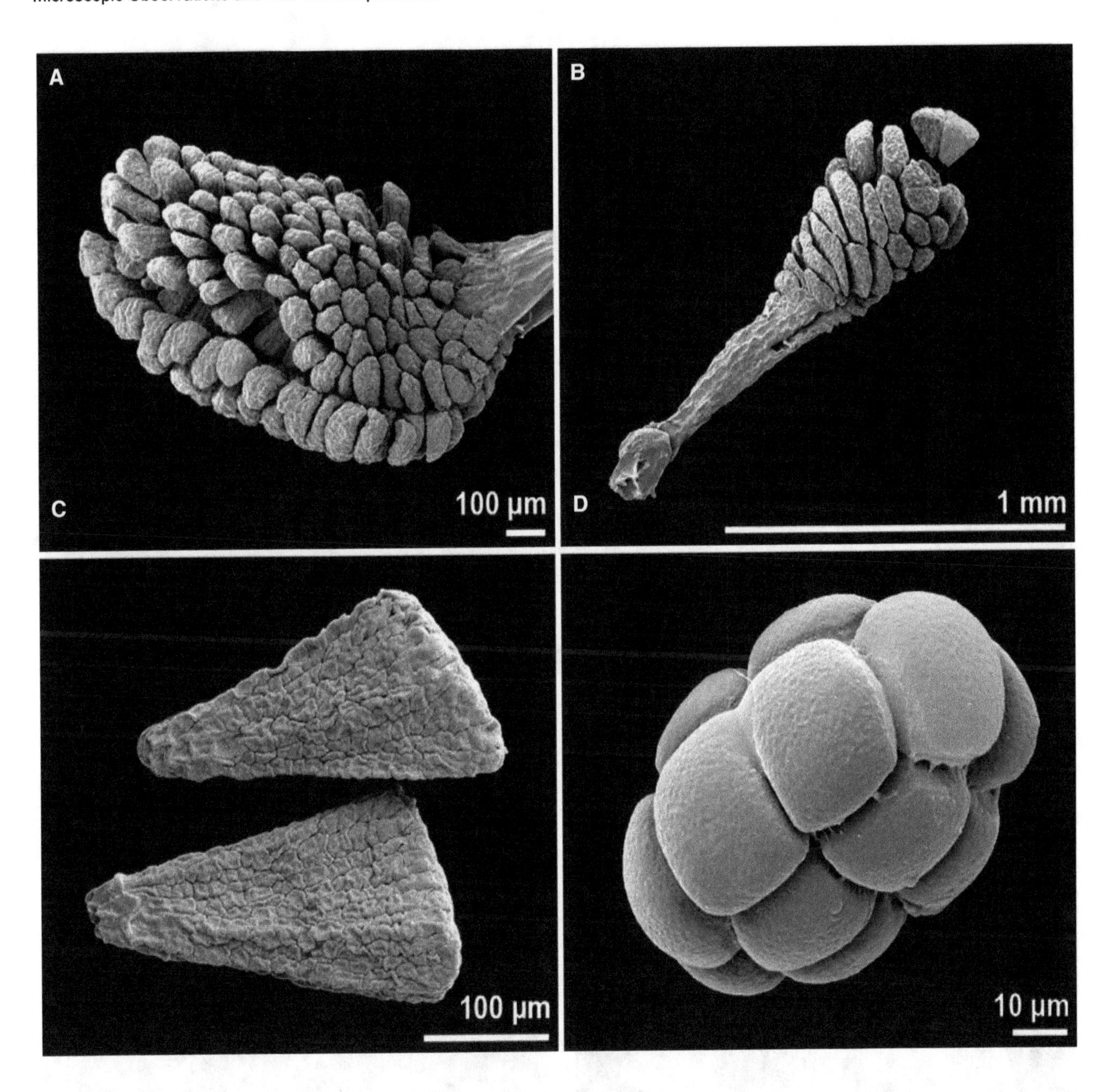

Fig. 18 Massula vs. polyad. A. *Habenaria* sp., Orchidaceae, pollinium composed of numerous massulae (massula highlighted). **B.** *Orchis ustulata*, Orchidaceae, pollinium composed of numerous massulae, two massulae partly segregated (massula highlighted). **C.** *Ludisia discolor*, Orchidaceae, 2 segregated massulae. **D.** *Albizia julibrissin*, Fabaceae, polyad (monad highlighted)

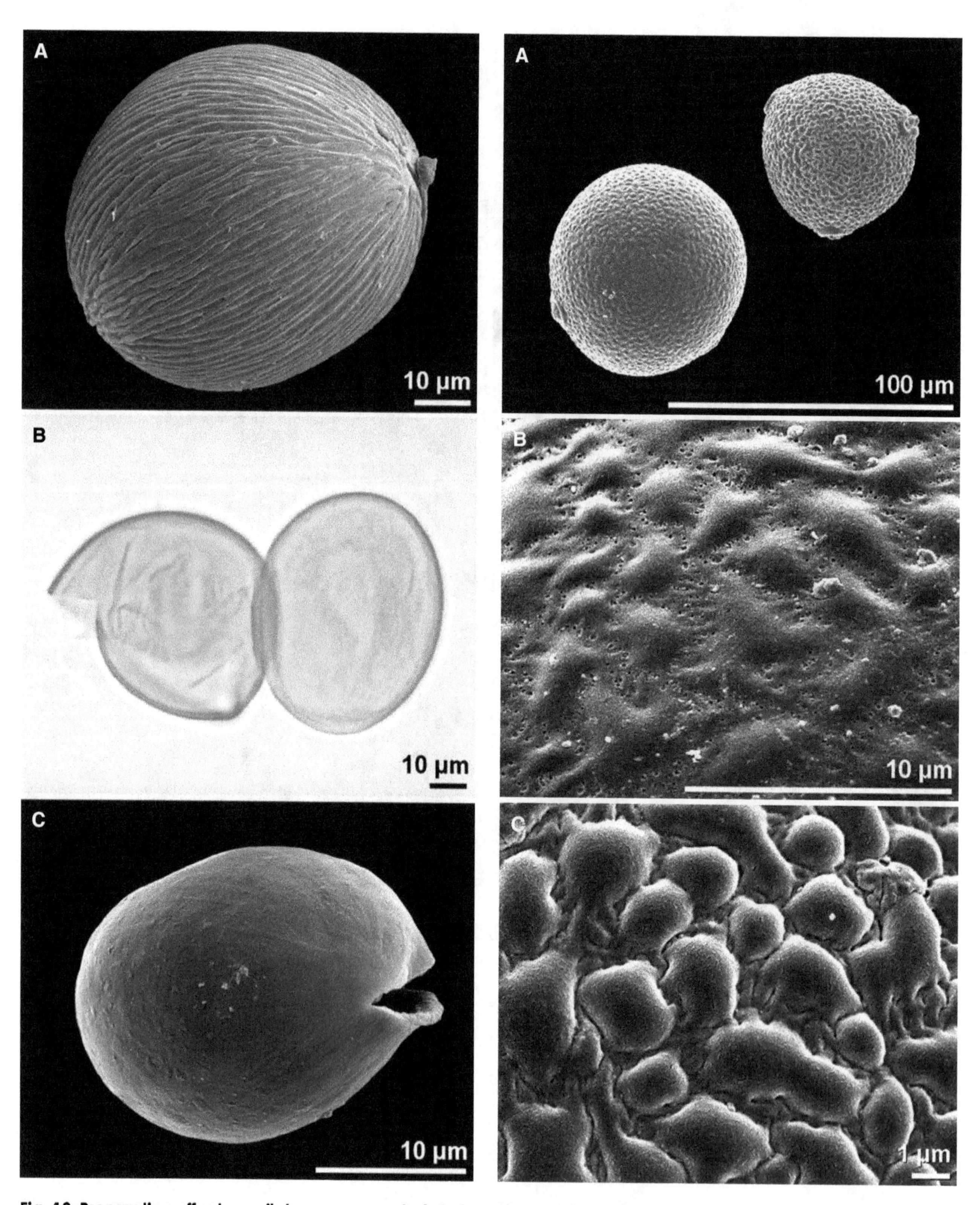

Fig. 19 Preparation effect — psilate vs. ornamented. A-C. *Amorphophallus krausei*, Araceae, pollen striate in hydrated condition (**A**), psilate after acetolysis, LM (**B**) and SEM (**C**)

Fig. 20 Preparation effect on ornamentation. A-C. *Trichosanthes anguina*, Cucurbitaceae. **A.** Pollen at different state of hydration: fully hydrated (left), less hydrated (right). **B.** Hydrated pollen, surface detail, verrucate, perforate. **C.** Less hydrated, surface detail, areolate-fossulate

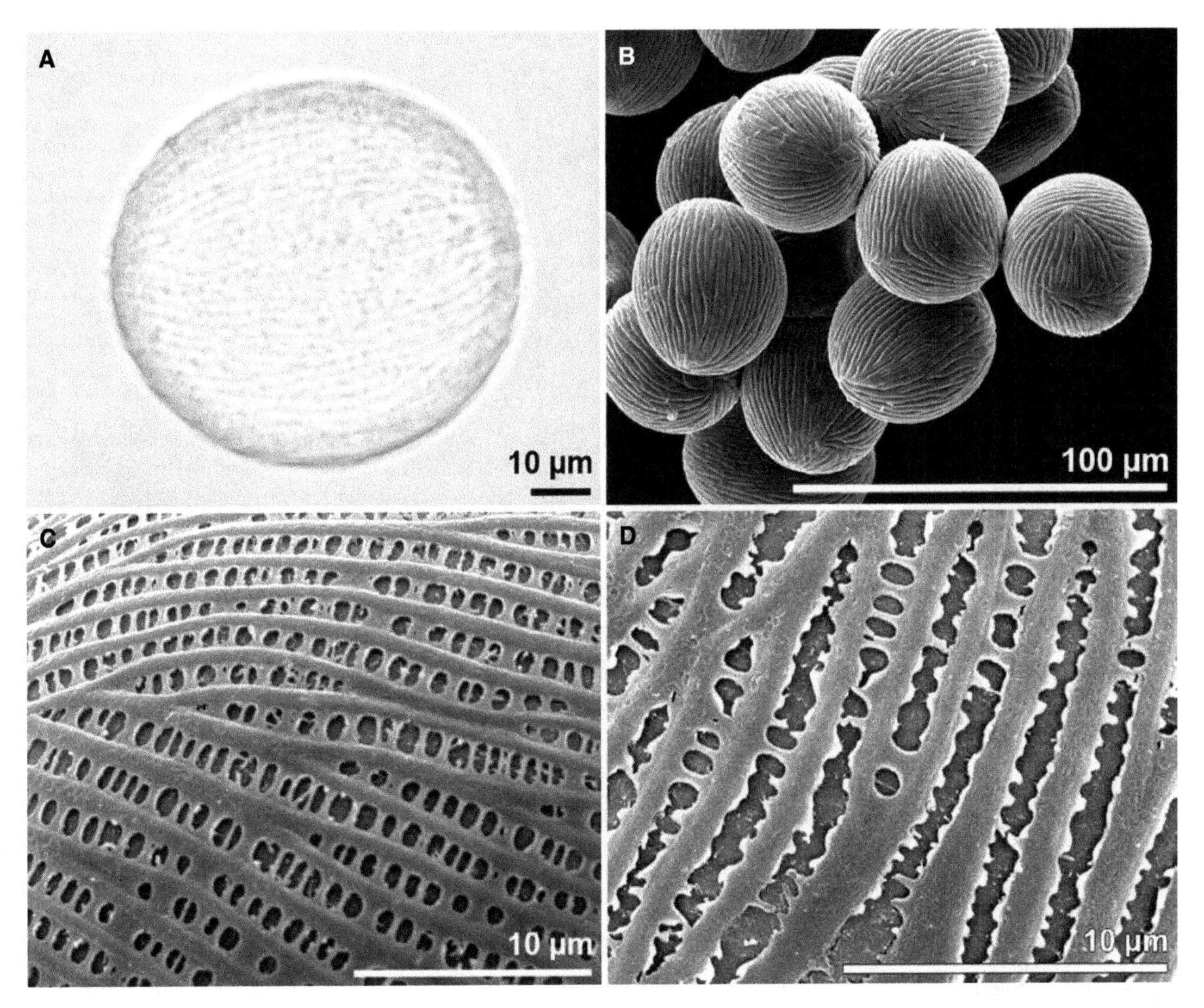

Fig. 21 Preparation effect on ornamentation. A-D. *Amorphophallus longituberosus*, Araceae, hydrated pollen in water with striate ornamentation, LM (**A**), dry pollen in SEM, striate (**B**), hydrated pollen in SEM, striate to reticulate (**C**), hydrated pollen in SEM, ornamentation striate with expanding thin surface layer (**D**)

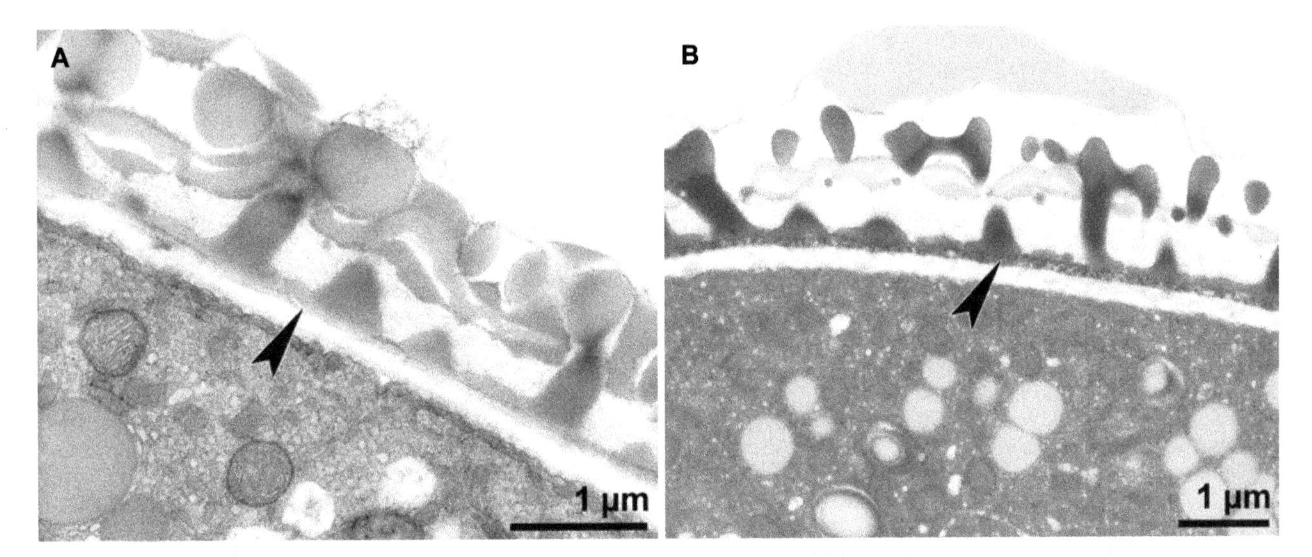

Fig. 22 Absence or presence of endexine. A-B. *Thymus odoratissimus*, Lamiaceae, U + Pb staining, endexine (arrowhead) not clearly visible (**A**), potassium permanganate staining, endexine (arrowhead) clearly visible (**B**)

References

Erdtman G (1952) Polelen Morphology and Plant Taxonomy. Angiosperms. Almqvist & Wiksell, Stockholm

Feuer SM (1977) Pollen morphology and evolution in the Santalales sen. str., a parasitic order of flowering plants. Thesis, University of Massachusetts

Feuer SM, Kuijt J (1985) Fine structure of mistletoe pollen VI. Small flowered neotropical Loranthaceae. Ann Missouri Bot Gard 72: 187–212

Gabarayeva NI, Rowley JR (1994) Exine development in *Nymphaea colorata* (Nymphaeaceae). Nordic J Bot 14: 671–691

Grayum MH (1992) Comparative external pollen ultrastructure of the Araceae and putatively related taxa. Monogr Syst Bot Missouri Bot Garden 43: 1–167

Grímsson F, Grimm, GW, Zetter R (2018) Evolution of pollen morphology in Loranthaceae. Grana 57: 16–116

Halbritter H (2016) Pontederia cordata. In: *PalDat* - A palynological database. https://www.paldat.org/pub/Pontederia_cordata/300430; [accessed 2018-08-07]

Halbritter H, Hesse M (1993) Sulcus morphology in some monocot families. Grana 32: 87–99

Harley MM (1999) Tetrad variation: its influence on pollen form and systematics in the Palmae. In: Kurmann MH, Hemsley AR (eds) The Evolution of Plant Architecture. Royal Botanic Gardens, Kew, p. 289–304

Harley MM (2004) Triaperturate pollen in the monocotyledons: configuration and conjecture. Plant Syst Evol 247: 75–122

Hesse M, Halbritter H, Weber M (2009) *Beschorneria yuccoides* and *Asimina triloba* (L.) Dun: examples for proximal polar germinating pollen in angiosperms. Grana 48: 1151–159

Hesse M, Zetter R (2005) Ultrastructure and diversity of recent and fossil zona-aperturate pollen grains. Plant Syst Evol 255: 145–176

Huynh KL (1976) Arrangement of some monosulcate, disulcate, trisulcate, dicolpate and tricolpate pollen types in the tetrads, and some aspects of evolution in the angiosperms. In: Ferguson IK, Muller M (eds) The evolutionary significance of the exine. Academic Press, London, p. 101–124

Iversen J, Troels-Smith J (1950) Pollenmorfologiske definitioner og typer. Pollenmorphologische Definitionen und Typen. Danm Geol Unders, ser 4, 3: 1–54

Kremp GOW (1968) Morphologic Encyclopedia of Palynology. 2nd edition, Arizona Press, Tucson

Punt W, Hoen PP, Blackmore S, Nilsson S, Le Thomas A (2007) Glossary of pollen and spore terminology. Rev Palaeobot Palynol 143: 1–81

Richard LC (1817) De Orchideis Europaeis annotationes, praesertim ad genera dilucidanda spectantes. Belin, Paris

Schols P, Furness CA, Merckx V, Wilkin P, Smets E (2005) Comparative pollen development in Dioscoreales. Int J Plant Sci 166: 909–924

Teppner H (2007) Notes on terminology for Mimosaceae polyads, especially in Calliandra. Phyton 46(2): 231–236

Traverse A (2007) Paleopalynology. 2nd ed, Springer, Dordrecht

Ulrich S, Hesse M, Bröderbauer D, Bogner J, Weber M, Halbritter H (2013) *Calla palustris* (Araceae): New insights with special regard to its controversial systematic position and to closely related genera. Taxon 62: 701–712

Ulrich S, Hesse M, Weber M, Halbritter H (2017) *Amorphophallus*: New insights into pollen morphology and the chemical nature of the pollen wall. Grana 56: 1–36

Wagenitz G (2003) Wörterbuch der Botanik. – 2nd edition, Spektrum, Heidelberg

Walker JW (1971) Pollen morphology, phytogeography, and phylogeny of the Annonaceae. Contributions from the Gray Herbarium 202: 1–131

Weber M, Halbritter H, Hesse M (1999) The basic pollen wall types in Araceae. Int J Plant Sci 160: 415–423

Weber M, Ulrich S (2010) The endexine: a frequently overlooked pollen wall layer and a simple method for detection. Grana 49: 83–90

Wettstein R (1907) Handbuch der systematischen Botanik 2(2/1): 161–394

Zetter R, Ferguson DK (2001) Trapaceae pollen in the Cenozoic. Acta Palaeobot 41: 321–339

5

Proper Presentation of Palynological Data

For the description of a pollen grain, a number of features are used including size, polarity and shape, aperture condition, ornamentation, and pollen wall structure. Additional and often more specialized features depend on the group of plants under study, Gymnosperms (Cycadales, Ginkgoales, Pinales, Gnetales) vs. Angiosperms (magnoliids, monocots, commelinids, eudicots). These features can only be obtained by the application of a combined analysis with LM, SEM, and TEM (Fig. 1). In order to compare and categorize pollen, a common language and understanding of technical terms is necessary.

The description and illustration of a pollen grain depends on how the material is going to be presented and if one is describing a single fossil pollen grain, pollen of a particular extant species, pollen representing several species, a whole genus, several related genera, a complete family, or even a number of families. For future work it is important to provide both LM and SEM micrographs (even TEM), including incorporated scale bars, showing each taxon and close-ups of what are considered diagnostic features of pollen. When documenting the sculpture of pollen grains in SEM it has to be made sure that the magnification is high enough to distinguish the shape and outline of sculpture elements larger than 0.1 μm in diameter. LM- and SEM-diagnosis may be different from each other, due to the methods and techniques used. The methods used to prepare pollen grains for LM, SEM, and TEM must be mentioned along with the pollen descriptions, preferably in a material and method section.

Pollen from a Single Extant Taxon: Online Publication in *PalDat*

Pollen grains from single extant species have rarely been accepted by scientific journals. There is now a new online venue *PalDat*, for publishing pollen from a single species. *PalDat* is the world's most comprehensive pollen database (www.paldat.org) and contains tools for pollen identification as well as global, free online submission and publication with review and editorial process (Weber and Ulrich 2017). *PalDat* already provides a large amount of pollen data on a variety of plant families. Each taxon entry (online publication) ideally includes a detailed description and micrographs (LM, SEM, and TEM) of the pollen, as well as images of the plant/inflorescence/flower and information on relevant literature (Fig. 2). *PalDat* is freely accessible and following a free registration it is open for contributions from all those willing to publish their pollen descriptions and micrographs online. Registered authors may also contribute as co-authors to existing publications by submitting new images and/or new data to pollen diagnosis (with review and editorial process). All changes are recorded in the database history as links to previous versions of the publication. Each contribution is citable and accessible for all users. Registered users can download publications in pdf form. The terminology used in *PalDat* follows this book.

Groups of Extant Pollen

Many of the classical papers on pollen morphology and ultrastructure, covering a large number of extant taxa, provide only a general description of pollen types with pollen of different species lumped together. Furthermore, micrographs are showing selected taxa and usually not the same taxon photographed in both LM and SEM. This makes the data unreliable and not very useful for among others paleopalynologists that want to compare their fossil pollen grains very precisely to particular extant taxa. The decision on particular potential modern analogues of the fossil pollen grain can have major effects on the paleovegetation reconstruction and paleoecological and paleoclimate interpretations of the fossil assemblage, as well as on the paleophytogeographic signal of the taxon. It is recommended, disregardless of the description, that all species be fully illustrated by LM and SEM (and TEM when possible) and their basic and diagnostic morphological features compiled in a table so they can be easily compared (Table 1). The example shown here are Winteraceae pollen tetrads. When portraying tetrads it is useful to show their basal-, lateral-, as well as apical view, in both LM (Fig. 3) and SEM (Fig. 4). Pollen grains should be portrayed in polar and equatorial view. Illustrating pollen from different taxa together on a plate/figure with the same magnification makes it easier to realize size differences. The SEM close-ups are then used to highlight the main sculpture features or the dissimilarities of the taxa. Ideally all close-ups showing sculpture elements should have the same magnification for an easy comparison (Fig. 5).

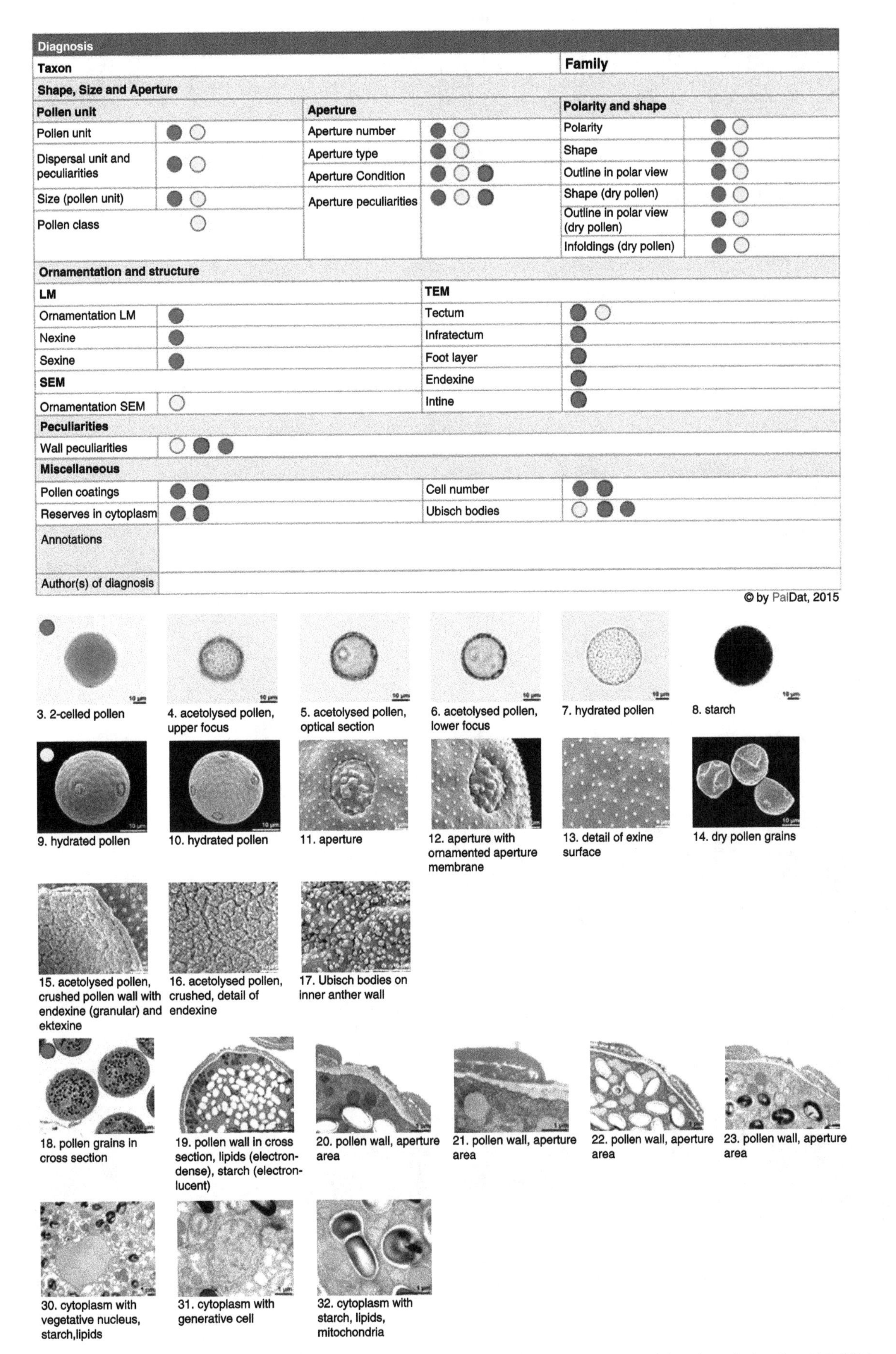

Fig. 1 Diagnosis worksheet. *PalDat* worksheet with all pollen features obtained by a combined analysis using LM, SEM, and TEM. Blue dots indicate LM-, yellow dots SEM-, and red dots TEM-based analyses. *PalDat* pictures showing *Plantago maritima*

PalDat - Palynological Database

an online publication on recent pollen

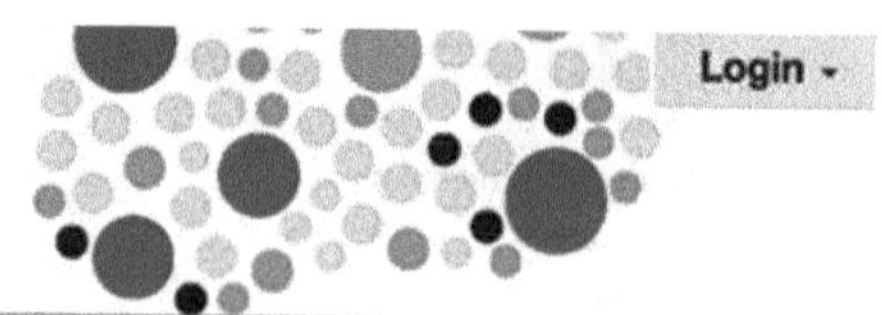

Login -

Home Search Data Submit Data Terminology Information Get Pictures Register

Alphabetical Search Combined Search

Betula pendula

Taxonomy: Angiospermae, Fagales, Betulaceae, *Betula*

Published: 2016-03-25

Pollen Description

Shape, Size and Aperture

pollen unit: **monad,** dispersal unit and peculiarities: **monad,** size (pollen unit): **small (10-25 µm),** pollen class: **porate,** polarity: **isopolar,** shape: **spheroidal,** outline in polar view: **circular,** shape (dry pollen): **irregular,** outline in polar view (dry pollen): **irregular,** infoldings (dry pollen): **irregularly infolded, interapertural area sunken,** aperture number: **3,** aperture type: **porus,** aperture condition: **porate, triporate,** aperture peculiarities: **annulus, operculum, oncus**

Ornamentation and Structure

LM ornamentation LM: **psilate,** nexine: **-,** sexine: **-,** **SEM** ornamentation SEM: **rugulate, microechinate,** **TEM** tectum: **eutectate,** infratectum: **columellate,** foot layer: **continuous,** endexine: **absent,** intine: **monolayered,** wall peculiarities: **-**

Miscellaneous

pollen coatings: **absent,** reserves in cytoplasm: **starch,** cell number: **2-celled,** Ubisch bodies: **present**

Annotations
tectum very mighty

Author(s) of diagnosis: Halbritter, Heidemarie; Diethart, Bernadette

Pictures

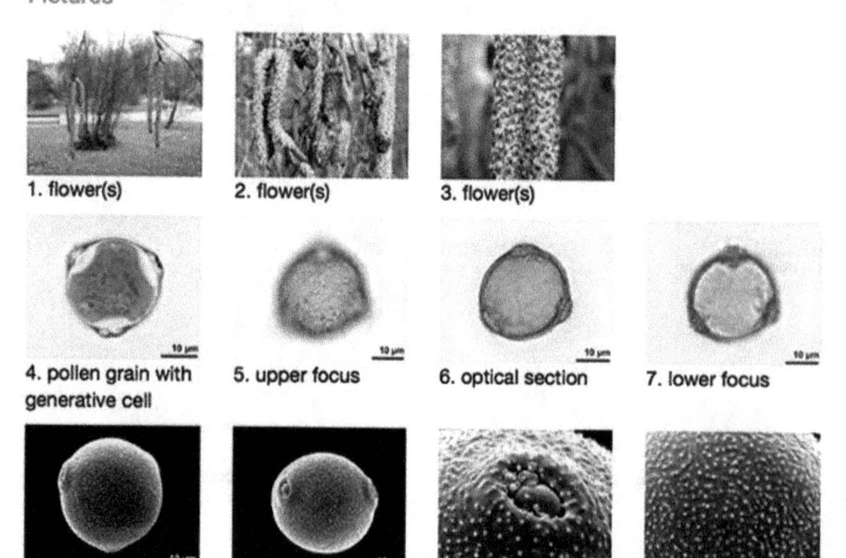

Fig. 2 Online publication in *PalDat*. Screenshot showing part of the online publication of *Betula pendula* (Halbritter and Diethart 2016)

Table 1 Winteraceae pollen tetrads

	Takhtajania perrieri	*Exospermum stipitatum*	*Tasmannia insipida*
Tetrad diameter (LM; µm)	58–65	32–38	28–33
Apertures surrounded by an annulus-like rim (width; µm)	Yes, 2.5–6	No	No
Width of aperture region (µm, longest axis)	12–17	5–6	7–11
Exine thickness (LM; µm)*	Max. 5.5	Max. 3	Max. 3.2
Nexine thickness (LM; µm)*	Max. 0.9	Max. 0.7	Max. 0.9
Sexine thickness (LM; µm)*	Max. 4.2	Max. 1.7	Max. 2.3
Sculpture (SEM)	Reticulate	Perforate to nanoreticulate	Reticulate
Muri	Broad and rounded	(Broad and rounded)	Narrow and crested
Diameter of (largest) lumina (µm; longest axis)	7–11	≤1	5–6
Number of lumina/perforations (one grain in lateral view)	c. 15/20	c. 120	c. 15–20
Height ratio columellae vs. muri	~1–1.5:1	?	~1:1
Columellae per µm	2 per 5 µm	2–3	1–2
Free-standing columellae	Frequent, mostly ≤1 µm; gemmae, bacula, and clavae	Absent	Rare, mostly ≤0.5 µm; verrucae, gemmae, and clavae
Ulcus membrane (SEM)	Granulate to microverrucate	Granulate to nanoverrucate	Granulate, nano- to microclavate

Main features of three different Winteraceae pollen tetrads

Annotation: Measurements like exine, nexine, and sexine thickness provided in Table 1 (asterisks) are commonly used in (paleo) palynological literature. Scientists should be aware that such measurements (e.g., 0.7 or 0.9 µm) vary highly, up to 30%, depending on the methods and tools used. Therefore, the measurements should not be overrated or used for taxonomic discrimination.

Fossil Pollen

From the birth of paleopalynology this branch of science has been plagued by the lack of taxonomic foundation when interpreting paleoenvironments. It is very unfortunate that numerous new "scientific" publications dealing with the subjects of paleoecology, paleovegetation, paleoclimate and various aspects of paleophytogeography still present only a list of taxa observed in LM. Some publications include LM micrographs of the most "common" taxa, but only in exceptional cases the LM micrographs are accompanied by SEM micrographs. The absence of illustrations makes it impossible for any reader to verify, or later revise, the taxonomic background and to conclude if the modern living relative or potential modern analogue of the fossil taxon is justified. Every proper scientific journal should make it a mandatory request that all pollen types are represented by at least one LM micrograph. Furthermore, all taxa that suggest some sort of different, abnormal or exceptional paleo-parameters, in an otherwise "homogeneous" assemblage, or taxa that are used to set any sort of boundaries (temperature, precipitation, biozone, time, etc.), should be illustrated using both LM and SEM (in some cases even TEM). These contrasting taxa might include a dry element in an otherwise humid assemblage, a tropical element in an otherwise temperate assemblage, or an African element in an otherwise North American-Eurasian assemblage. Even though the journal would not allow these illustrations in the printed version most of them now offer the possibility to archive online supplementary files where the pollen can be illustrated.

For those who want to produce a taxonomically valid study based on fossil material are advised to use the single-grain method when investigating fossil pollen and make sure not to sieve the sample

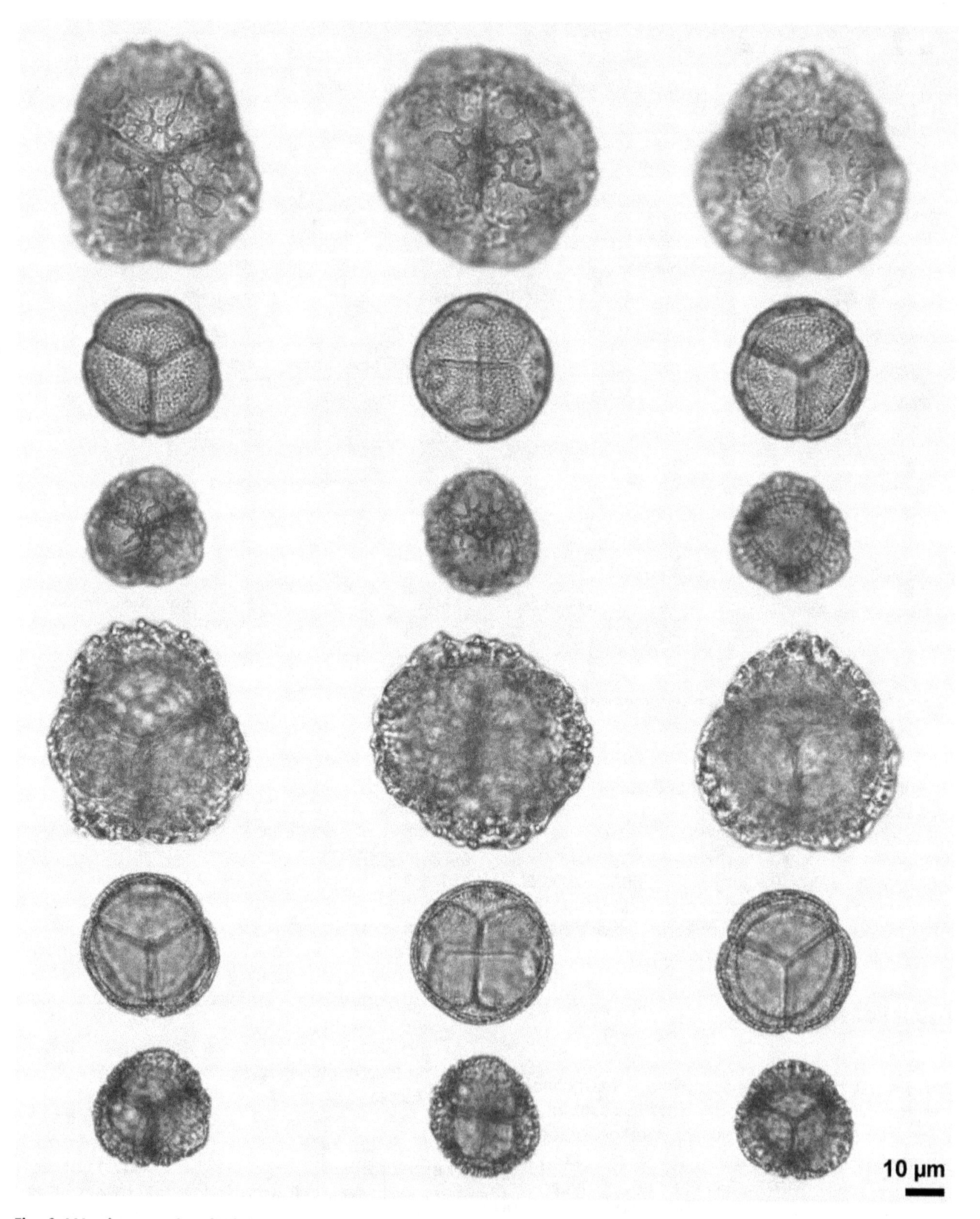

Fig. 3 LM micrographs of Winteraceae pollen tetrads. Tetrads shown in basal-(left), lateral-(middle), and apical (right) view at high focus (upper three rows) and in optical cross section (lower three rows). *Takhtajania perrieri* (first and fourth row), *Exospermum stipitatum* (second and fifth row), *Tasmannia insipida* (third and sixth row)

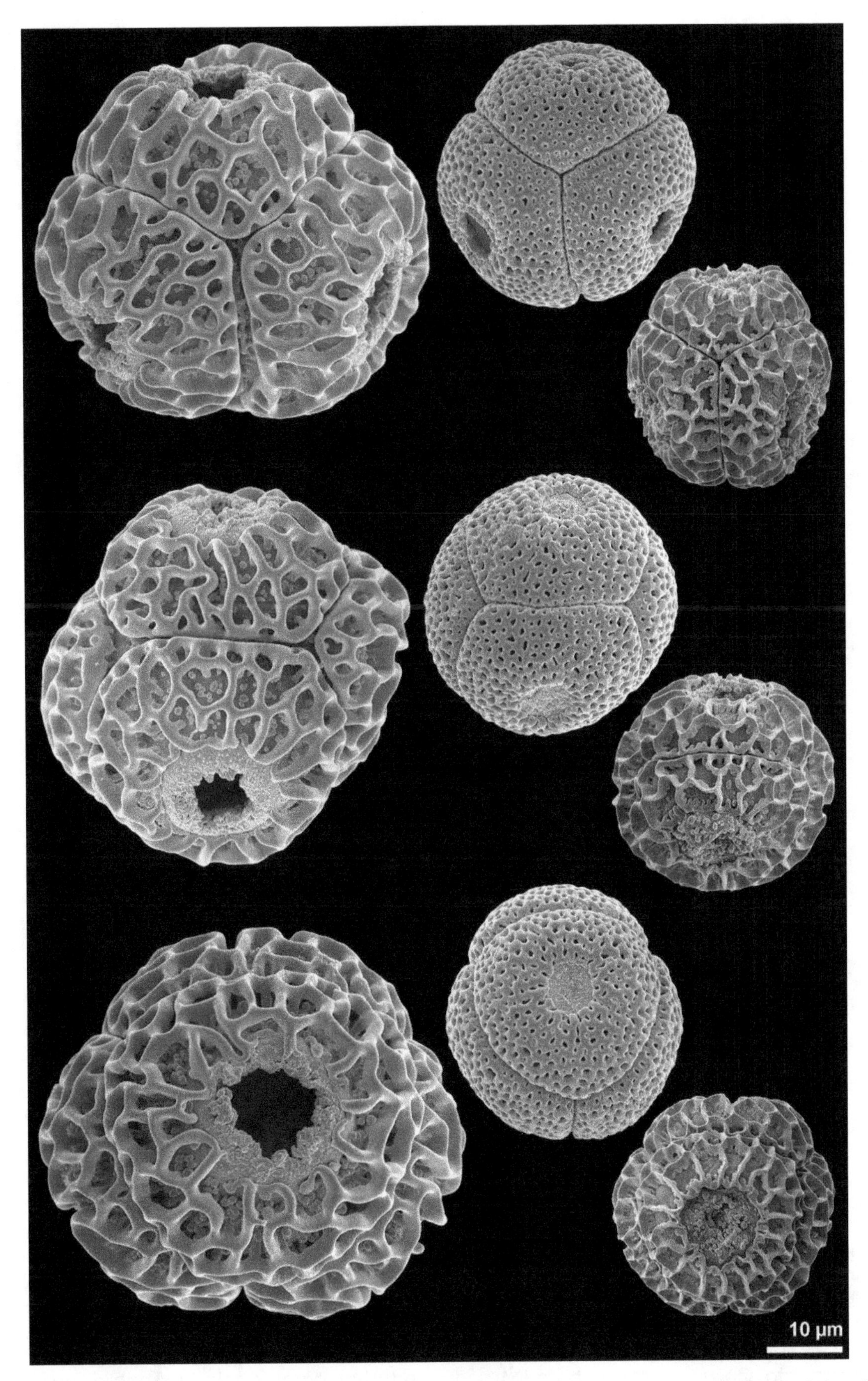

Fig. 4 SEM micrographs of Winteraceae pollen tetrads. Tetrads shown in basal view (upper row), lateral view (middle row) and apical view (lower row). *Takhtajania perrieri* (left), *Exospermum stipitatum* (middle), *Tasmannia insipida* (right)

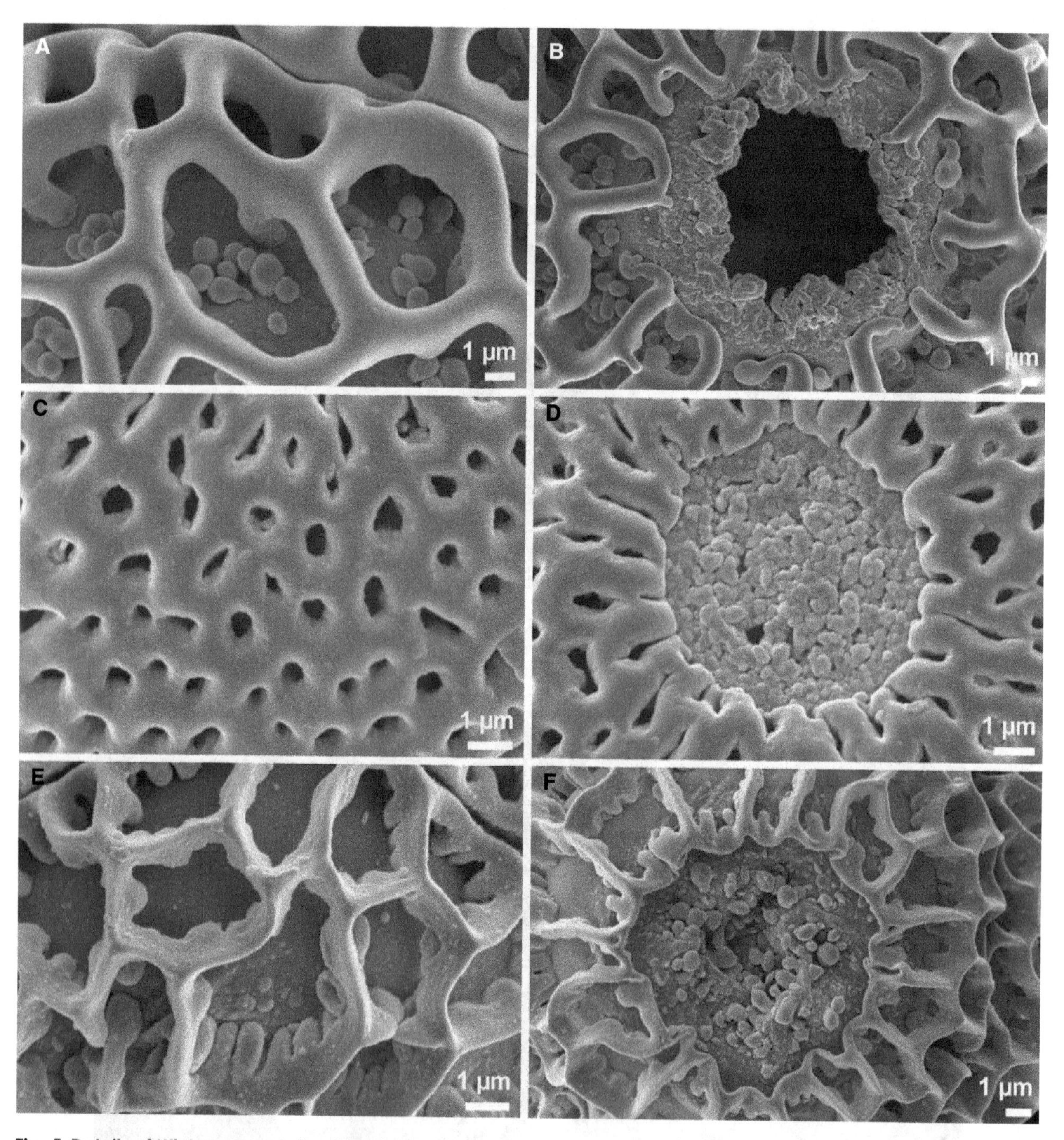

Fig. 5 Details of Winteraceae pollen tetrads. SEM close-ups of *Takhtajania perrieri* (**A-B**), *Exospermum stipitatum* (**C-D**) and *Tasmannia insipida* (**E-F**), showing sculpture on distal face of pollen (**A, C, E**) and the aperture region and ulcus membrane (**B, D, F**)

during preparation (see "Methods in Palynology"). This allows the researcher to study all elements occurring within a sample using both LM and SEM and to investigate even very small and/or rare pollen grains. The small and/or rare pollen (Fig. 6) would otherwise be overlooked during the old-fashion routine LM observation, where the researcher usually counts 300–600 grains. When illustrating fossil pollen it is important to show the grain in both LM and SEM. Close-ups taken with the SEM should have magnification high enough so all sculpture elements larger than 0.1 µm become distinguishable. Sculpture and suprasculpture elements smaller than 1 µm are not observed or hard to distinguish using LM only, but will be revealed using high magnification SEM (Fig. 7). Many pollen grains that look similar or the same in LM can be distinguished using SEM. In some cases it is beneficial to turn the pollen grain once it has been photographed in SEM, re-sputter and photograph again. This applies especially to heteropolar pollen grains (Fig. 8) as well as pollen dispersed in permanent tetrads. When single pollen

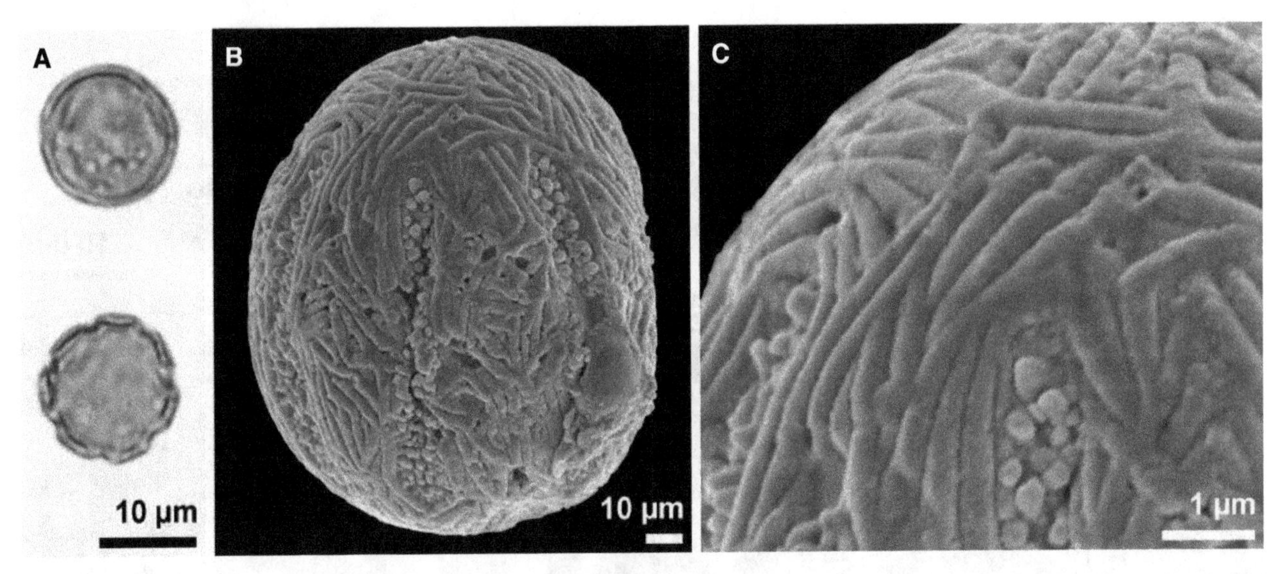

Fig. 6 Small and rare pollen, Paleocene, Western Greenland. A. small fossil grains (≤10 µm in diameter) usually absent in samples after sieving. LM micrographs (left) in equatorial (upper) and polar view (lower). **B.** pollen in equatorial view, SEM. **C.** striate sculpture not seen under low magnification LM

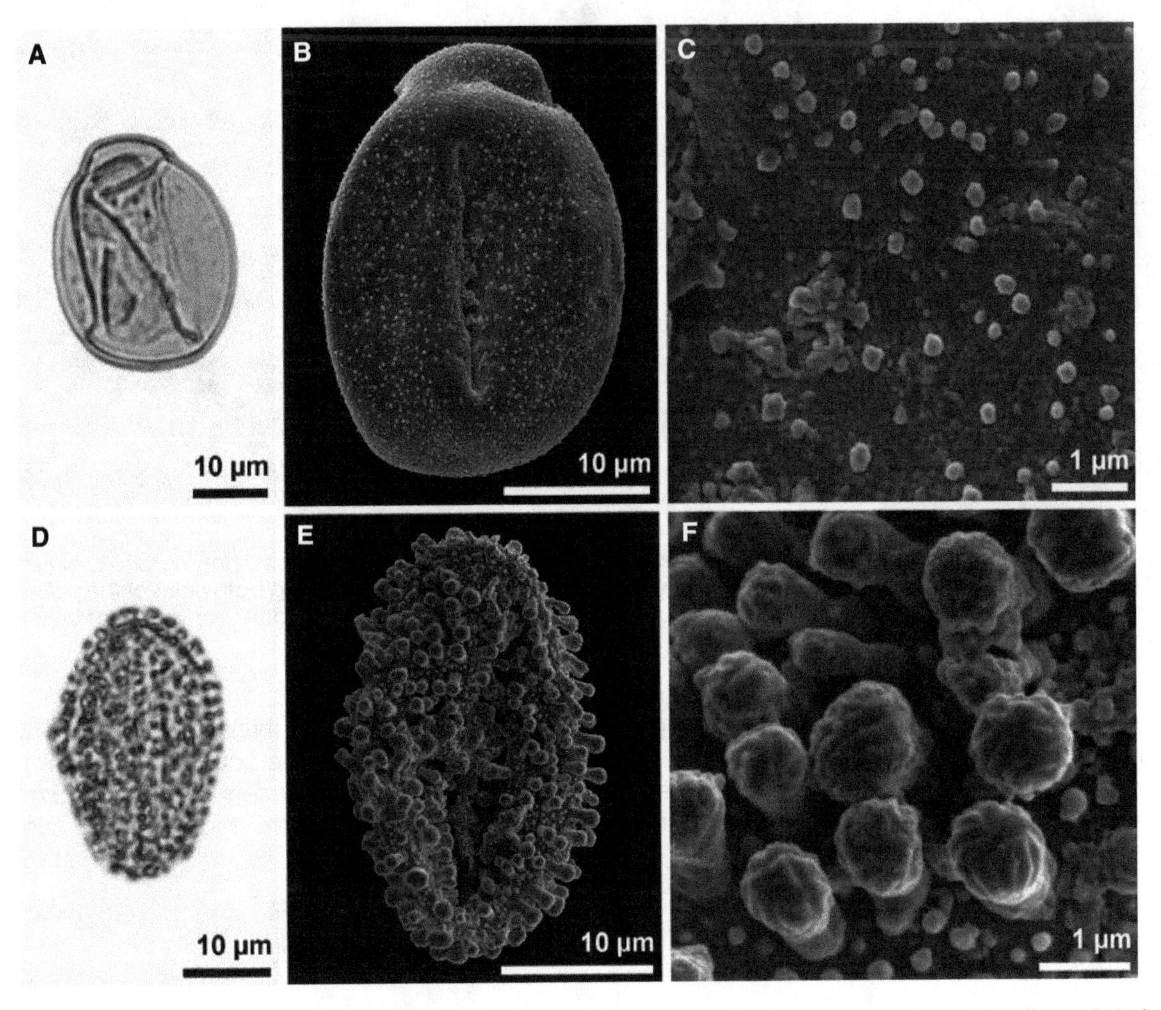

Fig. 7 Ornamentation LM vs. SEM, fossil, Middle Eocene, Western Greenland. A-C. *Eucommia* sp. **A.** Pollen psilate in LM. **B.** Pollen in SEM, equatorial view, note sculpture. **C.** Ornamentation nanoechinate (≤0.5 µm) and granulate. **D-F.** *Ilex sp.*, **E.** LM and SEM overviews show the typical clavate sculpture known for this genus. **F.** Microrugulate suprasculpture present on the distal part of the clavae, only observed using high magnification SEM

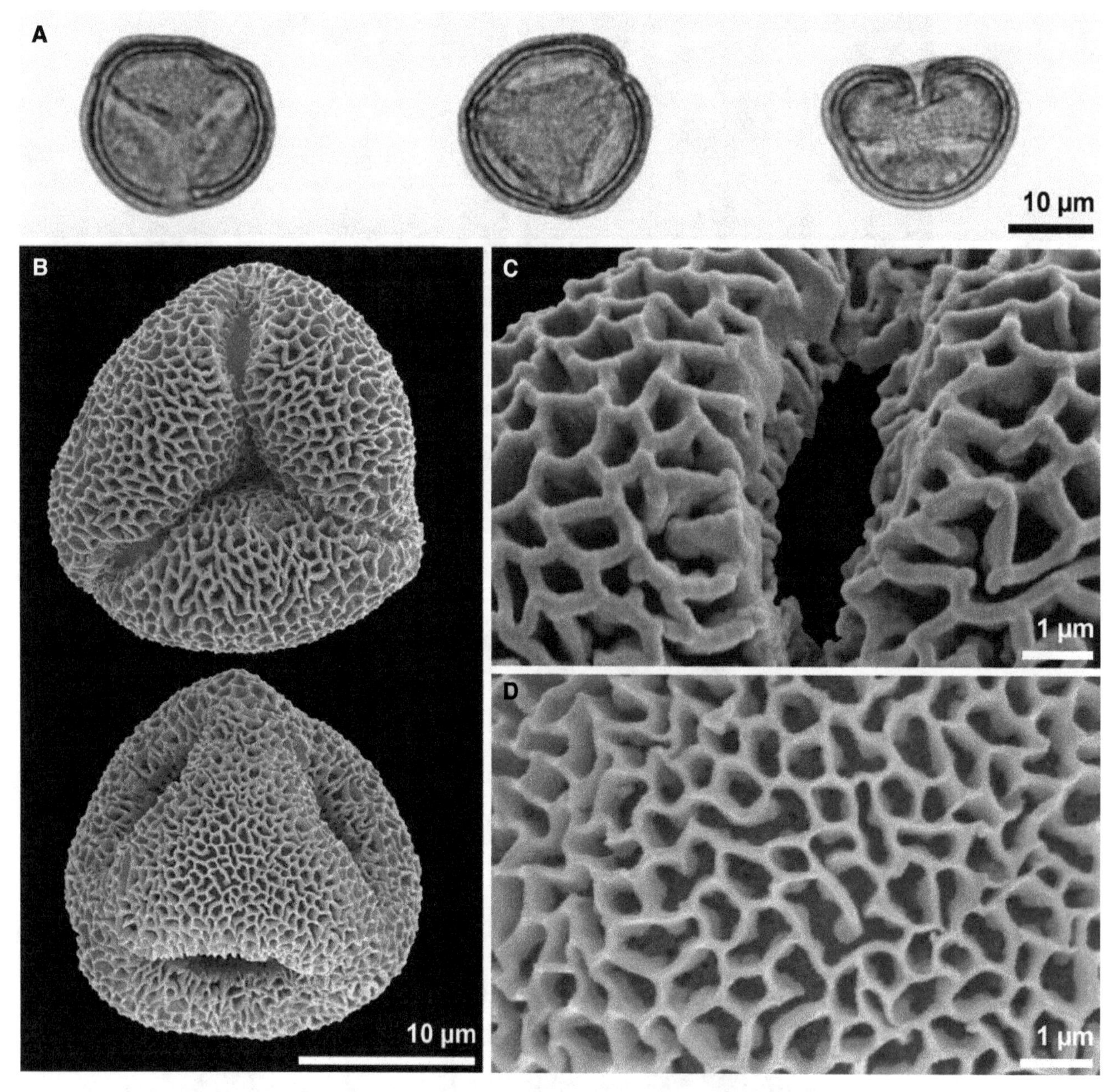

Fig. 8 Fossil heteropolar pollen grain, Paleocene, Western Greenland. A. LM micrographs showing proximal (left) and distal (middle) poles of pollen grain and equatorial view (right). **B.** SEM overviews showing both poles of the pollen grain and the different aperture arrangements. **C-D.** SEM close-ups of proximal (**C**) and distal poles (**D**) show that the muri are much broader on the proximal pole

grains or tetrads are studied using SEM, changes in sculpture over the pollen surface are often observed, for example polar vs. equatorial region, mesocolpium vs. aperture region vs. aperture membrane (Fig. 9). Some pollen or tetrads also have Ubisch bodies or viscin threads (Hesse et al. 2000). These differences in the sculpture of fossil pollen need to be documented and it is therefore often necessary to show more than a single close-up taken with the SEM.

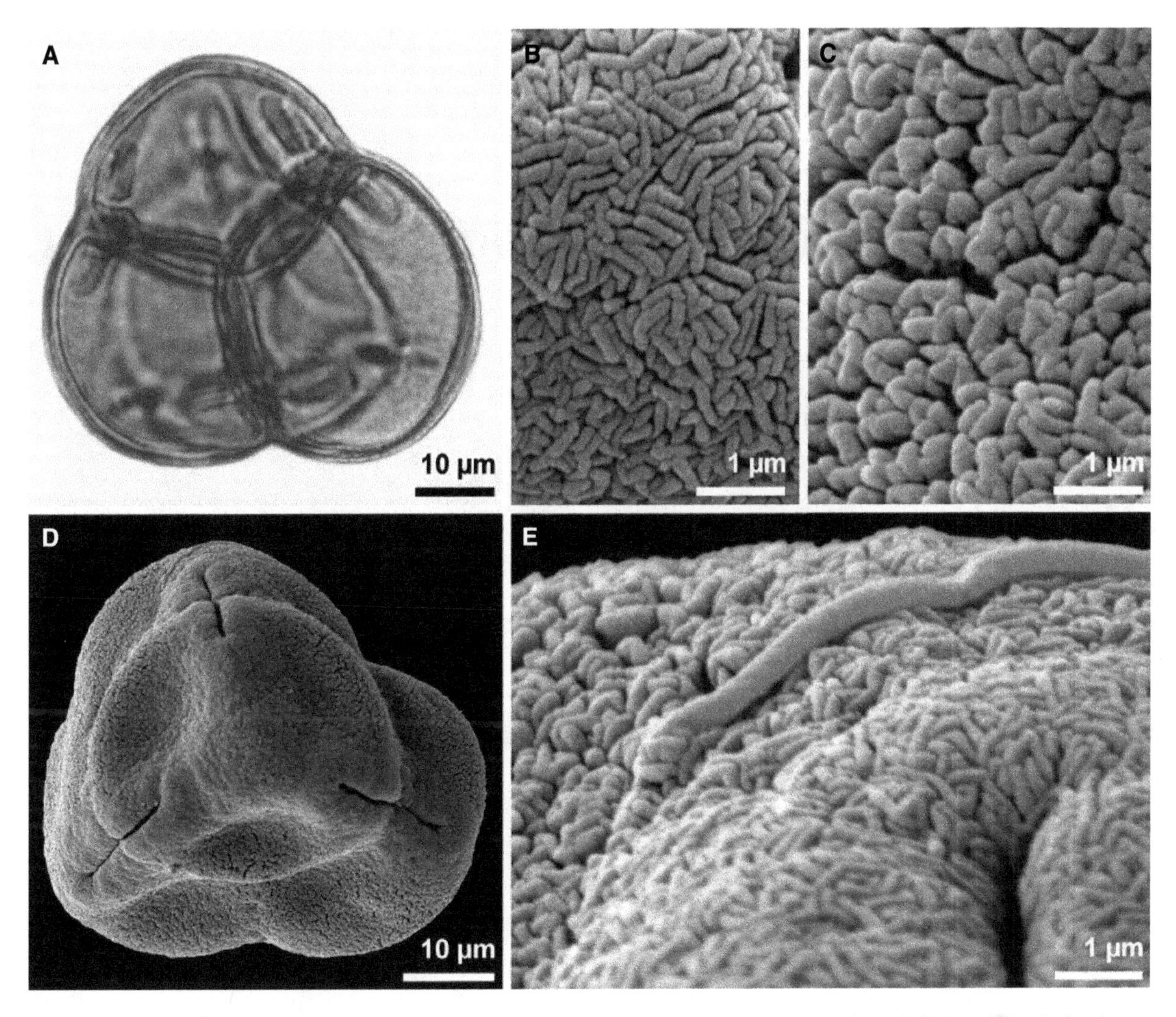

Fig. 9 Fossil tetrad, *Rhododendron* sp., Miocene, North-east China. A, D. Tetrad, overviews in LM vs. SEM. **B-C.** close-ups at same, magnification show difference in sculpture at polar region of pollen grain (**B**) vs. interapertural area (**C**). **E.** exine surface with viscin thread, SEM

References

Halbritter H, Diethart B (2016) *Betula pendula*. In: PalDat – a palynological database. Published on the Internet https://www.paldat.org/pub/Betula_pendula/300732 [accessed 2017–04–28]

Hesse M, Vogel S, Halbritter H (2000) Thread-forming structures in angiosperm anthers: their diverse role in pollination ecology. Plant Syst Evol 222: 281–292

PalDat – a palynological database (2000 onwards, www.paldat.org)

Weber M, Ulrich S (2017) PalDat 3.0 – second revision of the database, including a free online publication tool. Grana 56: 257–262

6

Palynology Research: Techniques and Methods

Preparation of Recent and Fossil Material for LM, SEM, and TEM

Light Microscopy

Scanning Electron Microscopy: Preparation of Recent Pollen

Transmission Electron Microscopy: Pollen Wall Stratification and Ultrastructure

Staining Methods

Preparation of Fossil Pollen

Preparation Method: From Rock to Palynomorphs

The Single-Grain Method

Recipes

References

Preparation of Recent and Fossil Material for LM, SEM, and TEM

Multiple methods and techniques should be used when investigating pollen grains in order to provide comprehensive and accurate information about pollen morphology and ultrastructure (see also "Misinterpretations in Palynology"). The preparation methods used depend on the material to be studied, if the pollen grains are to be obtained from recent flower material (herbarium sheets, newly collected) or from various sedimentary rocks, sediments or soils (fossil to subfossil pollen). Recent and fossil pollen grains are easily studied using both LM and SEM, but recent pollen grains are also more often studied using TEM.

For an accurate description of any taxonomic value, it is important to study pollen grains in both LM and SEM. The LM will provide, among others, information on the endoaperture that cannot be obtained using SEM. Likewise the SEM will provide detailed information on the sculpture of the pollen grain that is not visible under the low magnification provided by the LM. For example, terms with "micro-" (like microreticulate) or "nano-" (like nanoechinate) can only be observed using SEM (Fig. 1).

Annotation: The methods described in this section are the standard palynological techniques applied by the authors of this book and may differ in other working groups/labs around the world. All LM, SEM, and TEM micrographs in this book are produced following these standard protocols. Recipes for preparations are included at the end of this section.

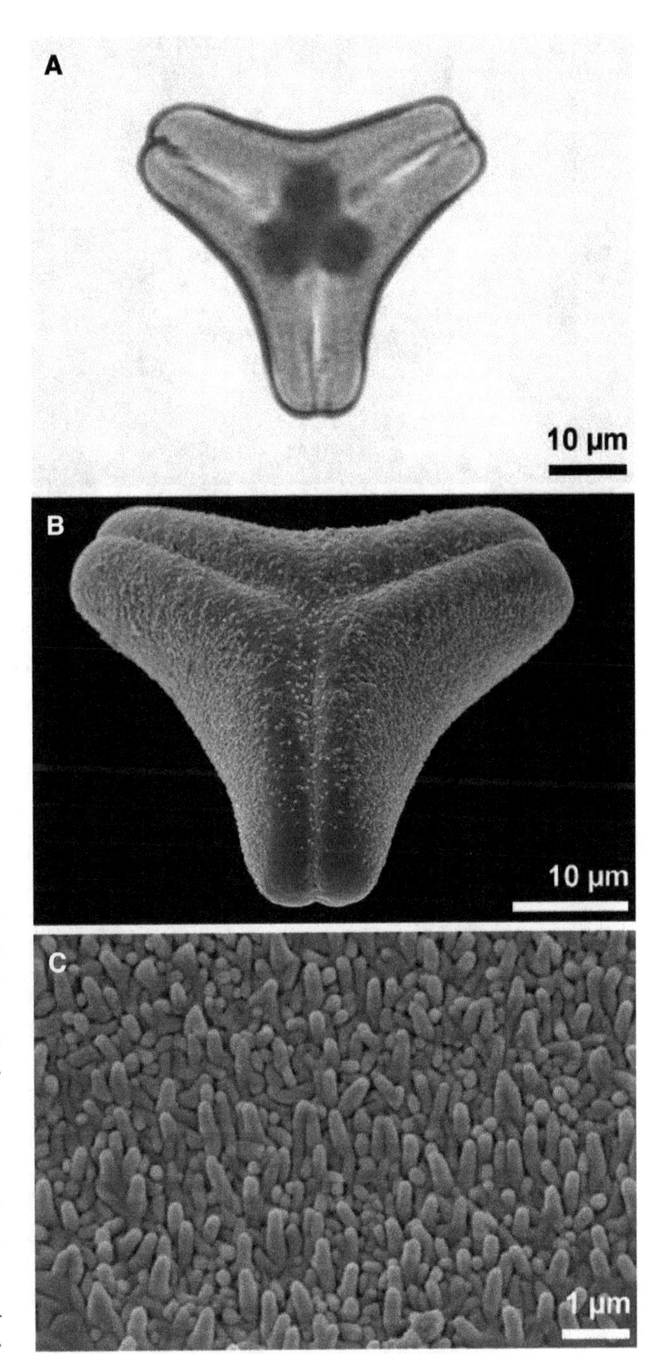

Fig. 1 LM vs. SEM. A-C. *Aetanthus coriaceus*, Loranthaceae. **A.** Pollen grain looks psilate or scabrate in LM. **B.** Sculpture elements become visible under SEM. **C.** The sculpture elements are nano- to microbaculate and only identifiable using high magnification

Light Microscopy

Pollen Hydration Status at Dispersal

To clarify the dehydration status of pollen grains at anthesis, pollen must be collected from newly opened anthers (Fig. 2). Fresh pollen grains are transferred immediately into a drop of pure glycerine and should be observed as soon as possible, as pollen grains expand in glycerine (within days or

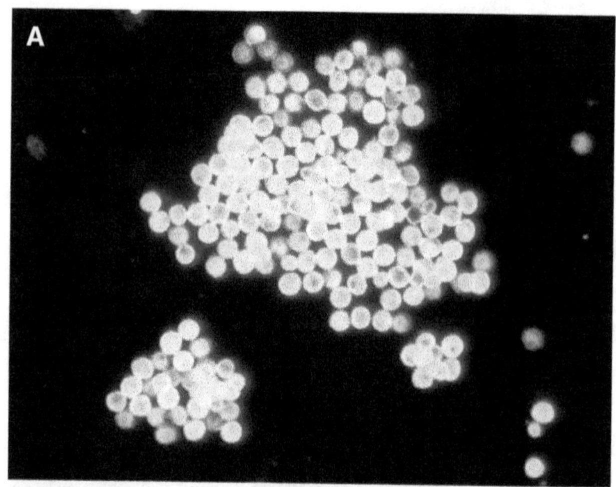

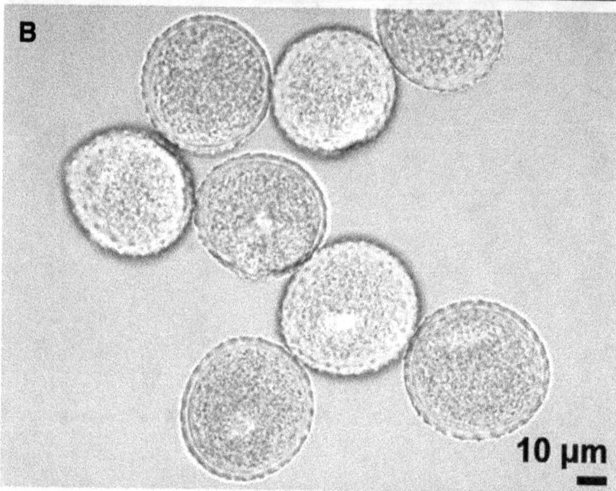

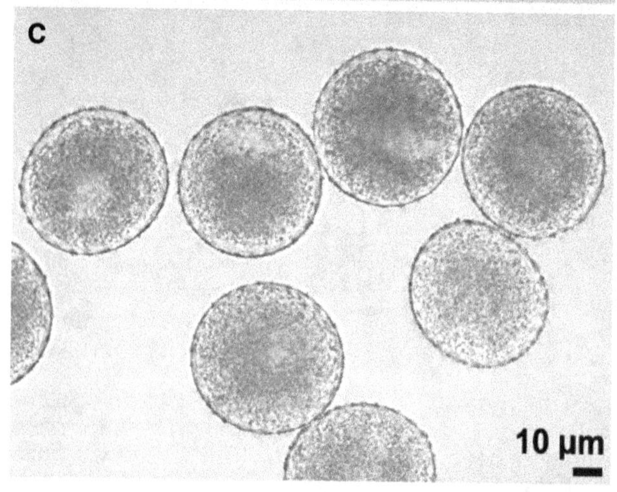

Fig. 2 Pollen hydration status at dispersal. A-C. *Alocasia* sp., Araceae. **A.** Pollen grains fully hydrated at anthesis, binocular microscope. **B.** Pollen in glycerine, LM. **C.** Pollen hydrated in water, LM

weeks). The water content of pollen grains at the time of dispersal varies and pollen can be fully hydrated, partially hydrated, or partially dehydrated (Heslop-Harrison 1979; Nepi et al. 2001; see also **harmomegathic effect** in "Pollen Morphology and Ultrastructure").

Pollen Hydrated in Water

Fresh or dry pollen grains are hydrated in a drop of water on a glass slide and observed in LM. This should be the first step before preparing pollen for SEM to get an impression about the quality of the collected material, to make sure that the material is not degenerated or contaminated by fungi (Fig. 2). Observations on pollen hydrated in water with the LM can reveal interesting aspects. One example is *Montrichardia* (Araceae), where a drop of water triggers a massive expansion of the thick intine resulting in an explosive opening of the pollen wall (Weber and Halbritter 2007).

Clarify the Pollen Polarity and Aperture Type

To clarify the pollen polarity and the aperture type, anthers with pollen tetrads must be collected before anthesis (usually found in flower buds). Pollen tetrads can be released from the anthers in a drop of water or in glycerine. Quite often different developmental stages can be found in one anther: microspores in early and late tetrad stages (with or without callose wall), but also young microspores (before first pollen mitosis) released from the tetrad as well as mature pollen grains (Fig. 3; see also Fig. 1 in "Pollen Development"). For the investigation it might be useful to stain the material, e.g. with toluidine blue or basic fuchsin (Siegel 1967).

Acetocarmine Staining: Detection of the Cellular Condition

For the detection of the cellular condition of pollen grains, fresh pollen are put into a drop of acetocarmine and warmed on a heating plate (up to

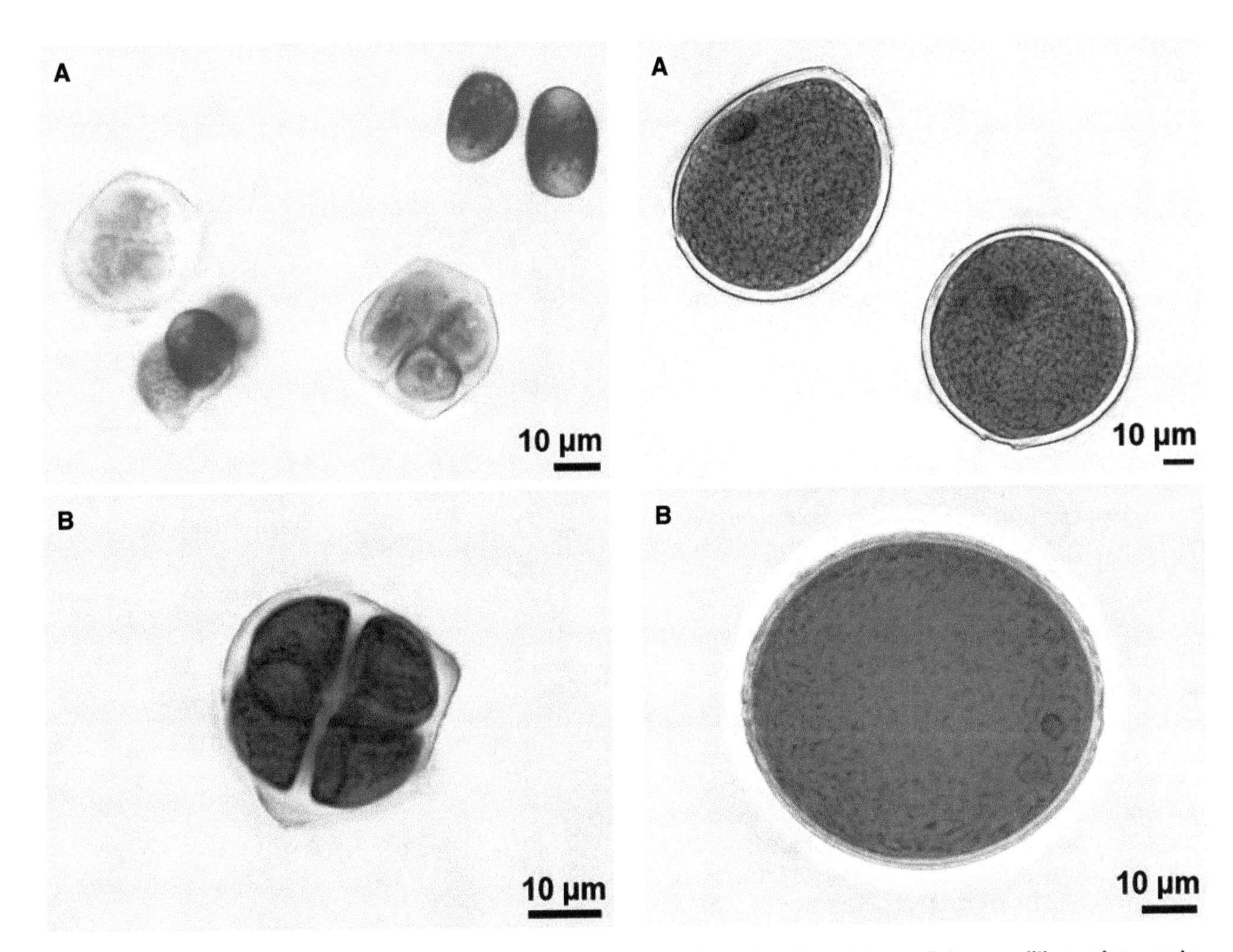

Fig. 3 Clarify the pollen polarity. A-B. *Calla palustris*, Araceae, tetrads in different stages as well as free microspores stained with toluidine blue (**A**) and basic fuchsin (**B**)

Fig. 4 Clarification of the cellular condition using aceto-carmine. A. Binucleate pollen of *Anchomanes welwitschii*, Araceae, generative nucleus stains intensive red. **B.** Trinucleate pollen of *Amorphophallus krausei*, Araceae, sperm nuclei stain intensive red

70 °C), for a few seconds to several minutes (species dependent), and observed under the LM (Gerlach 1984). The generative nucleus in binucleate pollen grains and the sperm nuclei in trinucleate pollen stain intensively red with aceto-carmine (Fig. 4). The generative nucleus usually stains less intensive.

Potassium Iodine: Detection of Starch

For the detection of starch as reserves in the cytoplasm, pollen grains are stained with aqueous potassium iodine (Gerlach 1984). Fresh or dry pollen grains are transferred into a drop of staining solution on a glass slide. Starch present in pollen grains will stain dark brown to black (Fig. 5).

Acetolysis: Visualizing Pollen Ornamentation and Aperture Number in Recent and Fossil Pollen

Acetolysis (Erdtman 1960) is a standard palynological preparation technique and an indispensable method for illustrating pollen grains with the LM. Untreated or stained pollen grains will hide much of the important information for the description of a pollen grain. The acetolysis treatment should remove the cellular content and the intine, but can also destroy the aperture membrane. Moreover, it cleans pollen surfaces and colors pollen grains brown, which makes it easier to observe all details of the pollen wall.

The normal preparation procedure is a combination of two steps, chlorination and acetolyzation

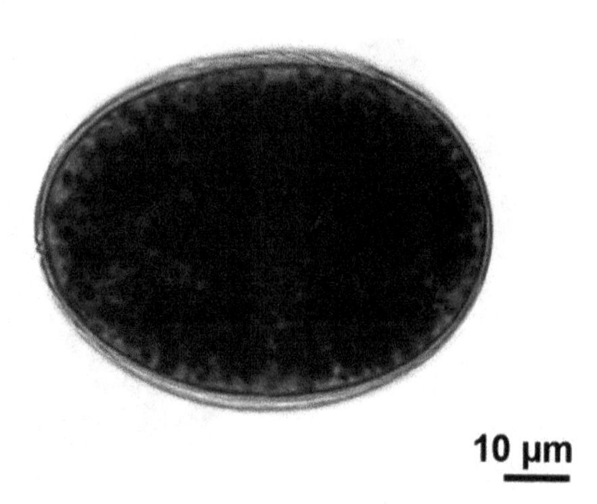

Fig. 5 Detection of starch using potassium iodine. *Amorphophallus interruptus*, Araceae, starch (in amyloplasts) stained with potassium iodine

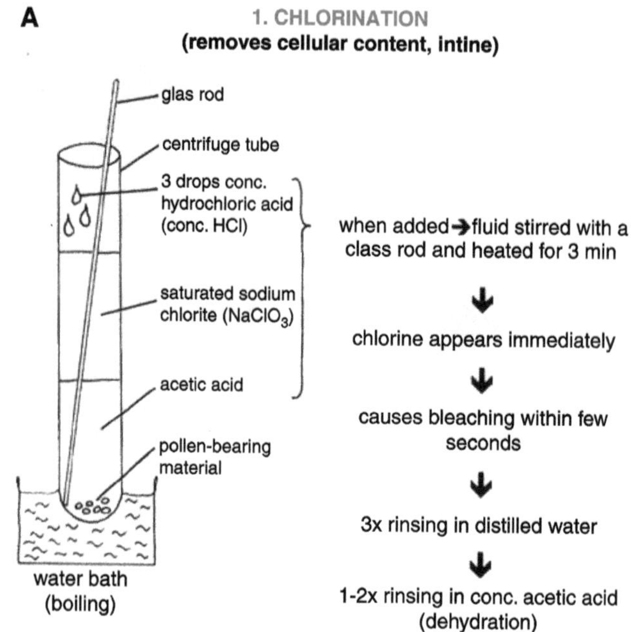

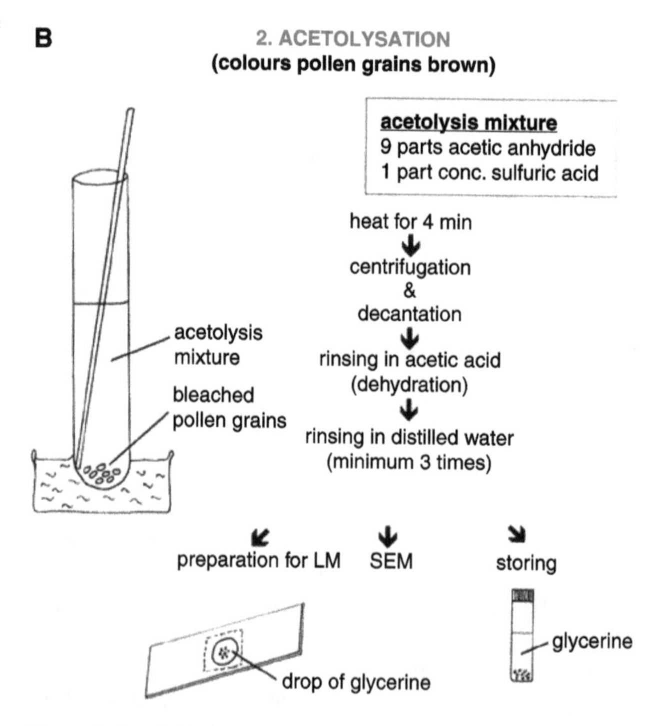

Fig. 6 Acetolysis treatment. Chlorination (**A**) and acetolyzation (**B**), the two steps of acetolysis

(Fig. 6). For **chlorination**, the sample is transferred to a test tube and covered with a layer (1.5 cm) of glacial acetic acid and a layer (ca. 3 cm) of a freshly prepared solution of saturated sodium chlorate. After adding 3 or 4 drops of concentrated HCl, the mixture is stirred with a glass rod, heated in a bath of boiling water for 3 min, centrifuged, and the liquid fraction decanted. The residue is carefully rinsed to eliminate any remaining chemicals and then finally washed in concentrated acetic acid or acetic anhydride to remove the water. For the **acetolyzation**, the sample is put into a mixture of 9 parts acetic anhydride and 1 part concentrated sulfuric acid and heated to 100 °C (at least 80 °C) for approximately 4 min (up to 10 min). The samples are ideally acetolyzed in an ultrasonic bath to avoid boiling retardation and to reduce water condensation. After the mixture has been centrifuged and the liquid fraction decanted, the residue is washed in acetic acid and 3 times with water. After washing, test tubes are turned upside down and the content dried. Glycerine is then added to the sample. For fossil pollen material both steps (chlorination and acetolyzation) are usually applied.

When **preparing recent material** (Fig. 7) it is routine to apply only the second step (acetolyzation). Traditionally, the term "acetolysis" is also used even when pollen grains have been acetolyzed only and not previously chlorinated. For acetolysis of recent pollen fresh or air dried pollen/anthers are transferred into test tubes and can be acetolyzed directly. For the analysis of soil, dust, honey, or any other samples, the material has to be washed in a beaker with about 200 ml distilled water (and detergent, e.g., Tween) and can be sieved to remove bigger parts (leaves, branches) from the sample. In order to prevent pollen loss, it is important to use sieves with big mashes (E-D-quick sieve "260 µm"). The material is then concentrated in test tubes by centrifuging at 3000 rpm and the water decanted. The residue is washed in concentrated acetic acid to remove the remaining water and

Fig. 7 Acetolyzation treatment of recent material. A. Washing the sample in a beaker. **B.** Washing with a detergent "Tween". **C.** Sieving the sample. **D.** Decanting water from the test tube after centrifuging; the organic fraction remains at the bottom. **E.** Fresh acetolysis mixture is added to the sample in the test tube. **F.** Samples are heated in an ultrasonic bath. **G.** During acetolyzation the solution turns brown. **H.** Residue washed in acetic acid followed by water. **I.** Drying of the acetolyzed sample. **J.** Acetolyzed material in glycerine stored in cryo tubes. **K.** acetolyzed pollen from honey in LM

subsequently acetolyzed (see description "acetolyzation" above). For light microscopy one part of the acetolyzed material is transferred into glycerine. For scanning electron microscopy, acetolyzed pollen is transferred into a drop of anhydrous ethanol on a SEM stub and sputter coated with gold (see also below "Preparation of fossil material").

Annotation: After rehydration or washing of the material (pollen/anthers) use acetic acid before and after the use of the acetolysis mixture, as it reacts intensively with water. Fresh acetolysis mixture is light yellow colored and highly reactive. Over time the mixture obtains a dark brown color and becomes less reactive.

Heavy Liquid Separation

Samples (recent and fossil) that still contain a very high mineral content after acetolysis should be treated with heavy liquid (e.g., zinc bromide solution; e.g., Eyring 1996, Traverse 2007; Fig. 8). Add ca. 2 cm of zinc bromide solution into the centrifuge tube and mix with the organic residue. Distilled

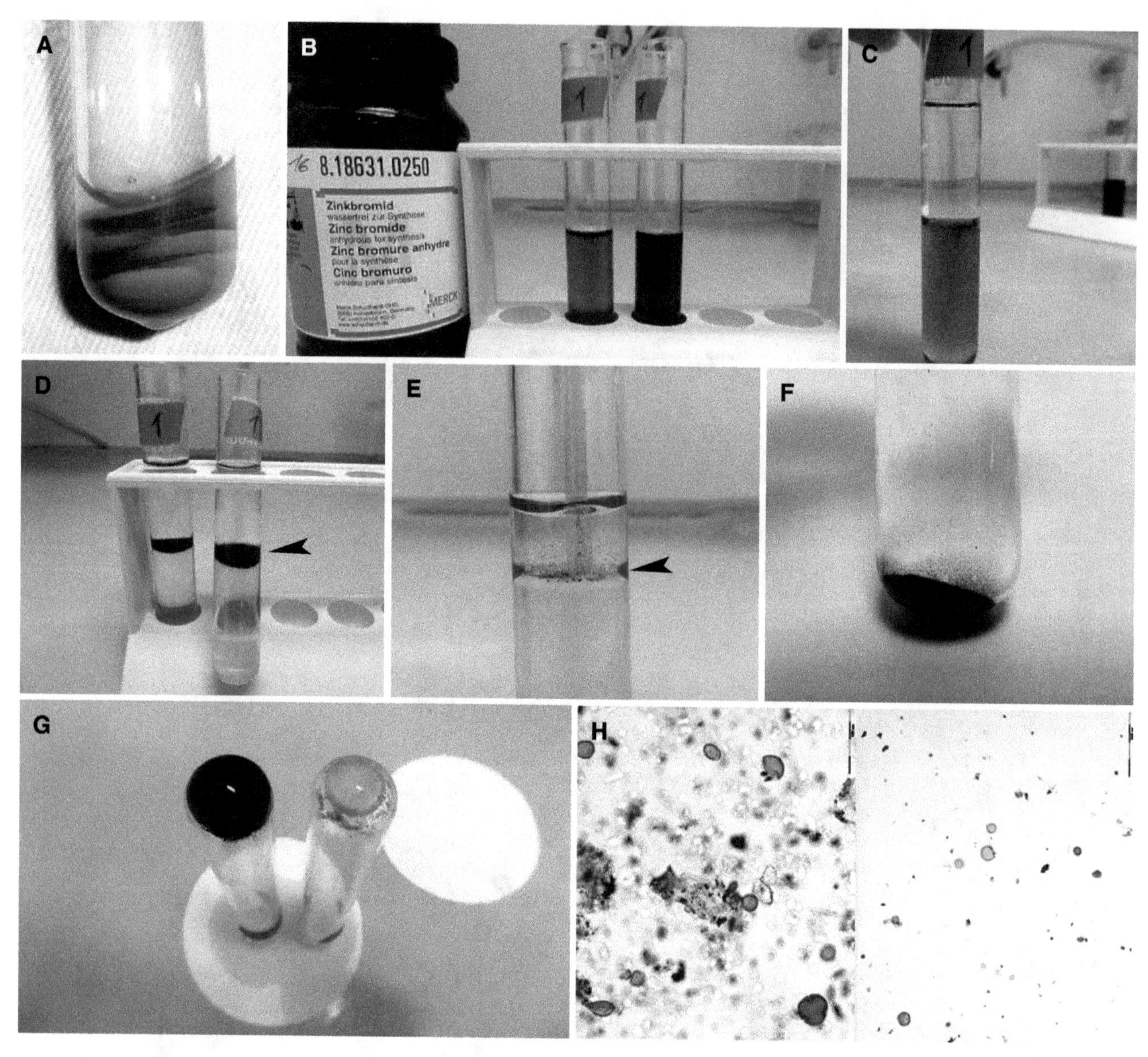

Fig. 8 Heavy liquid separation. A. Sample with high mineral content (light grey layers) after acetolysis. **B.** Mixing the sample with heavy liquid. **C.** Distilled water added without intermixing the liquids. **D.** Organic fraction (arrowhead) floating on the heavy liquid. **E.** Organic fraction (arrowhead) pipetted to a new test tube. **F.** Washing the organic fraction with water. **G.** Drying the acetolyzed sample (left) and the mineral fraction (right). **H.** Sample untreated (left) and treated with heavy liquid separation (right)

water is then carefully poured into the test tube (ca. 2 cm) and make sure that the two liquids do not intermix. After centrifuging for about 5–8 min at 3000 rpm the organic material is floating on the heavy liquid and below the distilled water. The organic material can then be transferred with a pipette into a new test tube for further washing. The inorganic parts remain at the bottom of the solution.

Acetolysis the Fast Way

A fast and easy way to prepare recent pollen grains for LM and SEM is to have a small glass bottle with a readymade acetolysis fluid (nine to one mix of 99% acetic anhydride and 95–97% sulfuric acid) at hand. Place a drop(s) of the acetolysis fluid on a glass slide. Remove anthers from the flowers and place them into the fluid on the glass slide (Fig. 9). To soften up the material let it lay in the liquid for some time and break the anther/flower material by squeezing and pressing it with the tip of a teasing needle. The slides are then heated over a candle flame for a short time to soften up the anthers, release the pollen grains from the anthers, dissolve extra organic material on pollen grain surfaces, "rehydrate" pollen grains and release their cell contents, and finally, to stain the pollen grains for LM photography. Make sure not to hold the slide over the flame for too long since it will make the pollen grains too dark. Best is to

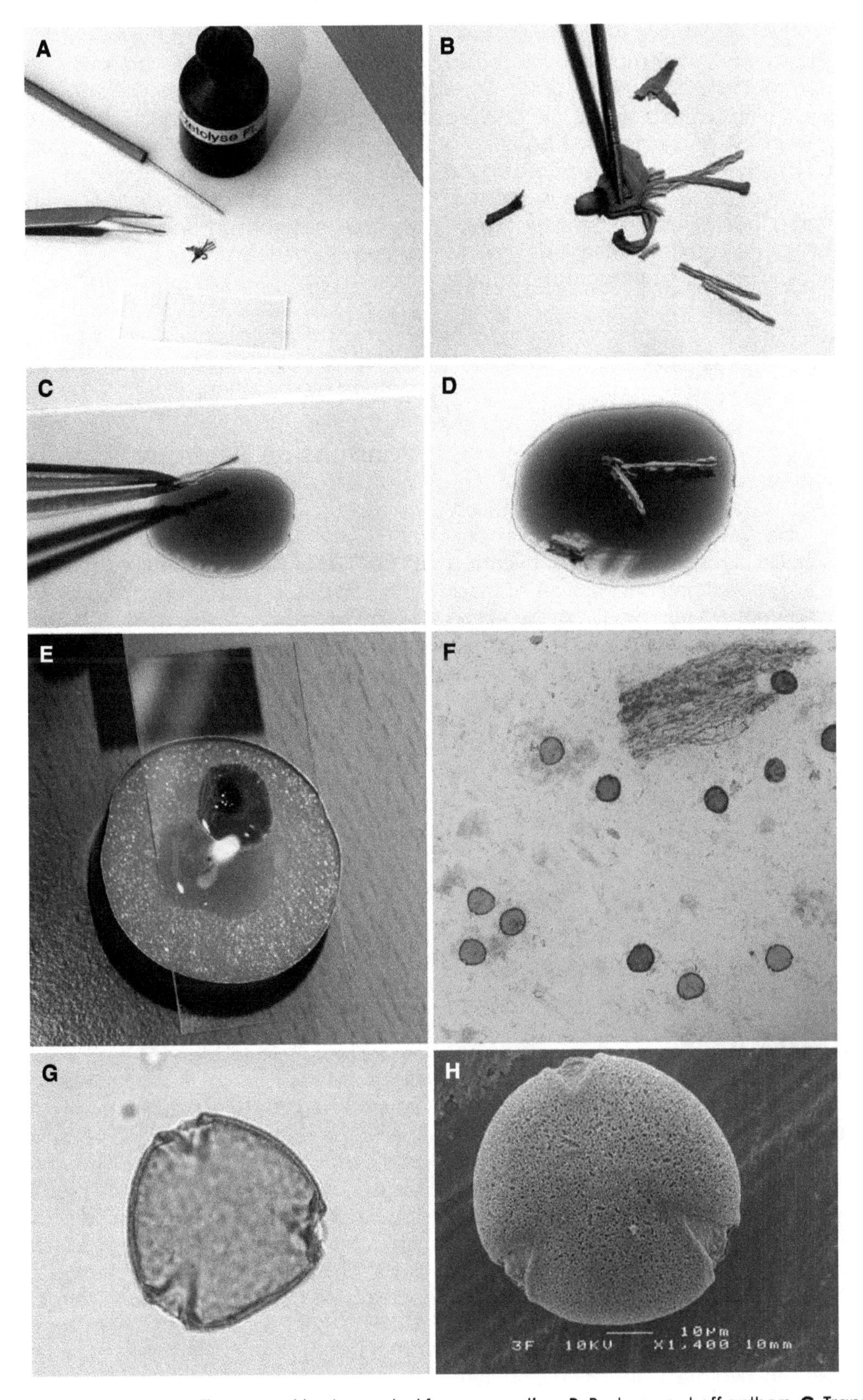

Fig. 9 Acetolysis the fast way. A. Flower and tools needed for preparation. **B.** Brake or cut off anthers. **C.** Transfer anthers into acetolysis fluid on glass slide. **D.** To soften up the material it can lay in the fluid for some time. **E.** Carefully heat the slides over a candle light. **F.** Readymade pollen grains in the acetolysis fluid. **G.** Transfer pollen grains to fresh drops of glycerine on new glass slides and photograph in LM. **H.** Same grain photographed in SEM using the "single-grain method"

heat the slides shortly and then use the teasing needle to break down the anther material. This should be repeated until the pollen have gained the required color. Using a micromanipulator (see below) selected pollen grains are then transferred into fresh drops of glycerine on new glass slides and photographed under LM. Some pollen grains can also be transferred to SEM stubs using the technique of the "single-grain method" described below, sputter coated with gold and photographed under the SEM.

Scanning Electron Microscopy: Preparation of Recent Pollen

SEM techniques cannot substitute LM, but they can provide a great deal more information, especially about ornamentation. Samples prepared for SEM should ideally reflect the fully hydrated condition of a living pollen grain. In addition, all types of pollen coatings must be removed from the pollen surface, not to obscure details of the pollen wall.

For scanning electron microscopy dehydration and drying techniques are of great importance. The principle of critical point drying (CPD) is to avoid any damaging to the pollen due to surface tension forces occurring during transition from the liquid to the vapor phase. Due to the slow penetration time of DMP, large samples (e.g., large anthers, whole parts of flowers) should be dehydrated in a series of alcohol (70–85–96%, each about 20 min) and acetone or dehydrated in 70% ethanol (3 days) and formaldehyde dimethyl acetal (FDA, 1 day or overnight).

The DMP Direct Method: Dimethoxypropane

With the DMP direct method (Halbritter 1998) important details of hydrated pollen grains, which may be lost by conventional methods (alcohol), are well preserved without shrinkage, distortion, or dissolution (Fig. 10). The best results are obtained using acidified dimethoxypropane (DMP) for dehydration. Anthers should be collected at anthesis. Take whole or parts of anthers, or loose pollen grains and put them into a pouch made of filter paper. For analyzing pollen in hydrated condition, moisture the filter pouch with a droplet of water and wait for a few seconds before transferring them into acidified 2,2-dimethoxypropane. After 20–30 min (or up to 24 h) in DMP samples are transferred into pure acetone for a few minutes and critical-point dried in CO_2 using acetone as the intermediate fluid. The CPD-pollen samples are then mounted on stubs using double-sided adhesive tape, sputter coated with gold and observed with an SEM. CPD samples can be stored, e.g., in a sealed plastic box to protect them from humidity.

This method can be used for fresh material as well as for herbarium samples (after rehydration in water). The chemical dehydration of unfixed plant material with DMP is a simple and fast method and can be applied to small samples only.

Unless stated otherwise, the pollen grains shown in this book are prepared using the DMP direct method by Halbritter (1998).

Transmission Electron Microscopy: Pollen Wall Stratification and Ultrastructure

For TEM studies of recent and fossil pollen, more than one protocol for fixation and staining may be needed.

Fixation and Embedding

Fixation of samples for TEM studies (Hayat 2000) is a time-consuming process that starts with fixation on the first day (Fig. 11), followed by dehydration and infiltration on the second and third day and ends with embedding on the fourth day (Fig. 12). For prefixation, the samples (closed anthers or pollen suspension) are placed in phosphate buffered glutaraldehyde (3%). In case of large specimens (flower/anther), the relevant parts of the sample are prepared/cut within the fixation solution under a binocular microscope (placed at the fume hood to prevent toxic substances from inhalation). Samples must be free of gaseous/air-bubbles. Transfer samples into Eppendorf tubes and make holes into the lid. Place the tubes into the vacuum desiccator and evacuate from air for 10–30 min. For pre-fixation the evacuated samples are then placed for 6 h in a specimen rotator (at room temperature). After rinsing in buffer and distilled water, samples are postfixed in 2% osmium tetroxide plus 0.8% phosphate-buffered potassium ferrocyanide (2:1) for 8–12 h at 6 °C (for osmium storage see also Fig. 28). On the second day osmium tetroxide is removed and samples are washed in distilled water (3 times for 5 min each) followed by dehydration in 2,2-dimethoxypropane (3 times, for 10 min each) and finally by pure acetone (2 times for 15 min each). The infiltration process starts by adding a few drops of the embedding media (1:2) to the samples

Fig. 10 The DMP direct method. A. Pollen collected at anthesis, *Fuchsia magellanica*, Onagraceae. **B.** Filter pouches for pollen preparation; moisture filter pouches (pollen samples) with a droplet of water (asterisk) before dehydration in DMP (arrowhead). **C.** Critical point dryer (CPD) with closed chamber and upper view on open chamber (arrowhead). **D.** CPD-pollen samples mounted on stubs using double-sided adhesive tape. **E.** Sputter coater. **F.** Samples sputter-coated with gold. **G.** SEM. **H.** Open chamber. **I.** Pollen in hydrated condition, SEM. **J.** Pollen in dry condition, SEM

and swirl the mixture. Repeat the procedure in 6–7 h, then let samples infiltrate overnight. This process has to be repeated on the third day. On the fourth day, acetone has to be removed before embedding the material: extract half of the acetone-resin-mixture with a pipette and wait for 2–3 h until the remaining acetone evaporates. After the fixation process the material should be stained intensive black (due to osmium), if not start from the beginning with new material.

The fixed material can now be transferred into embedding forms filled with fresh **embedding media** (Agar low-viscosity resin, see section "Recipes for TEM"). Polymerization takes place in an oven for about 12 h at 70 °C. After polymerization the specimen blocks can be stored in small plastic bags and are ready for ultrathin sectioning.

Annotation: For fixation of pollen, the material must be centrifuged after each step and the fixation mixture/water/DMP must be extracted with a pipette.

Ultramicrotomy

A lot of equipment and preliminary steps are involved in the ultramicrotomy process: **preparation of formvar film-coated grids, section-manipulators**

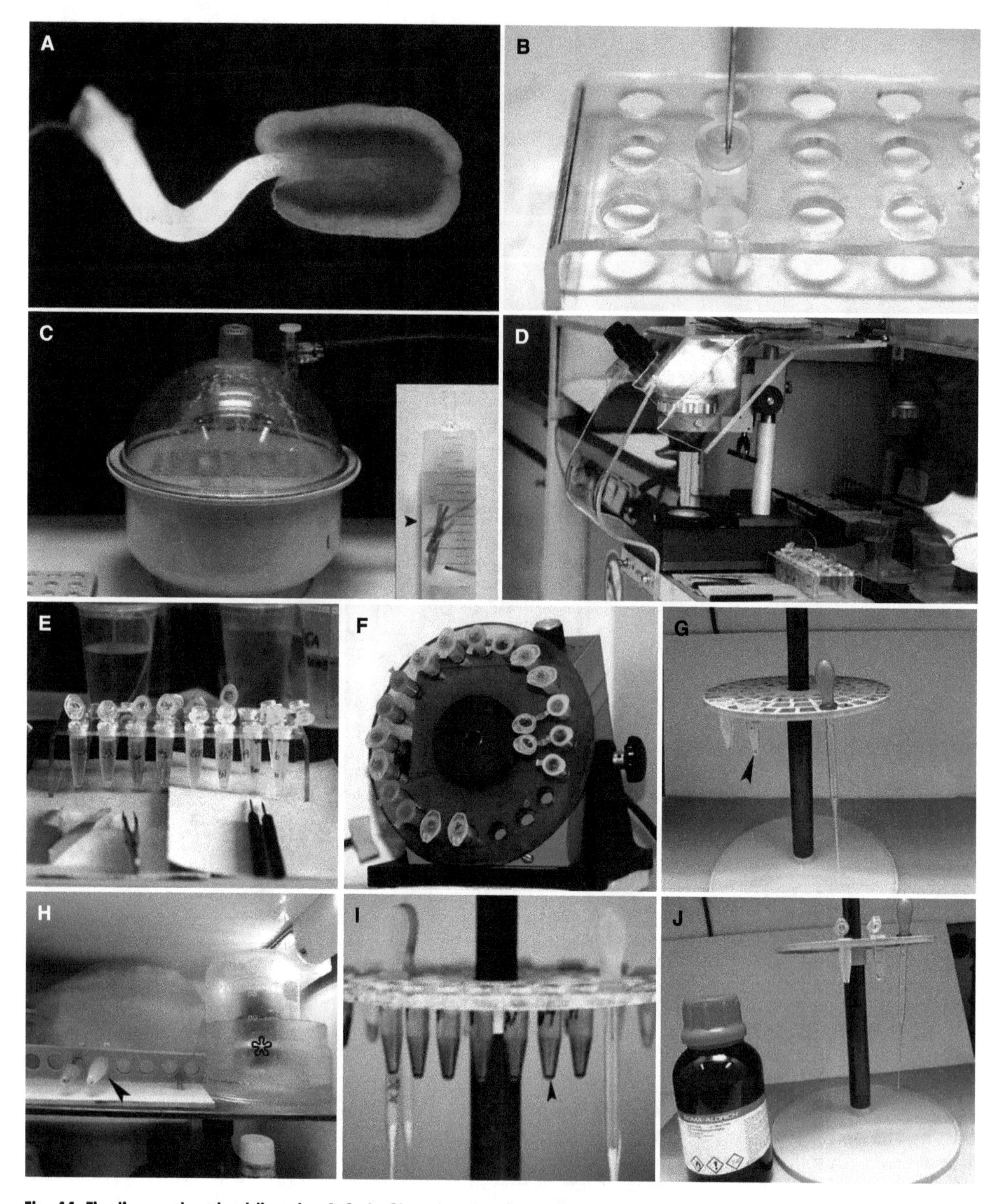

Fig. 11 Fixation and embedding day 1–2. A. Closed anther for pre-fixation. **B.** Material in Eppendorf tube with fixation solution, make holes in lid before evaporation. **C.** Evacuation in vacuum desiccator (left) or manually in a syringe (right). **D.** Preparation/cutting of samples within the fixation solution under a binocular microscope (placed at the fume hood). **E.** Transfer of selected parts of the sample into small Eppendorf tubes with fixation solution (3% GA). **F.** Samples in specimen rotator. **G.** Post-fixation; arrowhead indicates sample with osmium solution. **H.** Post-fixation of samples (arrowhead) for 8–12 h at 6 °C (fridge in a fume hood) in Eppendorf tubes; Note: osmium solution stored in fridge (asterisk). **I.** Samples after 8–12 h: material blackened due to osmium (arrowhead). **J.** After removal of osmium, samples are dehydrated, followed by pure acetone

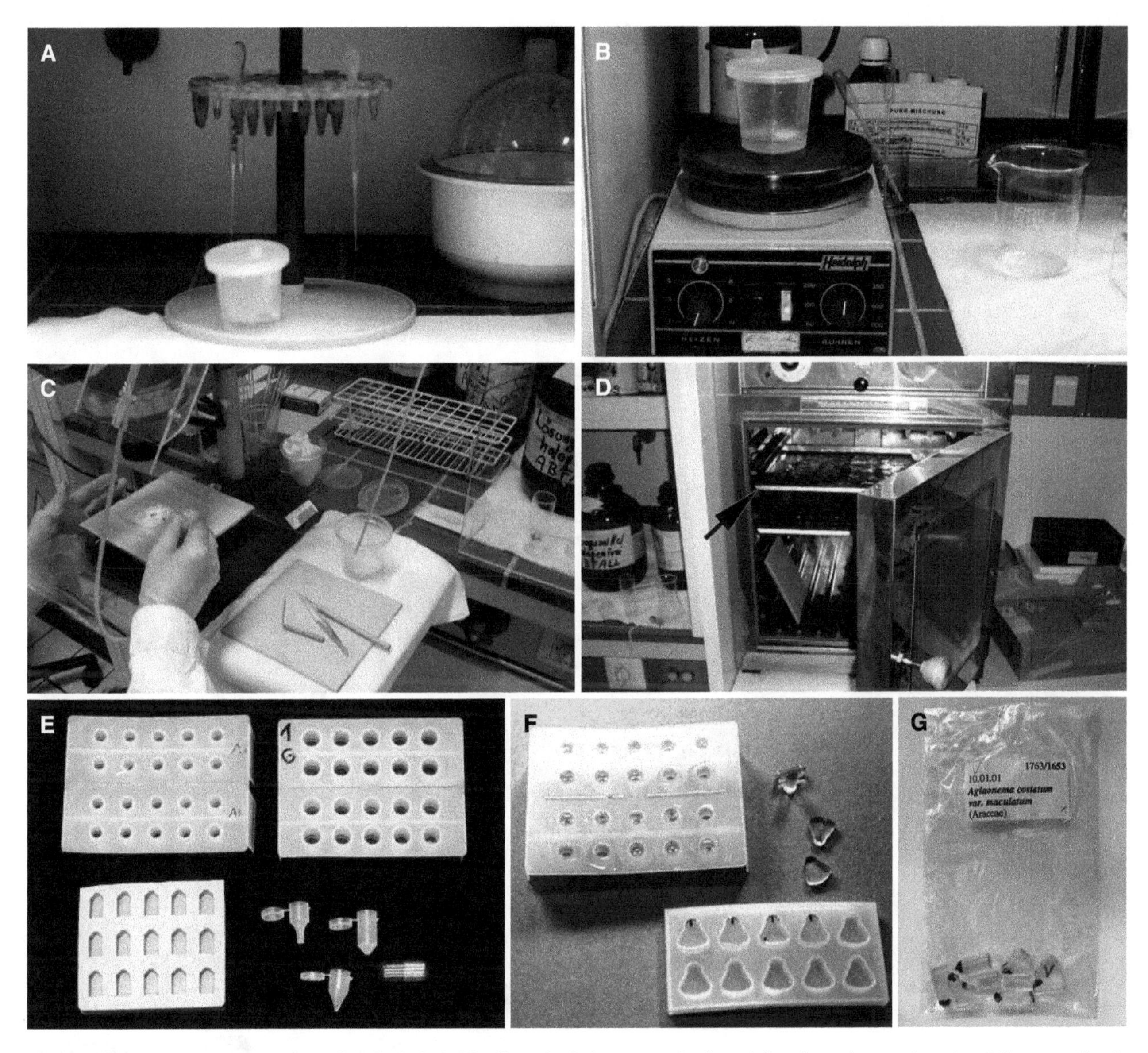

Fig. 12 Fixation and embedding day 2–4. A. Infiltration starts by repeatedly adding few drops of embedding media. **B.** Embedding solution (Agar low-viscosity resin) mixed using a magnetic stirrer. **C.** Final embedding into adequate embedding forms under binocular microscope. **D.** Polymerization at 70 °C in a thermostat oven (arrow). **E.** Examples of various embedding forms. **F.** Polymerized samples. **G.** Specimen blocks stored in small plastic bags

and preparation of loops, specimen block trimming, semi-thin sectioning, making of glass knives, diamond knives, and **ultra-thin sectioning.** Another indispensable equipment for ultramicrotomy are tweezers with an ultra fine pointed, curved, and angled precision tip.

Formvar Film-Coated Grids

Coated grids are made with a formvar solution (see "Recipes for TEM"; Fig. 13). New and cleaned glass slides are dipped with a special self-made "filming machine" into the formvar solution (minimize evaporation of the chloroform). The extraction speed of the slide influences the thickness of the formvar film: a thin film is produced by a slow, steady movement. After 1–2 min remove the glass slide steadily from the solution and dry for 2–3 min. The film can then be transferred onto a clean water surface (use distilled water in a clean staining cuvette). To loosen the film, cut the film with a scalpel along the edges of the slide and blow moist air (with a straw from your mouth) onto the film. In the same instance, dip the slide into the water at an angle of 45° to remove the film from the glass slide. When the film is floating on the water surface, don't pull out the slide, but let it slowly set into the cuvette. The quality of the film is indicated by the color: a thin film is grey to silver, whereas gold is too thick. Grids cleaned with chloroform are placed using fine pointed tweezers onto the film. To know which side of the grid is coated, always put one side (either

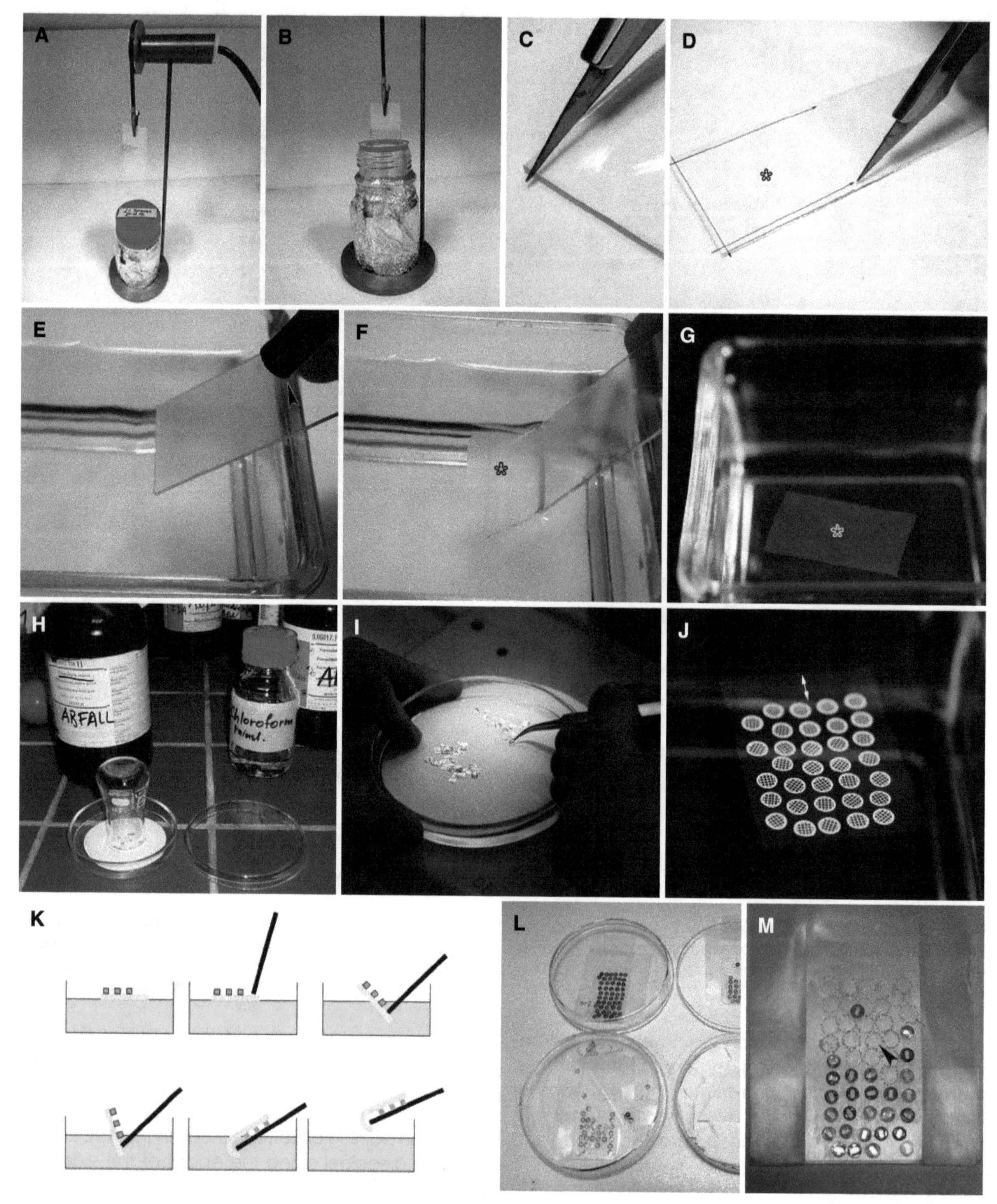

Fig. 13 Making formvar film-coated grids. A. Filming machine with holder for glass slide; filming solution should be protected from light. **B.** Glass slide dipped into formvar solution (under fume hood). **C-D.** Film cut along edges (arrowhead); arrows indicate cutting line on film (asterisk). **E.** Moisture film before dipping the slide under water; arrowhead indicates straw. **F.** Dipping the slide into the water at an angle of 45°; film partly floating on water (asterisk). **G.** Thin film floating on water surface (silver colored). **H.** Clean grids with chloroform. **I.** Shiny or dull side of the grid is visible under binocular. **J.** Grids on thin floating formvar film, arrow indicates space left for film extraction. **K.** Extraction of coated grids from the water surface using a parafilm-coated glass slide. **L.** Coated grids dried and stored in petri dish. **M.** Parafilm-coated slide with formvar coated grids, perforations (arrowhead) outline removed grids

shiny or dull side) of the grid down on the film. Make sure to leave enough space between grids and along one short margin to extract the film from the water surface. Use a parafilm-coated glass slide to extract the filmed grids: place the slide on free space of the film and dip with quick and steady motion at about 45° angle into the water and then pull out the slide again (Fig. 13K). Place the slide on a filter paper in a petri dish and let it dry. Formvar film-coated grids should be stored protected from light and dust-free (e.g., in the petri dish). To isolate the grids, use a needle to make perforations around the grids and remove them carefully with a forceps. Before ultra-thin sectioning, check the formvar film-coated grids for defects (e.g., holes, dust) under binocular microscope and place them with the filmed side up on a filter paper (see also Fig. 20 "Section pick up").

Section Manipulators (Eyelash or Other Adequate Type of Hair)

To separate and move semi-thin and ultra-thin sections floating on the water surface an **eyelash manipulator** is used. Usually a human eyelash (untreated) is fixed with glue or wax on a short glass pipette or wooden stick. The eyelashes should be cleaned with alcohol each time used and stored dust-free (e.g., covered with the back end of a bigger plastic pipette) (Fig. 14).

Loops

Loops are used to transfer ultra-thin sections onto formvar-coated grids (see Fig. 20 "Section pick up"). A loop should take up a droplet of water accurately and should fit exactly onto the grid. Therefore, two types of loops are produced (1) **circular loops**, that fit onto mesh-grids and (2) **oval** loops, used for slot-grids (Fig. 15).

Loops are made with wires from conventional electric cables (wires should not be too thick or thin). For making a circular loop a small piece of wire can be twisted around a circular object with appropriate diameter (e.g., screw driver). To produce an oval loop make a smaller circle and press it from two sides with a plier into an oval shape (fitting the grid slot). More ideally wrap the wire around a self-made model form fitting the grid size/slot. The wire of the loop is finally flattened with a hammer and the twisted (non-flattened) appendices fixed with glue or wax, e.g., on a short glass pipette. The loop should be cleaned before use with alcohol and stored free of dust.

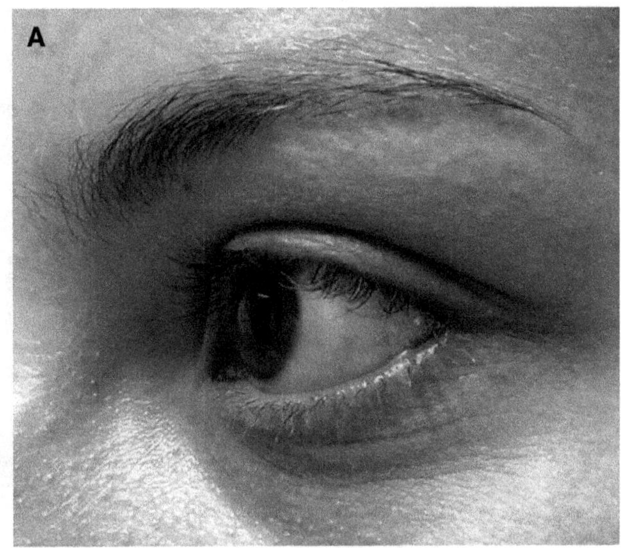

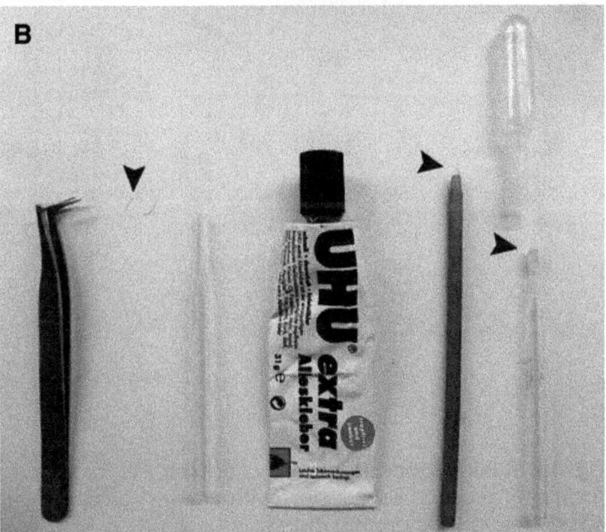

Fig. 14 Making a section manipulator. A. Human eyelash. **B.** Technical equipment for making a section manipulator; arrowheads indicate eyelashes. **C.** eyelash fixed with glue on wooden stick

Specimen Block Trimming

Criteria for block trimming are: (1) a small sample size, (2) the location of the sample should be in the center of the block-face (trapezoid) and surrounded by resin, (3) the straightness of the block-face edges (parallel edges).

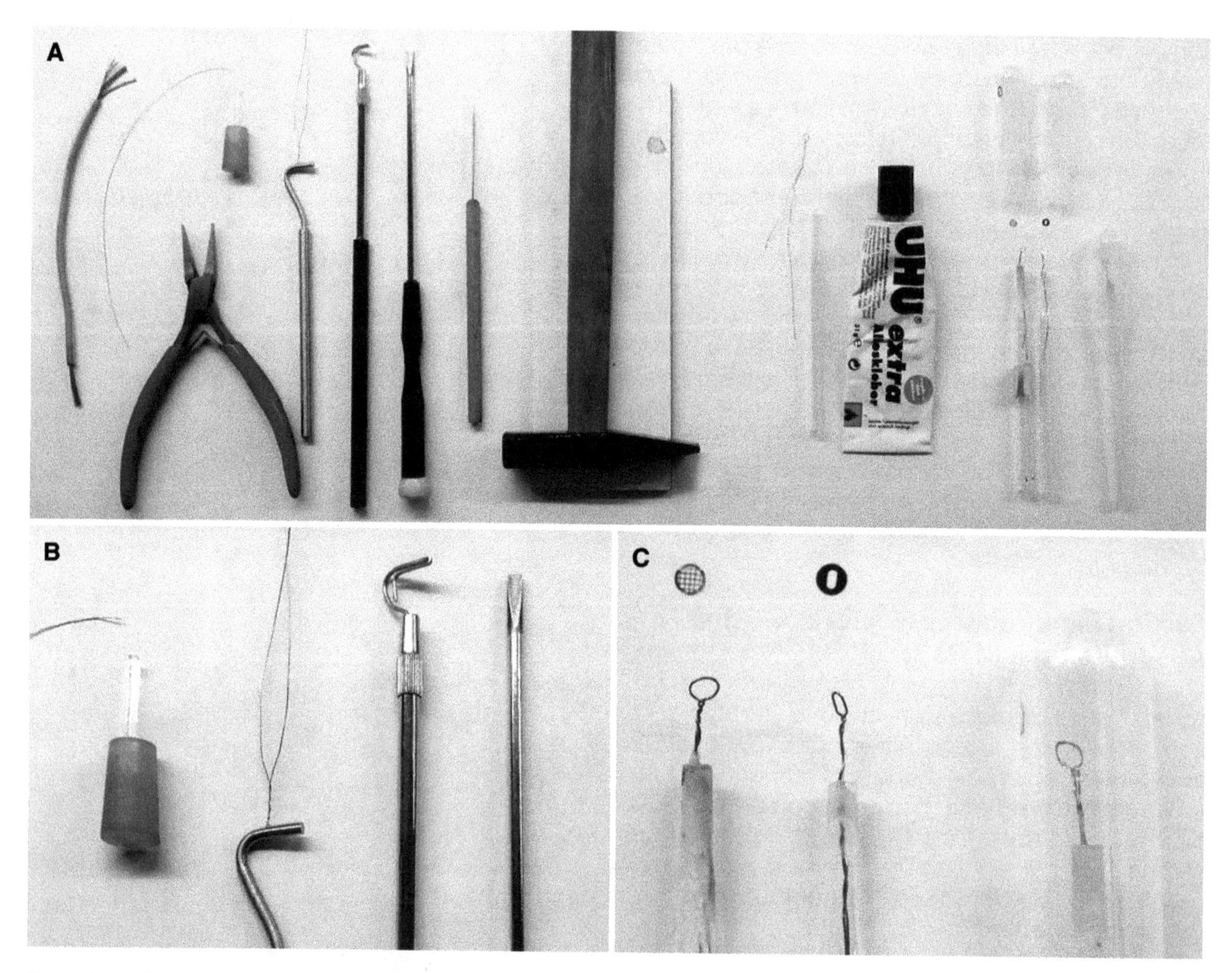

Fig. 15 Making loops. A. Technical equipment for making loops. **B.** Wire twisted around a circular object (model form) for mesh grids. **C.** Loops fixed with wax on glass pipettes (left), for storage loop covered with the back end of a disposable plastic pipette (right)

A specimen block must be trimmed (cut) to get small sections with a block-face of 4 mm by 4 mm in size (Fig. 16 O). A small block-face ensures good sectioning performance. Trimming is conducted with razor blades (for each block use a new razor blade). The block is fixed in a specimen holder and trimmed under a binocular microscope. The specimen block is trimmed into a pyramid with a trapezoid-shaped block-face. The tip of the pyramid should be cut away until you reach the appropriate level within the sample. A glass knife is used for initial cuts. If the specimen is rather big, the block-face can be larger for semi-thin sectioning (max. 4 mm^2) to ensure that the area of interest is preserved. Such a large block must be trimmed further to reach the final required block-face for ultra-thin sectioning.

Glass Knives

Glass knives are generally used for semi-thin sectioning and are replaced by diamond knives for ultra-thin sectioning. Glass knives are produced with a "knife-maker" (Fig. 17). Specially produced glass strips (e.g., 6.4 × 25 mm) are first cut into squares. The squares are then cut diagonally into two triangles, each with a knife edge (Fig. 17 E). The breaking line (stress line) indicates the quality of the knives. The left side of the glass knife is sharper and can also be used for ultra-thin sectioning, whereas the right side is used for semi-thin sectioning only. "**Glass knife boats**" (disposable plastic forms) are attached and sealed with hot melted dental wax (hot plate and ethanol burner) to the glass knife (see also Fig. 18). Glass knives should be stored dust-free and safe in a "glass knife box."

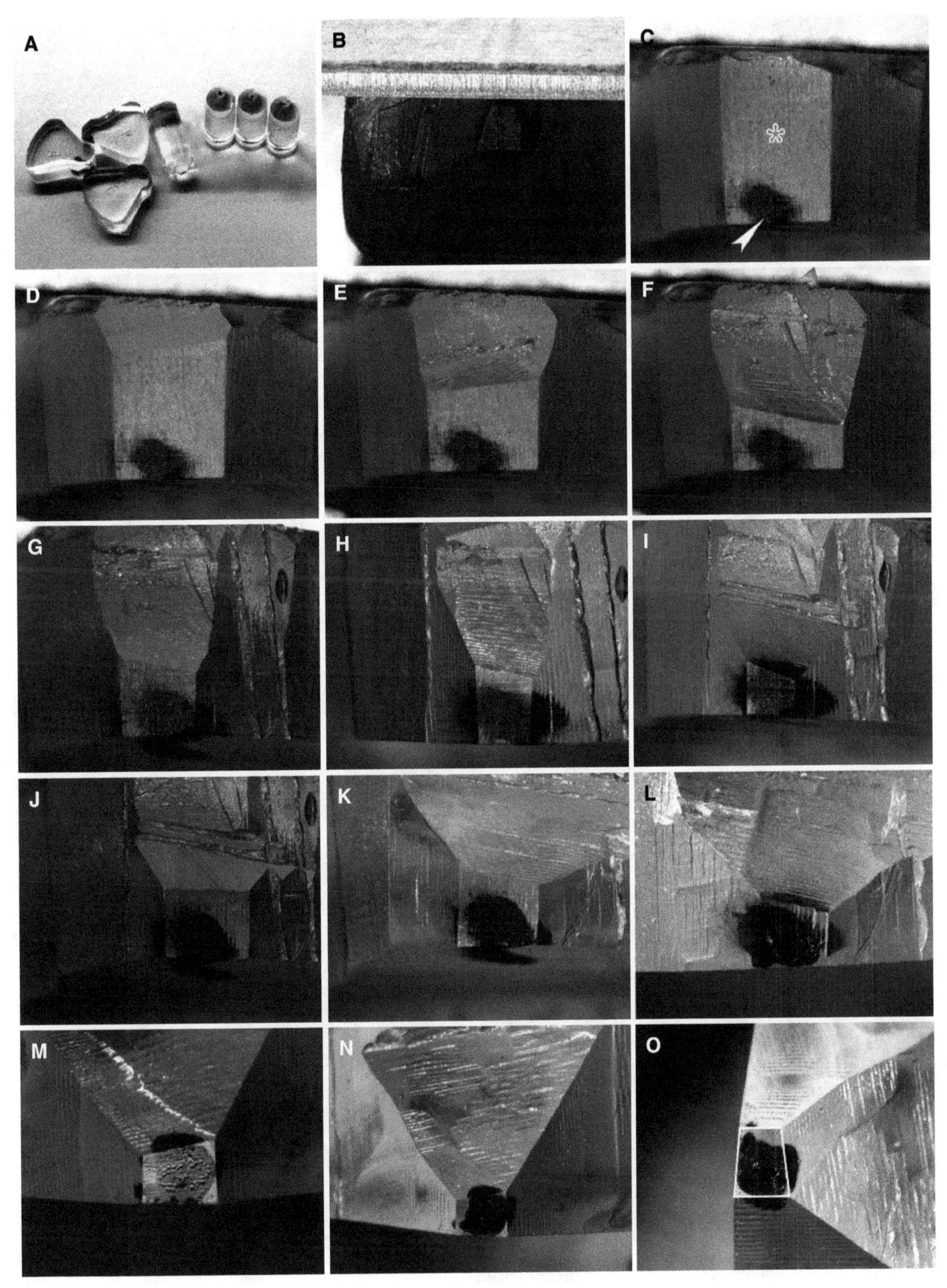

Fig. 16 Specimen block trimming. A. Specimen blocks of various shapes. **B.** Trimming is conducted with razor blades. **C.** Untrimmed specimen block with view on block-face (asterisk), arrowhead indicates position of specimen inside block. **D-N.** Blocks are trimmed into a pyramid with a +/– trapezoid shaped block-face and parallel edges. **O.** Final block-face with trapezoid form (white trapeze)

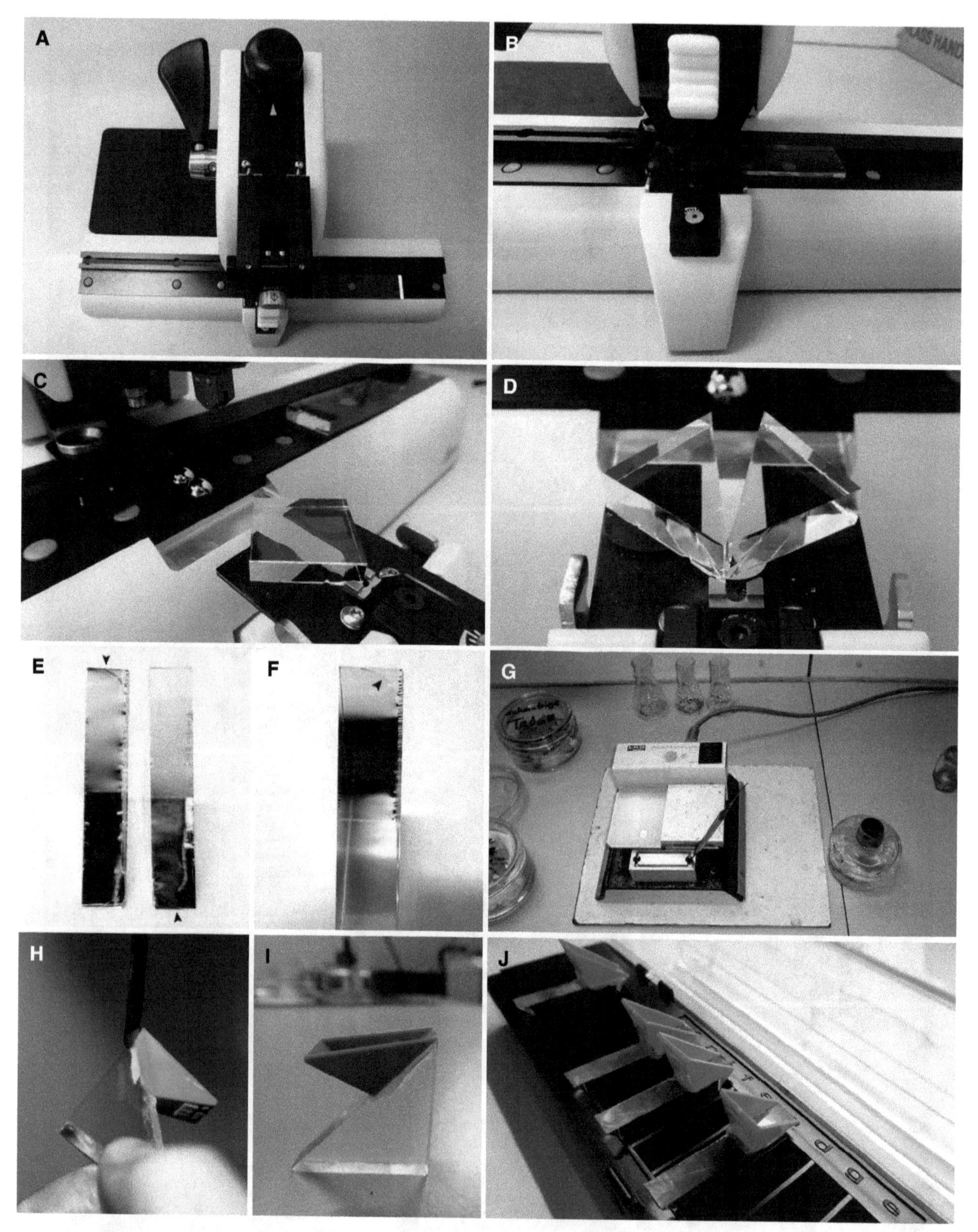

Fig. 17 Making glass knives. A. Knife maker. **B.** Glass stripes cut into squares. **C.** Squares are cut into two triangles. **D.** Two triangles (glass knives). **E.** Each triangle has a knife edge (arrowhead). **F.** Detail of triangle with knife edge, arrowhead indicating breaking (stress) line. **G.** Hot plate and ethanol burner for melting dental wax. **H.** Glass knife boats attached and sealed with hot wax using a spatula. **I.** Readymade glass knife. **J.** Knives stored in glass knife box

Semi-Thin Sectioning

Before selecting an area of the specimen block for ultra-thin sectioning, semi-thin sections are cut with an ultramicrotome, using a glass knife (Fig. 18). The settings for semi-thin sectioning are: section thickness between 0.5 and 2 μm (interference color purple to blue) and cutting speed 2 mm per second. Semi-thick sections are transferred with a loop into a drop of water on a glass slide. For a fast drying process put the slide on a hot plate (approx. 70 °C). While the water evaporates the sections will stretch. The dried sections are stained with toluidine blue on the glass slide, which can be sped up by placing the slide for max. 5 s on the hot plate. Carefully wash the slide with water and dry the glass slide in a filter paper block. The stained semi-thin sections are controlled with the LM to determine the quality of the fixation and to ensure that the appropriate area of the specimen is in the correct position for ultra-thin sectioning.

Ultra-Thin Sectioning

Ultra-thin sections between 60 and 90 nm (interference color silver to pale gold) are cut using an ultramicrotome (Fig. 19). **Diamond knives** are more suitable for cutting plant material, as e.g., crystals in cells destroy the cutting edge of glass knives, generating scratches within the sections or even splitting the sections.

The knife is placed in the knife holder and the knife boat filled with distilled water. The knife should be clean, free of dust and moistened with water. The specimen block has to be placed in the specimen arm in the upper position. Then the block has to be positioned parallel to the knife-edge by rotational or lateral adjustments of block as well as the knife. By moving the block up and down in front of the knife a slit of reflected light helps to adjust the block to the knife. A narrow slit of light indicates that the block is close to the knife and a constant thickness of the slit, along the whole block-face, indicates that the block face and the knife-edge are parallel. This is the ideal position for sectioning. The settings for ultra-thin sectioning are: section thickness between 60 and 90 nm and cutting speed 1 mm per second. The section settings can be adjusted while cutting until pale gold to silver sections are produced. Sections are floating on the water surface and can be manipulated with an eye-lash. Before the ultra-thin sections can be transferred to grids, sections must be stretched to remove compressions due to cutting. For stretching a solvent (e.g., xylol, chloroform, acetone vapor) or a hot pen can be used. For the vapor method use a thin, wedge-shaped piece of filter paper moistened with a drop of solvent, hold it closely above the sections while moving it back and forth.

Section Pick-Up

The stretched sections are picked up from the water surface with a loop (Fig. 20). Depending on the size of the sections between 3 and 10 sections can be picked up at once. Center the loop above the selected sections, dip it on to the water surface, lift the sections up within a droplet of water and transfer onto a grid under a binocular microscope. Center the loop above the grid and lower it onto the grid surface. Lift up the loop and the attached grid. The water is removed slowly with a filter paper touching the first twist by the loop (Fig. 20 D). Transfer the grid with a forceps into a grid-box (sections should face the same side). Make a section protocol. Store the grid box away from light.

Staining Methods

The application of different TEM staining techniques for one and the same sample is very important and highly recommended to avoid misinterpretations of the pollen wall structure. Therefore, sections of pollen grains are routinely stained using the several different staining methods (Figs. 21 and 22). Most staining solutions are harmful or even toxic and therefore applied under fume hood.

Annotation: In electron microscopy there is no grey-scale terminology from white to black. Use "electron dense" for black or darkly colored structures and "electron translucent" for white to light grey colored.

Uranyl Acetate-Lead Citrate Staining: U + Pb

Uranyl acetate-lead citrate staining is a conventional staining method (Hayat 2000; Figs. 21 and 22). Ultra-thin sections are usually collected on copper grids. Sections are stained in uranyl acetate solution (Leica Ultrastain-1) for 45 min followed by lead citrate staining (Leica Ultrastain-2) for 1–5 min at room temperature. Use of sodium hydroxide pellets for lead citrate staining prevents crystalline precipitation by absorbing moisture and carbon dioxide from the air. Sections are thoroughly washed in distilled water after each staining step (3 times for 5 min in a row of water drops).

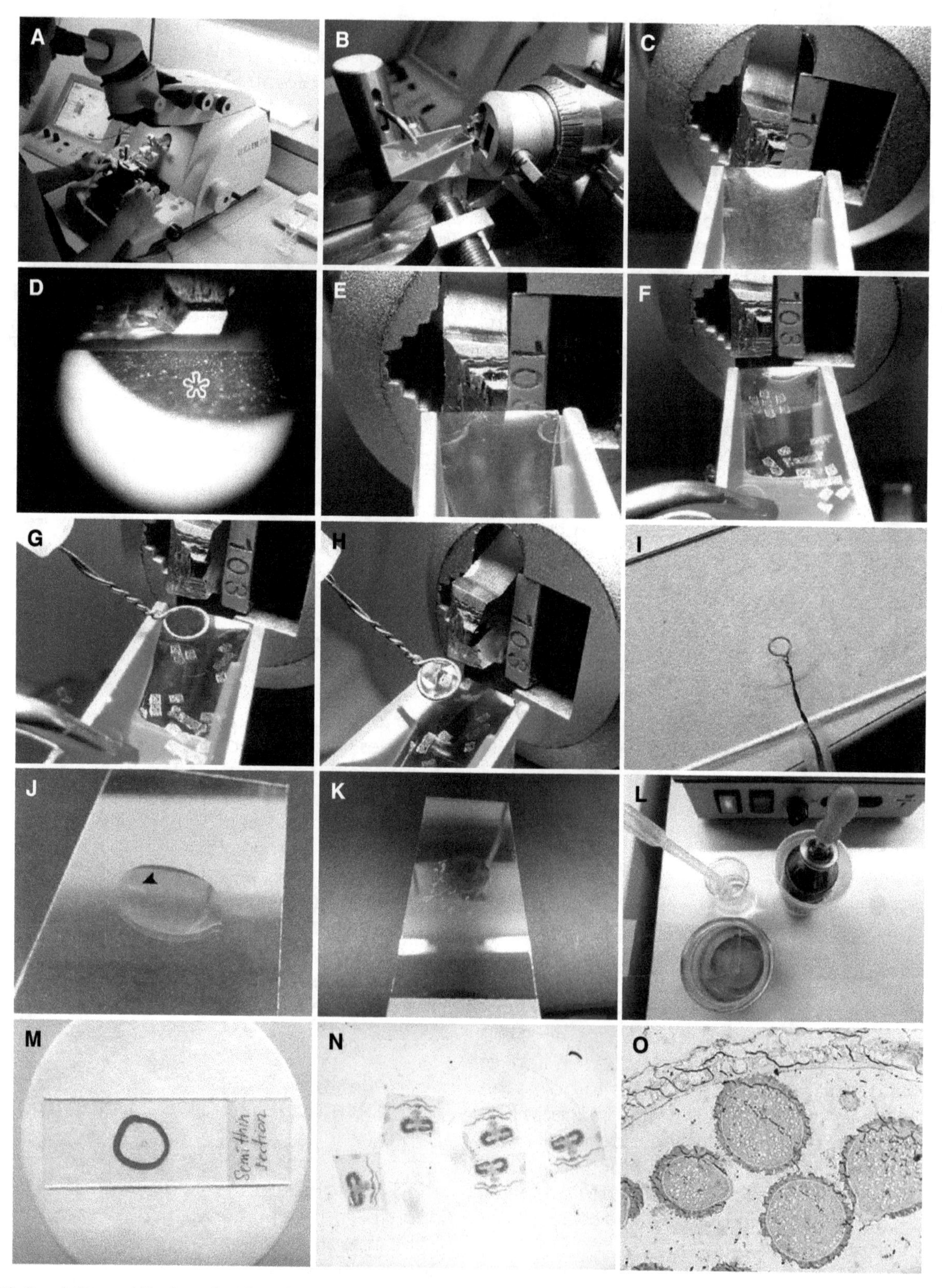

Fig. 18 Semi-thin sectioning. A. Ultramicrotome. **B.** Glass knife positioned in the knife holder and knife boat filled with distilled water; specimen block fixed within the specimen holder. **C.** Block adjustment parallel to knife-edge by use of reflecting light. **D.** Knife should be clean, free of dust and moistened with water, asterisk indicates slightly lowered water level at knife edge for sectioning, but still moistened. **E.** Block must be close enough to knife (until slit of light almost disappears) to start sectioning. **F.** Semi-thin sections between 0.5 and 2 μm (interference color purple to blue) floating on water. **G-H.** Section pick-up with a loop (see "Section pick-up"). **I.** Transfer of sections in a drop of water on a glass slide. **J.** Slide on a hot plate (arrowhead indicates semi-thin sections). **K.** Staining sections with toluidine blue on hot plate. **L.** Rinsing the stained sections with water. **M.** Stained semi-thin sections ready for LM. **N.** Toluidine blue sections seen under LM. **O.** Final quality check before ultra-thin sectioning

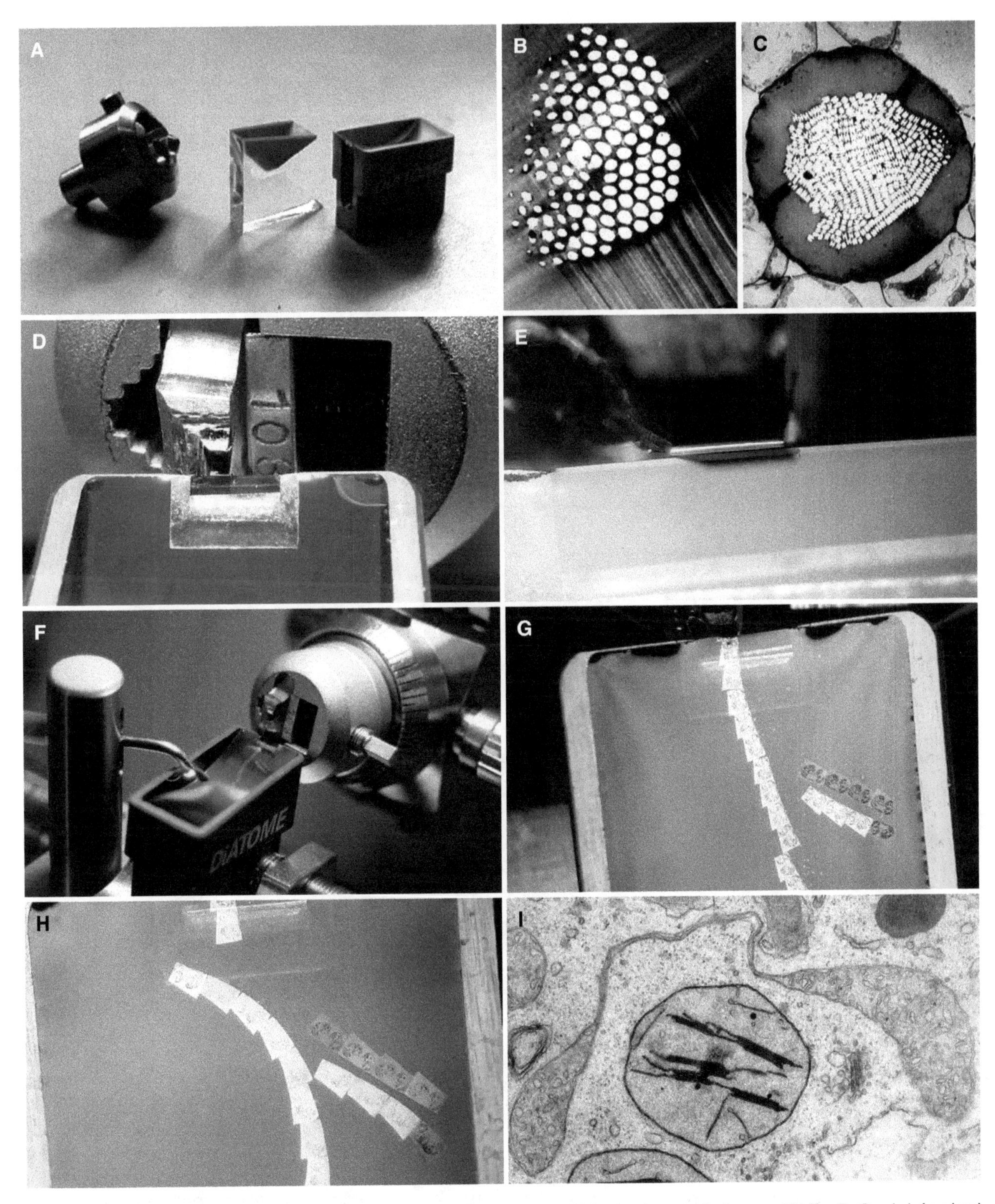

Fig. 19 Ultra-thin sectioning. A. Specimen block holder with trimmed block, glass and diamond knife. **B.** Crystals in plant cells cut with glass knife, note scratches. **C.** Crystals in plant cells cut with diamond knife. **D-E.** Block adjusted parallel to knife-edge by use of reflecting light. **F-G.** Sections between 60 and 90°nm (interference color silver to pale gold). **H.** Stretched sections, note the change in size and thickness (for color change compare to picture **G**). **I.** Ultrastructure of a plant cell showing high quality fixation of several organelles in TEM

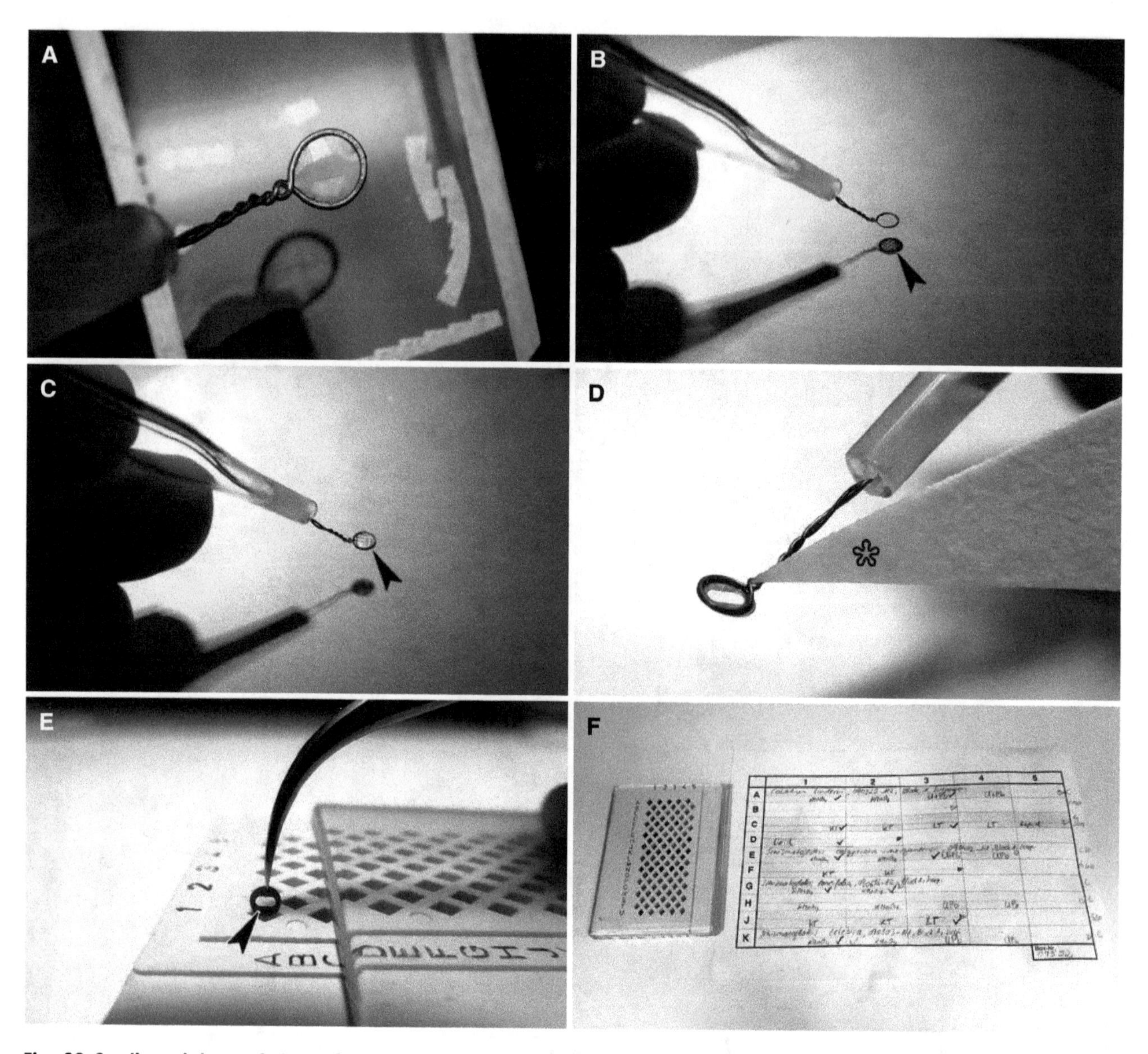

Fig. 20 Section pick-up. A. Loop for section pick-up. **B.** Loop centered above grid (arrowhead). **C.** Grid attached to loop (arrowhead). **D.** Water removed from grid, asterisk indicates wet filter paper. **E.** Dry grid placed into grid-box, sections on the left side (arrowhead). **F.** Grid-box and section protocol with color code used for different staining methods

The Lipid Test for the Detection of Unsaturated Lipids: TCH + SP

The endexine can be differentiated from the ektexine and the intine by thiocarbohydrazide-silver proteinate (TCH+SP) staining in osmium-fixed material. The endexine stains electron dense after the lipid test, indicating lipidic compounds (Fig. 22 B).

Ultra-thin sections on gold grids are treated with 0.2% TCH for 8–15 h and 1% SP for 30 min and thoroughly washed in water (3 times for 5 min in a row of water drops) (Rowley and Dahl 1977; Weber 1992).

Thiéry-Test: PA + TCH + SP

The Thiéry-test is used for the detection of neutral polysaccharides in osmium-free material (Thiéry 1967). Ultra-thin sections from osmium-free material are placed on gold grids and treated with 1% periodic acid (PA) for 45 min, 0.2% thiocarbohydrazide (TCH) for 8–15 h, and 1% silver proteinate (SP) for 30 min (Thiéry 1967). The polysaccharide intine and starch grains in amyloplasts stain electron dense (Fig. 22 C). For control samples leave out the thiocarbohydrazide step. If osmium fixed material is

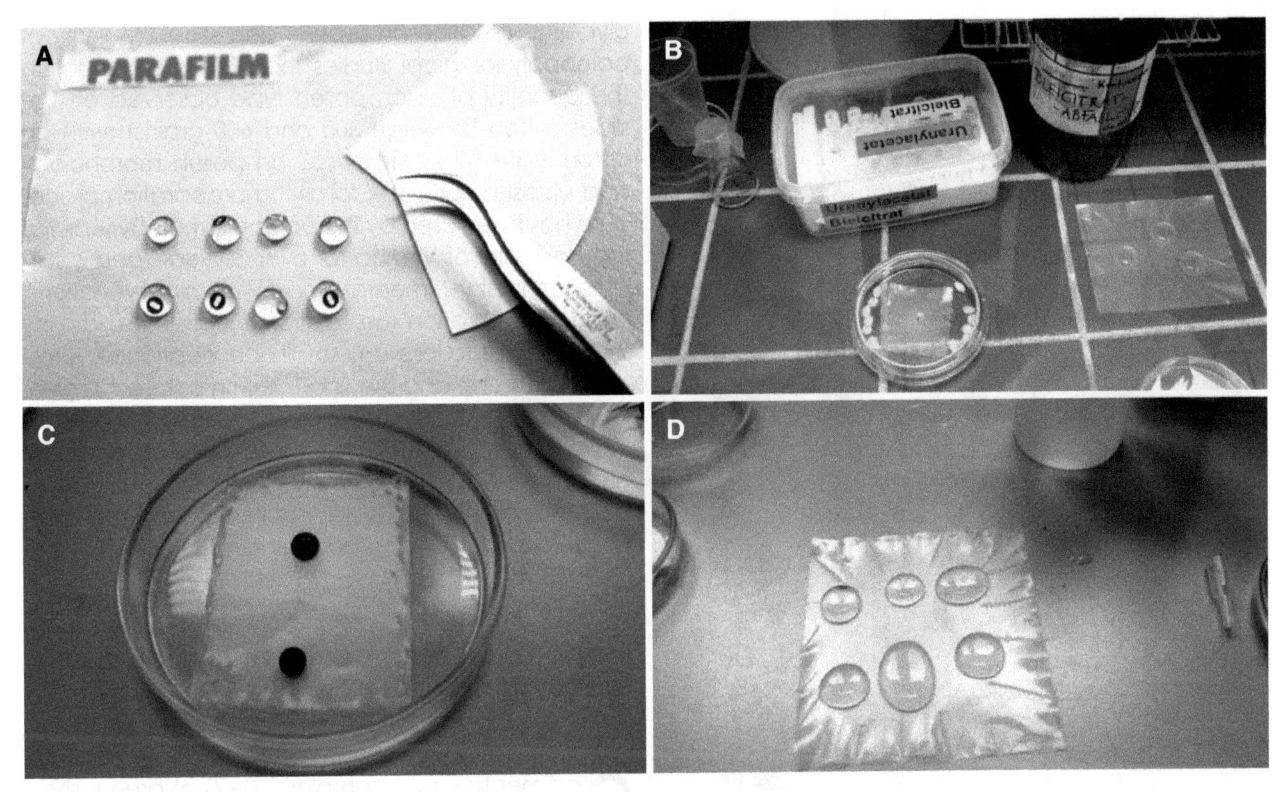

Fig. 21 Staining methods for ultra-thin sections. A. Ultra-thin sections on copper or gold grids stained in a small drop of uranyl acetate on parafilm. **B.** Small drops of lead citrate on parafilm and sodium hydroxide pellets in a closed petri dish. **C.** Small drops of potassium permanganate on parafilm. **D.** Row of large water drops for washing placed on parafilm

used for the Thiéry-test, the staining time for 1% periodic acid has to be prolonged up to 60 min (instead of 30 min), to remove the osmium tetroxide from the material.

Modified Thiéry-Test: PA + TCH + SP (short)

The modified (short) Thiéry-test (Weber and Frosch 1995) is especially effective after fixation of specimens with osmium and potassium ferrocyanide and is a good method for general enhancement of contrast in the cytoplasm and the pollen wall (Fig. 22 D). Ultra-thin sections are collected on gold grids and stained with 1% periodic acid (PA) for 10 min, 0.2% thiocarbohydrazide (TCH) for 15 min, and 1% silver proteinate (SP) for 10 min (at room temperature). After all steps the sections are thoroughly washed in distilled water (3 times for 5 min in a row of water drops), and following the TCH first washed in 3% acetic acid.

Potassium Permanganate: $KMnO_4$

Potassium permanganate staining is a simple method for the detection of the endexine. Using uranyl acetate and lead citrate, ektexine and endexine may differ in their electron opaqueness in that the endexine is higher in electron density than the ektexine, or vice versa. When the endexine is thin and less compact or discontinuous, the differentiation of the two layers may be insufficient. Typical for the endexine is its increasing thickness close to the aperture. Potassium permanganate stains the endexine electron dense, producing a distinct contrast (Weber and Ulrich 2010; Fig. 22 E). Ultra-thin sections from osmified material on copper grids are treated with 1% aqueous potassium permanganate solution for 7 min and thoroughly washed in water (3 times for 5 min in a row of water drops).

Preparation of Fossil Pollen

There are numerous methods currently used to extract organic material, including fossil and subfossil pollen, from all different types of sediments (rocks) and soils. These methods have been summarized in detail by, e.g., Erdtman (1943), Brown (1960), Fægri and Iversen (1989), Moore et al. (1991), Wood et al. (1996), and Traverse (2007). Most of these preparation methods involve sieving of some sort and the final production of palynomorphs enclosed in

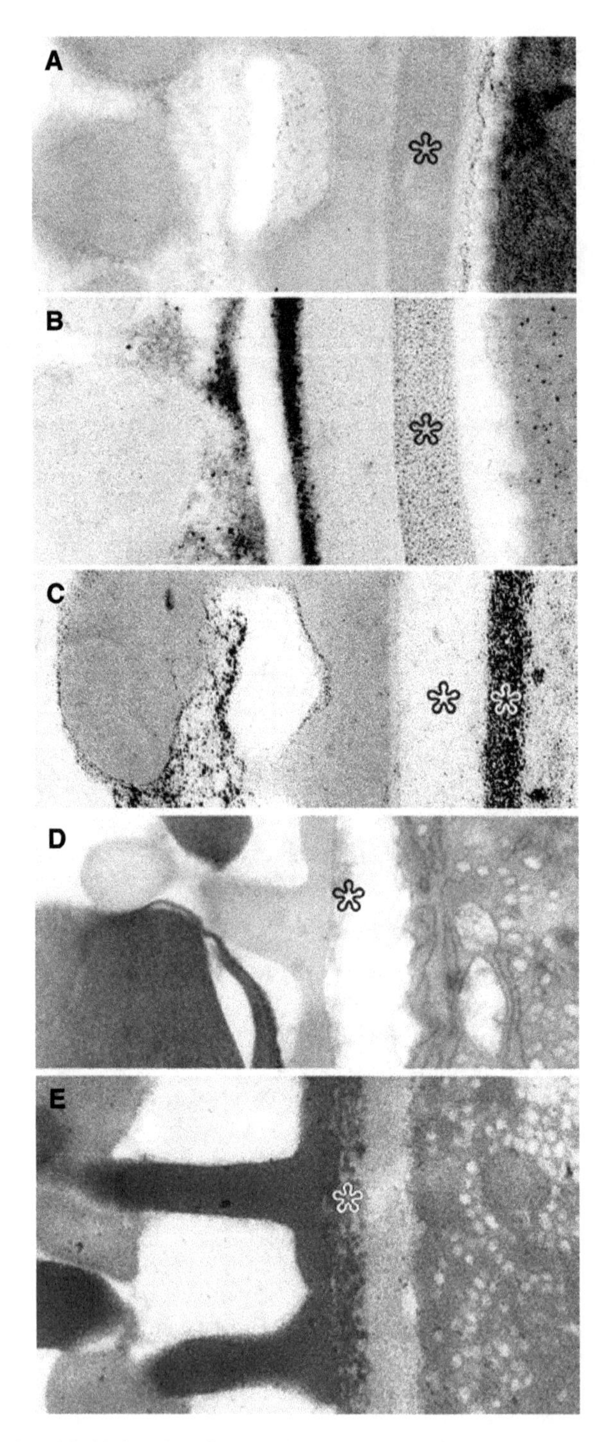

Fig. 22 Stained pollen walls and behavior of endexine (cross-section, TEM). A-C. *Apium nodiflorum*, Apiaceae. **A.** Uranyl acetate + lead citrate (U + Pb), compact-continuous endexine (asterisk). **B.** Lipid test (TCH+SP). compact-continuous endexine (asterisk) stains electron dense. **C.** Thiéry-Test (PA + TCH + SP), compact-continuous endexine (asterisk) stains electron translucent, intine electron dense (white asterisk). **D-E.** *Mentha aquatica*, Lamiaceae. **D.** Modified Thiéry-Test (PA + TCH + SP), thin compact-continuous endexine (asterisk) only slightly visible. **E.** Potassium permanganate ($KMnO_4$), thin compact-continuous endexine (asterisk) electron dense

glycerine gelatine on sealed glass-slides. Majority of paleopalynological studies then focus on counting the quantity of each pollen type observed on the slides (often between 300 and 600 grains), with an unfortunate minor emphasis on pollen morphology and ultrastructure. The following preparation procedure has been used by the paleopalynology team at the University of Vienna for over 30 years and is suitable for most sedimentary rocks with minor variations. During preparation the solution is not sieved at any stage, so not to lose any small or exceptionally large palynomorphs, and the final solution is stored in glycerine suspension in small sample tubes so the palynomorphs can be studied using the so-called "single grain method." This method has been evolved to able researchers to obtain pollen characters from single fossil grains using both LM and SEM and sometimes TEM.

Preparation Method: From Rock to Palynomorphs

Sedimentary rock samples (20–50 g) are washed and dried and hand ground in a mortar with a pestle (Fig. 23). Using a glass beaker the resulting powder is boiled in ≥200 ml of concentrated hydrochloride acid (HCl) for 5–10 min; this should remove all carbonates. Let the solution stand and when the residue has settled, decant most of the HCl liquid. Transfer the remainder of the solution into a copper pan or pot and add ≥150 ml of hydrofluoric acid (HF) and boil for approx. 10 min while stirring with a copper stick or spoon (or let stand in cold HF for 3–5 days, stir regularly, use acid-resistant plastic containers and tools); this should remove all the silicates. The solution is then poured slowly into a 4 L plastic beaker filled with water. After settling, the liquid is decanted and the remainder solution poured into glass beakers along with ≥200 ml of HCl and boiled again for 5–10 min; this prevents the formation of fluorite crystals. After cooling and settling decant most of the HCl and pour the remainder of the solution into two separate test tubes (glass centrifuge type). Wash the solutions 4 times with water and centrifuge and decant the liquid following each wash. Fill one large glass tube with cold water and add 1–2 teaspoons of sodium chlorate (pure crystalline powder; $NaClO_3$). Shake this large tube and when there are crystals that cannot be dissolved in the water the solution is ready. Pour ca 1 ml of acetic acid glacial (100%, CH_3COOH) and 3–4 ml of the sodium chlorate solution into the two original test tubes, then add five drops of HCl. Place the tubes in boiling water for at least 5 min and

Fig. 23 From rock to palynomorphs. A. Different types of sedimentary rocks: reddish, yellowish, and white-greyish samples usually contain few and/or badly preserved palynomorphs (back row), brown, dark-grey to blackish samples often contain well preserved pollen (front row). **B.** Sedimentary rock sample (ca 30 g) hand grounded in a mortar with a pestle. **C.** Sample boiling in ≥200 ml of HCl. **D.** Sample boiled in a copper pan with ≥150 ml of HF. **E.** The HF solution is poured slowly into a large plastic beaker filled with water. **F.** Organic material settled on the bottom of the beaker. **G.** Acetolysis, test tube in boiling water, note the stirring glass stick. **H.** Acetolyzed sample before decanting of water following the final wash

have a stirring glass stick in it at all times. The color of the sample solutions should change from dark blackish to brown or reddish. Centrifuge the test tubes and decant the liquid. Wash the residue 3–4 times with water and one last time with acetic acid glacial. Prepare a new solution in a clean and dry measuring glass-tube with 9 parts acetic anhydride (99%, $(CH_3CO)_2O$) and one part sulfuric acid (95–97%, H_2SO_4). Make sure to produce at least 10 ml of this solution for each original (fossil) test tube you process. Pour ca 10 ml of the new solution into each test tube. Direct tube away from your face and make sure no water comes into contact with the solution. Place the tubes again into the boiling water bath for at least 5 min. Then centrifuge and decant the liquid (again avoid contact with water). First wash the remaining residue once with acetic acid glacial, centrifuge and descant liquid, and then wash them 3–4 times with water. The remaining organic material in the test tube is finally mixed with glycerine and transferred, using pipettes, into small closable plastic test tubes. Test tubes are labelled accordingly.

The Single-Grain Method

A combined method for the investigation of fossil pollen grains was initiated by Daghlian (1982), suggesting that the same individual fossil grain should/could be observed in LM, SEM, and even TEM. This idea of how to properly investigate fossil pollen grains in a taxonomically valid way was taken further by Zetter (1989) who evolved a relatively easy method to investigate the same single fossil grain using the so-called "single-grain method," also described in Ferguson et al. (2007). To apply this method the following equipment and tools are necessary: samples prepared in the way described above, narrow glass-pipettes (see below, Fig. 24), teasing needle with an attached human nasal hair (see below, Fig. 25), an erect image compound microscope with a photographing unit, 10 and/or 20× objective lens with a minimum 10 mm working distance, glass slides, ethanol absolute, SEM stubs, sputter coater, and a functional scanning electron microscope.

Making Glass-Pipettes

It is important to have enough cheap and dispensable glass-pipettes to transfer parts of the sample from the storage tubes onto the glass slides for primary LM investigations. These pipettes are also used to make very small drops of ethanol on the surface of the SEM stubs when transferring pollen from glycerine drops using the micromanipulator (see below

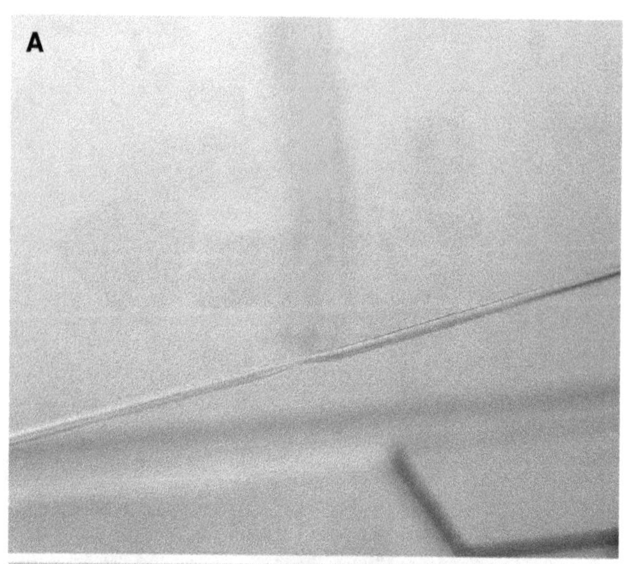

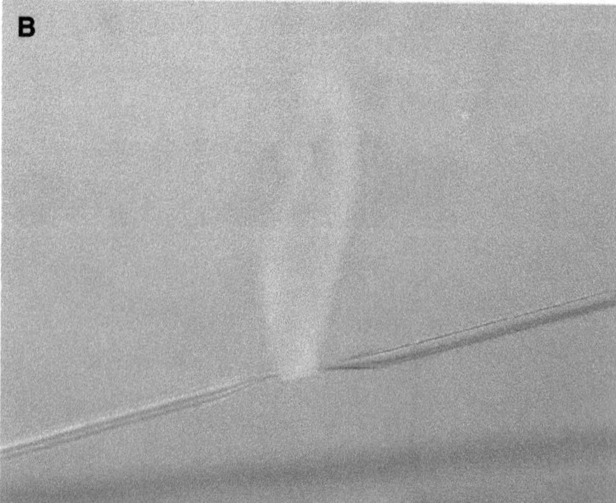

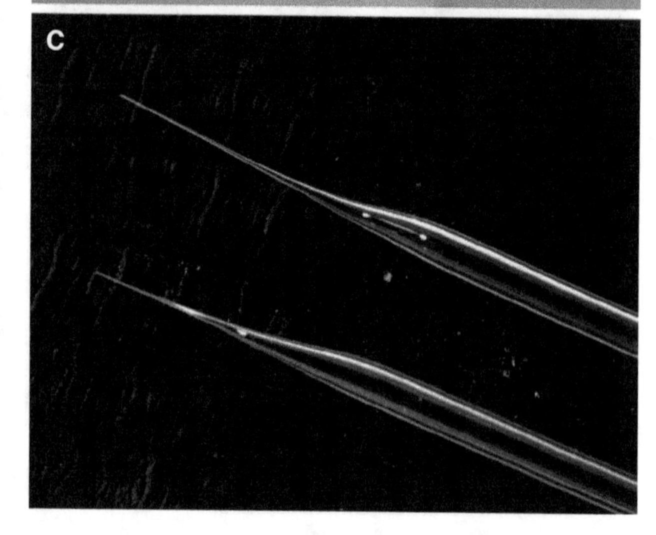

Fig. 24 Making glass-pipettes. A. Glass pipe held in burning gas flame starting to melt. **B.** Melting glass pulled very slowly and gently apart. **C.** Two freshly made pipettes ready for use

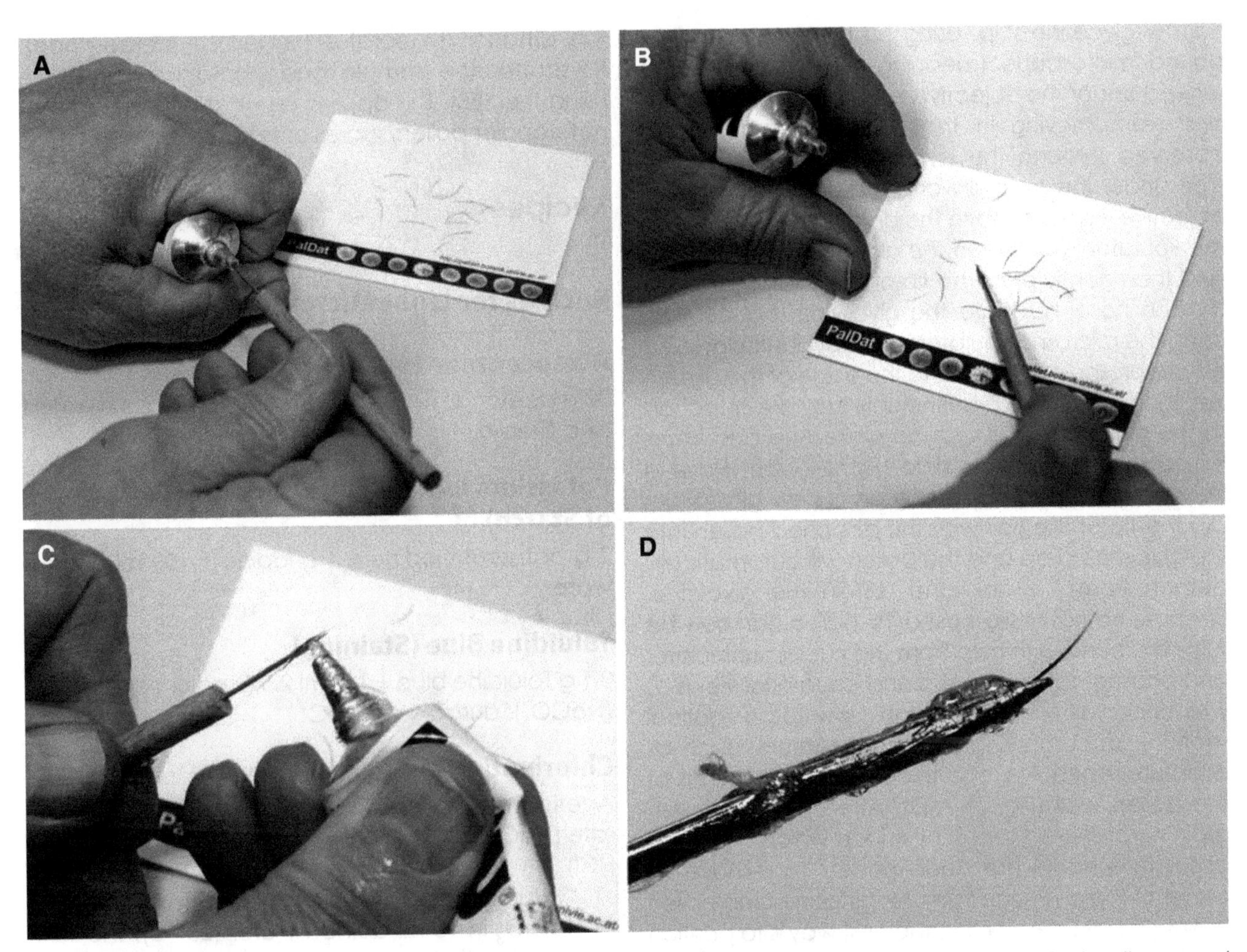

Fig. 25 Producing a micromanipulator. A. Needle pushed ca 1 cm into glue tube and turn in circles. **B.** Needle pressed onto a nasal hair. **C.** Extra glue added around the proximal part of the hair. **D.** Readymade micromanipulator

"Producing a Micromanipulator"). One possibility is to make your own pipettes (Fig. 24) by cutting down 4 mm wide glass pipes (cylinders) into ca 30 cm long units. The middle part of these is then held in/over a burning gas flame. While the glass starts to heat and melt you pull it apart from each end. The pipes will quickly give away in the middle as the glass melts. When pulled apart the glass will form two very long and narrow cones until they finally detach and one holds a perfect pipette in each hand.

Producing a Micromanipulator

The easiest way to make a really good and functional micromanipulator, that can be used to push around and pick out single pollen grains, is to attach a human nasal hair to a teasing needle (Fig. 25). Collect fresh nasal hairs from your professor or senior scientist (avoid the grey and white hairs) and lay them on a sheet of paper. Take a teasing needle and push it ca 1 cm into a glue tube while squeezing gently and turning the needle in circles. Pull out the needle and press onto one of the hairs already laying on the sheet of paper. Make sure that the distal end of the hair is facing the same way as the distal end of the needle and that it extends a few mm longer than the needle. When the hair is attached to the glue, add a little extra glue to cover the proximal part of the hair. Place the needle across the small opening of the glue tube, then press the tube gently for additional glue and at the same time turn the needle in circles while moving it back and forward.

Applying the Single-Grain Method

Use one of the self-made glass-pipettes to stir the sample and blow air through it to mix up the particles real good. Then suck up a tiny portion of the sample using the pipette and transfer onto a glass slide. When the tip of the pipette touches the glass slide drag the pipette along the middle section of the slide (left to right) to produce a long and relatively narrow glycerine strip. Using an erect image compound microscope (meaning when something is moved under the objective lens from left to right it is also seen moving in that same direction when observed through the eyepiece) place the glass slide under the special working distance 10× or 20× objective lens and move the distal end of the micromanipulator in-between the glass slide and the lens and then gently press the tip of the micromanipulator (the nasal hair) into the glycerine (Fig. 26 A-B). Using the micromanipulator grains of particular interest are brushed or pushed to the edge of the glycerine, then out of the glycerine until they are attached to the nasal hair and can be picked up and transferred to another glass slide (Fig. 26 C-H). Have a fresh drop of glycerine ready on a new glass slide. Dip the tip of the hair with the attached pollen into the glycerine drop and the pollen will automatically detach from the hair and rest in the glycerine. Because no cover slip is used this pollen can now be turned around with help from the micromanipulator and photographed in polar and equatorial views as well under different foci (high-, low focus, optical section), documenting important features such as sculpture, apertures, and thickenings or thinnings of the pollen wall (Fig. 27 A-D). After this, the pollen grain is transferred to a SEM stub to which a drop of absolute ethanol has been added to remove all traces of the glycerine from the surface of the pollen grains (Fig. 27 E-G). For this, the best way is to position the light microscope close beside a binocular stereoscope. Place a single SEM stub under the stereoscope and have a small container with fresh ethanol at your side as well as one of the glass-pipettes mentioned above. First pick out a pollen grain with the micromanipulator from the glycerine drop and slowly move over to the stereoscope. Dip the tip of the pipette into the ethanol container and it will automatically suck up a small portion of the ethanol. Press the tip of the pipette on the surface of the SEM stub to leave a tiny drop of ethanol. Then gently press the tip of the nasal hair with the attached pollen into the drop of ethanol and the pollen will be detached from the hair, float a bit in the drop and finally rest on the stub surface when the ethanol evaporates. Try to make the ethanol drops small and close to the center of the SEM stub. Up to 10 different types of grains can be placed on a single stub and additional ethanol drops can be added to clean the glycerine thoroughly off the pollen grains. The stub is then sputter coated with gold and the pollen photographed using a SEM (overviews and close-ups). Pollen of particular interests can be turned. Add a drop of ethanol to the sputtered sample and flip the grain over using the micromanipulator before the ethanol evaporates (under the stereoscope). Re-sputter the sample and photograph it again using the SEM. This applies especially to any kind of heteropolar pollen/spores or tetrads of some sort.

Recipes

Recipes for Light Microscopy (LM)

Acetocarmine (Staining)

30 g acetocarmine + 2 L 45% acetic acid, 4 h boiled and filtered.

Potassium Iodine (Lugol's Iodine, Detection of Starch)

2 g potassium iodine + 1 g iodine + 100 ml distilled water

Toluidine Blue (Staining)

0.1 g Toluidine blue + 100 ml 2.5% sodium carbonate ($NaCO_3$); durable at +4 °C

Chlorination Mixture

Acetic acid (CH_3COOH) + saturated sodium chlorate ($NaClO_3$)* + 3–5 drops hydrochloric acid (conc. HCl)

*Saturated sodium chlorate solution: about 10 g of $NaClO_3$ in 10 ml distilled water (25 °C); the solution is saturated when crystals are still present.

Annotation: solubility of sodium chlorate is depending on the temperature of water.

Acetolysis Mixture

Acetolysis mixture: 9 parts acetic anhydride (99%) are mixed with 1 part concentrated sulfuric acid (96%).

Zinc Bromide Solution (Heavy Liquid Separation for Samples with a High Mineral Content)

250 g zinc bromide (Merck 8.18631.0250) + 25 ml 10% HCl*, mix until all zinc bromide is solved (takes some time!), then add 100 ml distilled water.

*10% HCl-Lösung: 27 ml H_2O + 10 ml HCl (37%) = 37 ml 10% HCl

Recipes for Scanning Electron Microscopy (SEM)

Dimethoxypropane (Dehydration)

30 ml 2,2-dimethoxypropane (DMP) + 1 drop 0.2 n hydrochloric acid (HCl)

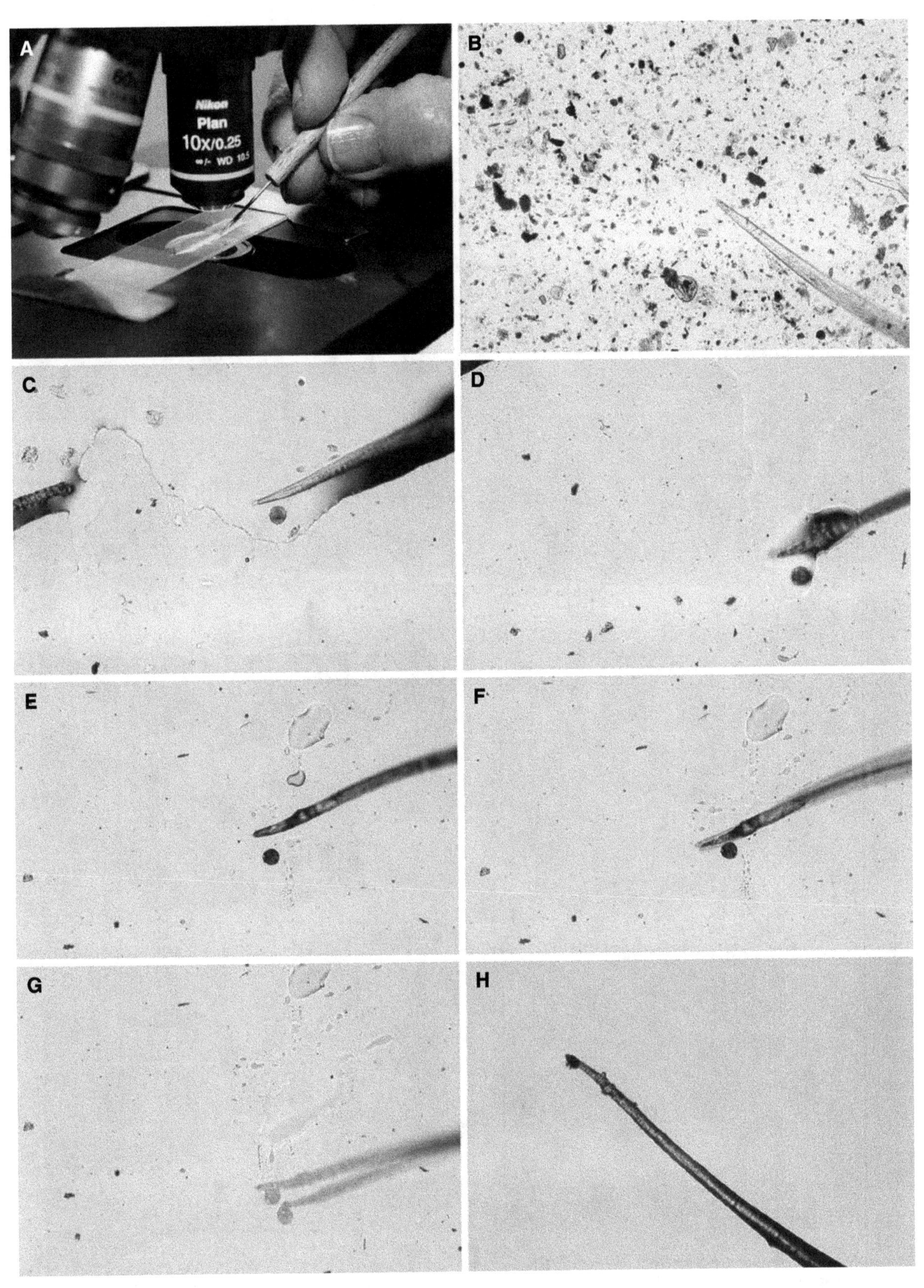

Fig. 26 Applying the single-grain method — Part 1. A. Sample on a glass slide under LM, working distance from sample to objective approx. 1 cm. **B.** Organic-rich sample and the tip of a nasal hair seen through the LM. **C.** Fossil pollen grain being brushed/pushed towards the margin of the glycerine. **D.** Fossil pollen grain pushed further away from the glycerine. **E.** Grain out of glycerine and ready to be picked up by the nasal hair. **F.** Pollen pushed a bit further. **G.** In a pushing or brushing motion the pollen is picked up from the glass slide. **H.** Single fossil pollen grain attached to tip of nasal hair

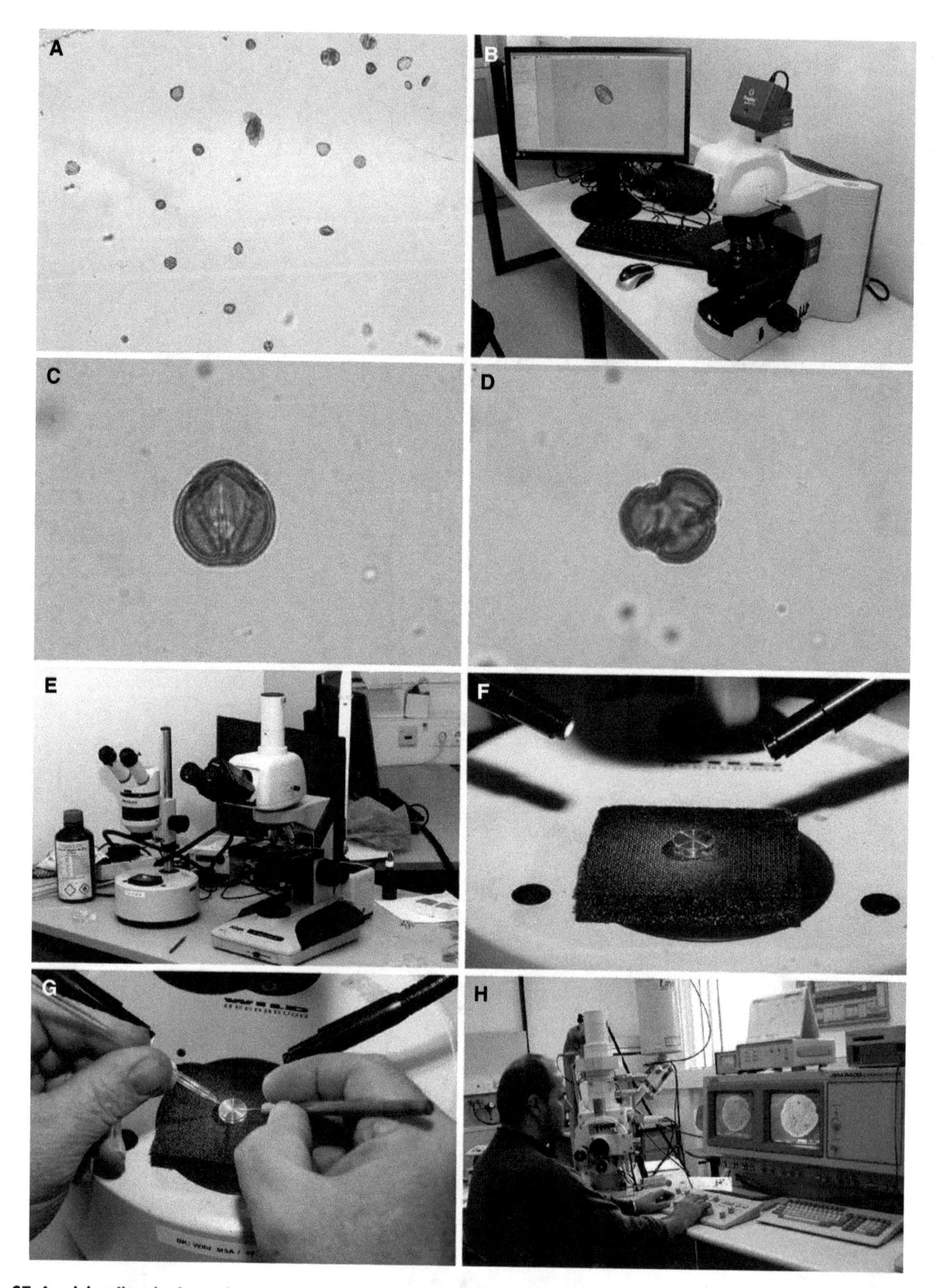

Fig. 27 Applying the single-grain method — Part 2. A. Selected well-preserved pollen grains in a fresh drop of glycerine. **B**. Light microscope equipped with a photographic unit to document pollen grains and their diagnostic features. **C**. Pollen turned and photographed in equatorial view. **D**. Same pollen grain turned and photographed in polar view. **E**. Arrangement of the light microscope and stereomicroscope along with a bottle of ethanol and other tools used when transporting pollen grains from glass slides over to SEM stubs. **F**. Cleaned SEM stub under a stereomicroscope waiting for fossil pollen grains. **G**. How to hold the pipette with the ethanol (left) and the micromanipulator (right) when transferring fossil pollen grains onto SEM stubs. **H**. Photographing fossil pollen using SEM

Recipes for Transmission Electron Microscopy (TEM)

3% Glutaraldehyde (Fixation)

100 ml glutaraldehyde: 12 ml glutaraldehyde (GA, 25%) + 88 ml phosphate buffer (pH 7.2).

1 % Osmium Tetroxide (Fixation)

0.1 g osmium tetroxide (OsO_4) + 10 ml distilled water.

Osmium can be acquired in crystalline form within glass ampullae. The osmium crystals usually adhere inside the ampulla and can be loosened by dipping the ampulla in liquid nitrogen (in a styrofoam box). The ampulla can then be opened and the osmium crystals transferred into distilled water in a vial. Close the vial and seal it with parafilm. For faster dissolution, place the vial in an ultrasonic bath. Mix the osmium solution and pipette it into a vapor-tight bottle. Store it at 6 °C.

Annotation: osmium is volatile and toxic, use in fume hood only; for storage, use oil with high percentage of unsaturated fatty acids (e.g. corn oil) to bind volatiles of osmium tetroxide (Fig. 28).

Phosphate Buffer pH 7.2 (Fixation)

1 phosphate buffer saline tablet (phosphate buffer saline tablets, $Na_2HPO_4{\cdot}2H_2O$, sodium hydrogen phosphate) + 200 ml distilled water (dispense tablet in ultrasonic bath).

0.8% Potassium Ferrocyanide (Accelerator for Osmium)

0.1 g potassium hexacyanoferrate (II) ($K_4Fe(CN)_6{\cdot}3H_2O$) + 12.5 ml distilled water (dispense in ultrasonic bath).

Annotation: the fresh solution is uncolored and becomes yellow after a few days.

Agar Low Viscosity Resin Kit (Embedding)

LV-resin (Agar Scientific): 48 g LV Resin + 8 g hardener VH1 + 44 g hardener VH2 + 2.5 ml accelerator.

Annotation: Mix the embedding solution in a disposable plastic beaker by using a magnetic stirrer. The first two components must be mixed first before adding the remaining ingredients, then mix well again. The mixture can be used immediately for infiltration and then for embedding. Embedding solution can be stored in a freezer.

Potassium Iodine (Staining)

3 g potassium iodide + 7 g iodine + 100 ml ethanol (92%).

Fig. 28 Osmium storage. A. Osmium solution stored at 6 °C (fridge placed in a fume hood). **B-D.** Osmium solution in a sealed bottle and stored in a plastic container, plastic container placed in glass vessel containing oil. **C.** Arrowheads showing osmium contamination from volatiles. **D.** Second glass vessel placed over the osmium containers, osmium vapor is bound to the oil and cannot escape into the atmosphere

1% Potassium Permanganate (Fixation and Staining)

1% potassium permanganate: 1 g potassium permanganate in 100 ml distilled water

1% Periodic Acid (Staining)

1 g periodic acid (PA, Firma Fluka) + 100 ml distilled water

0.2% Thiocarbohydrazide (Staining)

0.2 g thiocarbohydrazide (TCH, by *Serva*) + 100 ml 20% acetic acid (20 ml 100 % CH_3COOH + 80 ml distilled water)

1% Silver Proteinate (Staining)

0.25 g silver proteinate (SP, by *Merck*) + 25 ml distilled water

Uranyl Acetate (Staining)

Prefabricated solution: "Ultrostain 1" by Leica

Lead Citrate (Staining)

Prefabricated solution: "Ultrostain 2" by Leica; used with potassium hydroxide pellets

Formvar Filming Solution (Film-Coated Grids)

2 g formvar (15/45 E) + 100 ml chloroform (pure); mix with a magnetic stirrer

References

Brown CA (1960) Palynological techniques. Lousiana State University, Baton Rouge, La

Daghlian CP (1982) A simple method for combined light, scanning and transmission electron microscope observation of single pollen grains from dispersed pollen samples. Pollen Spores 24: 537–545

Erdtman G (1943) An introduction to pollen analysis. Chronica Botanica, Waltham, Mass

Erdtman G (1960) The acetolysis method. Svensk Bot Tidskr 54: 561–564

Eyring MB (1996) Soil pollen analysis from a forensic point of view. Microscope 44: 81–97

Fægri K, Iversen J (1989) Textbook of Pollen analysis. 4th edition, John Wiley & Sons, Chichester

Ferguson DF, Zetter R, Paudayal KN (2007) The need for the SEM in paleopalynology. C R Palevol 6: 423–430

Gerlach D (1984) Botanische Mikrotechnik. 3rd edition, Thieme, Stuttgart

Halbritter H (1998) Preparing living pollen material for scanning electron microscopy using 2,2-dimethoxypropane (DMP) and critical-point drying. Biotech Histochem 73: 137–143

Hayat MA (2000) Principles and techniques of electron microscopy: Biological applications. Cambridge University Press, Cambridge

Heslop-Harrison J(1979) An interpretation of the hydrodynamics of pollen. Am J Bot 66: 737–743

Moore PD, Webb JA, Collinson ME (1991) Pollen analysis. 2nd edition. Blackwell Scientific Publication, Oxford

Nepi M, Franchi GG, Pacini E (2001) Pollen hydration status at dispersal: cytophysiological features and strategies. Protoplasma 216: 171–180

Rowley JR, Dahl AO (1977) Pollen development in *Artemisia vulgaris* with special reference to Glycocalyx material. Pollen Spores 19: 169–284

Siegel I (1967) Toluidine blue O and naphthol yellow S; a highly polychromatic general stain. Stain Technol 42: 29–30

Thiéry J-P (1967) Mise en évidence des polysaccharides sur coupes fines en microscopie électronique. J Microscopie 6: 987–1018

Traverse A (2007) Paleopalynology. 2nd ed, Springer, Dordrecht

Weber M (1992) Nature and distribution of the exine-held material in mature pollen grains of *Apium nodiflorum* L. (Apiaceae). Grana 31: 17–24

Weber M, Frosch A (1995) The development of the transmitting tract in the pistil of *Haquetia epipactis* (Apiaceae). Int J Plant Sci 156: 615–621

Weber M, Halbritter H (2007) Exploding pollen in *Montrichardia arborescens* (Araceae). Plant Syst Evol 263: 51–57

Weber M, Ulrich S (2010) The endexine: a frequently overlooked pollen wall layer and a simple method for detection. Grana 49: 83–90

Wood GD, Gabriel AM, Lawson JC (1996) Chapter 3. Palynological techniques – processing and microscopy. In: Jansonius J, McGregor DC (eds) Palynology: principles and applications. American Association of Stratigraphy Palynologists Foundation, Vol. 1. Publishers Press, Salt Lake City, Utah, USA, p. 29-50

Zetter R (1989) Methodik und Bedeutung einer routinemäßig kombinierten lichtmikroskopischen und rasterelektronenmikroskopischen Untersuchung fossiler Mikrofloren. Cour Forsch-Inst Senckenberg 109: 41–50

Part II

Pollen Terminology: An Illustrated Guide

7

Diverse Pollen Grains and their Dispersal Units

monad

dyad

pseudomonad

tetrad

tetrad tetrahedral

tetrad decussate

tetrad planar

polyad

massula

pollinium

pollinarium

monad

unit consisting of a single pollen grain

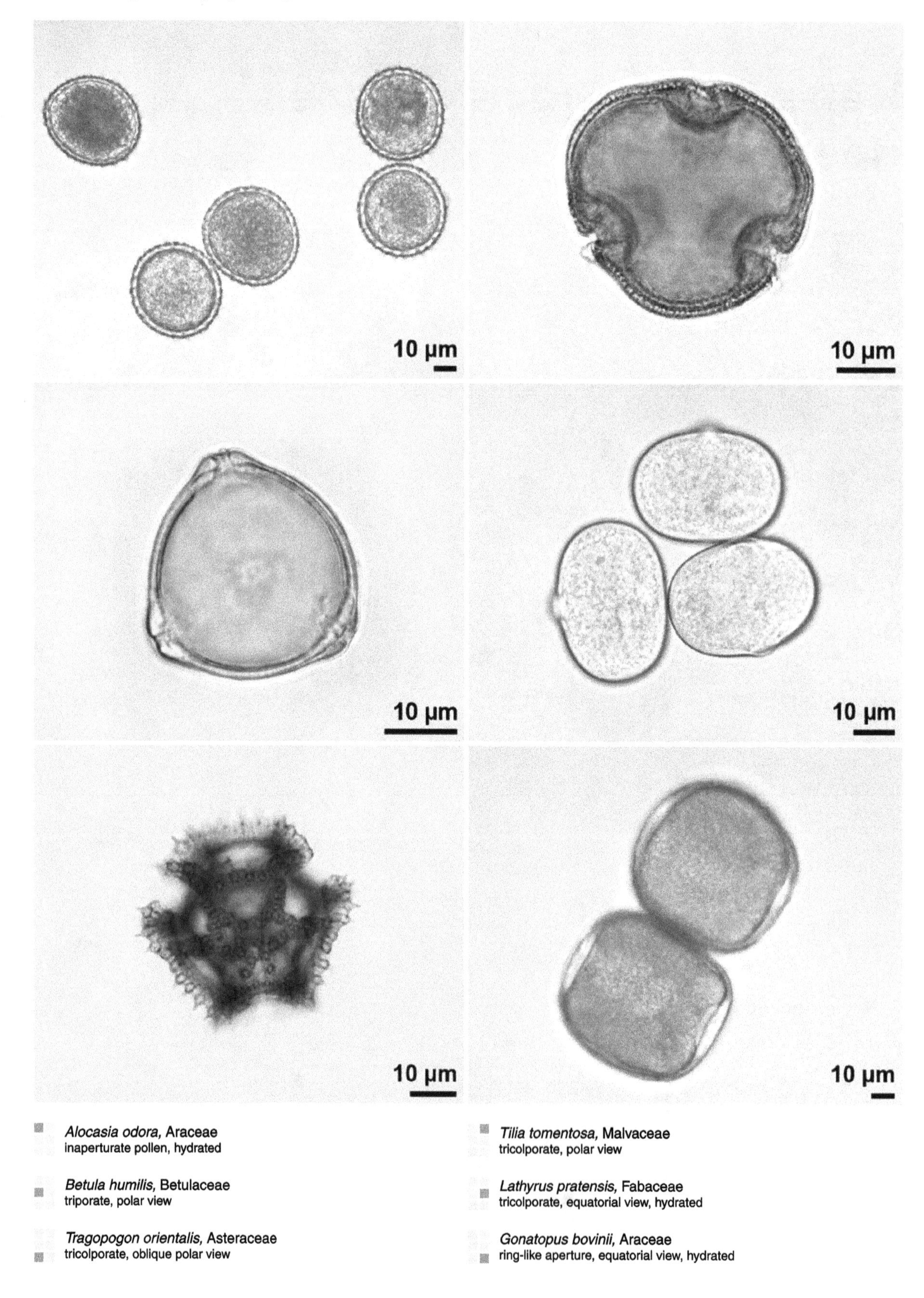

Alocasia odora, Araceae
inaperturate pollen, hydrated

Tilia tomentosa, Malvaceae
tricolporate, polar view

Betula humilis, Betulaceae
triporate, polar view

Lathyrus pratensis, Fabaceae
tricolporate, equatorial view, hydrated

Tragopogon orientalis, Asteraceae
tricolporate, oblique polar view

Gonatopus bovinii, Araceae
ring-like aperture, equatorial view, hydrated

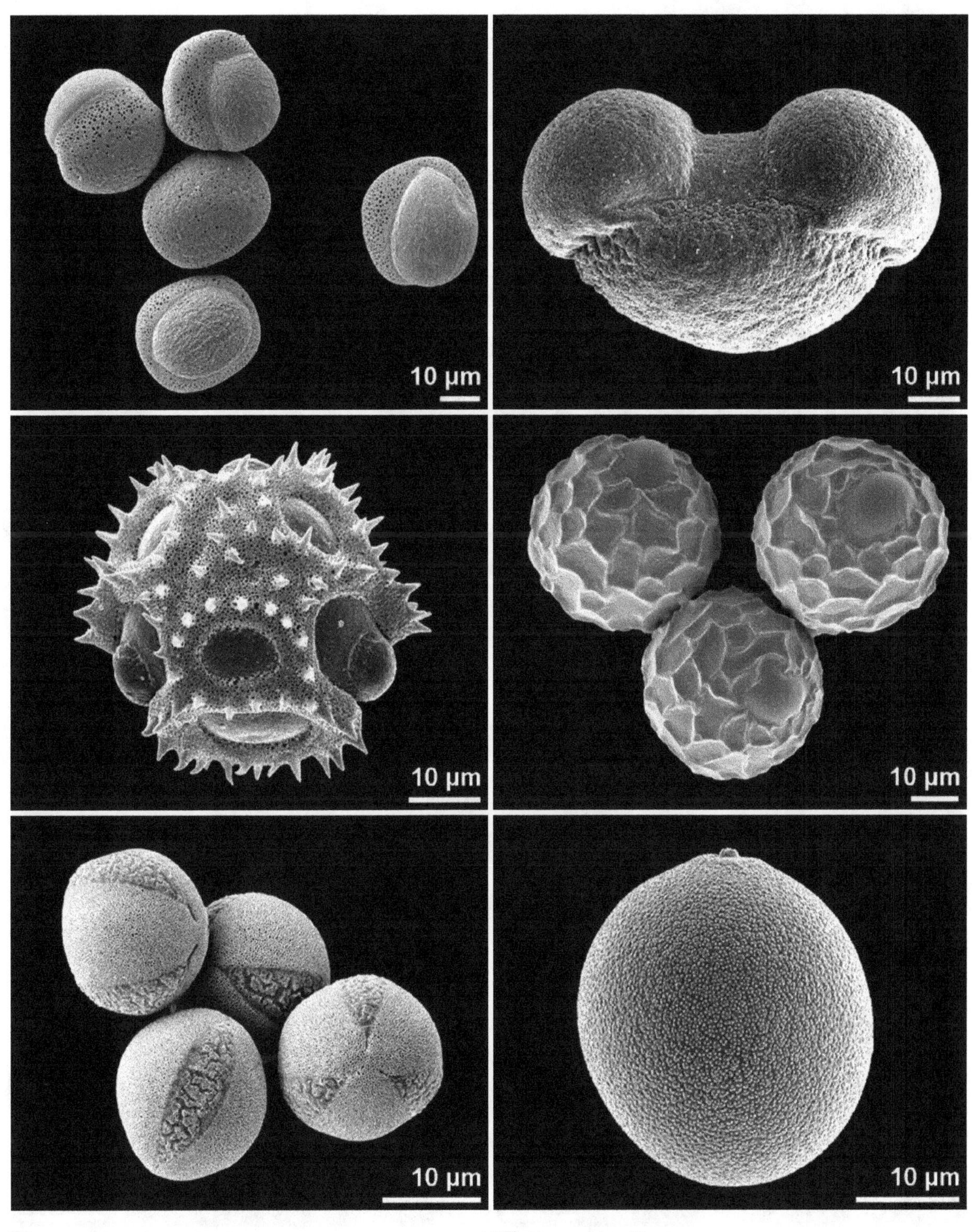

Haworthia herbacea, Xanthorrhoeaceae
sulcate

Tragopogon orientalis, Asteraceae
tricolporate, oblique polar view

Lamium galeobdolon, Lamiaceae
tricolpate

Pinus strobus, Pinaceae
bisaccate, equatorial view

Bituminaria bituminosa, Fabaceae
tricolporate, dry pollen

Dactylis glomerata, Poaceae
ulcerate, equatorial view

dyad

unit of two pollen grains

Polypleurum stylosum, Podostemaceae

Polypleurum stylosum, Podostemaceae
pollen collapsed

Zeylanidium olivaceum, Podostemaceae
equatorial view

Zeylanidium subulatum, Podostemaceae

Thelethylax minutiflora, Podostemaceae
equatorial view

Scheuchzeria palustris, Scheuchzeriaceae

pseudomonad

unit of a permanent tetrad with three rudimentary pollen grains

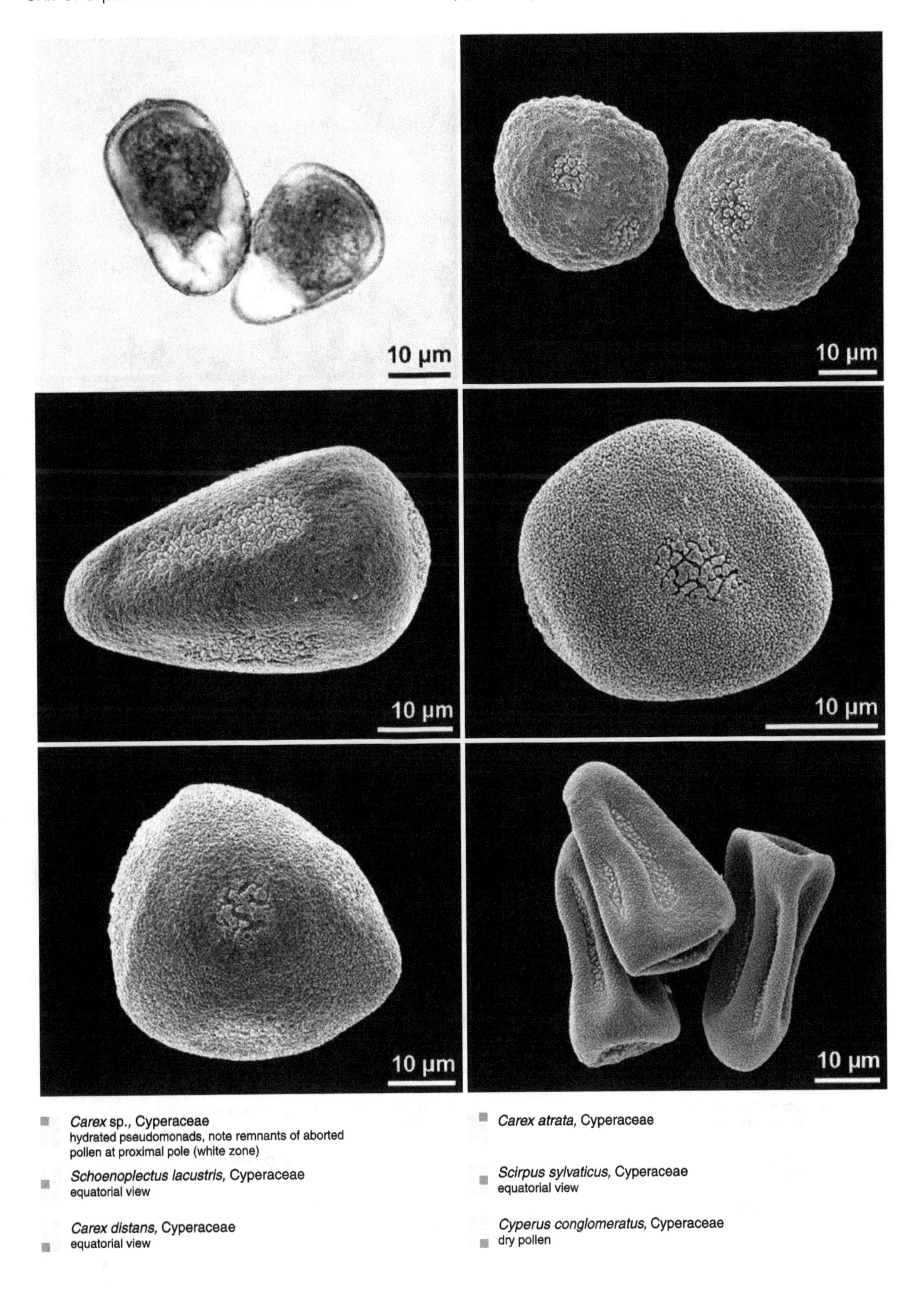

- *Carex* sp., Cyperaceae
  hydrated pseudomonads, note remnants of aborted pollen at proximal pole (white zone)
- *Carex atrata*, Cyperaceae
- *Schoenoplectus lacustris*, Cyperaceae
  equatorial view
- *Scirpus sylvaticus*, Cyperaceae
  equatorial view
- *Carex distans*, Cyperaceae
  equatorial view
- *Cyperus conglomeratus*, Cyperaceae
  dry pollen

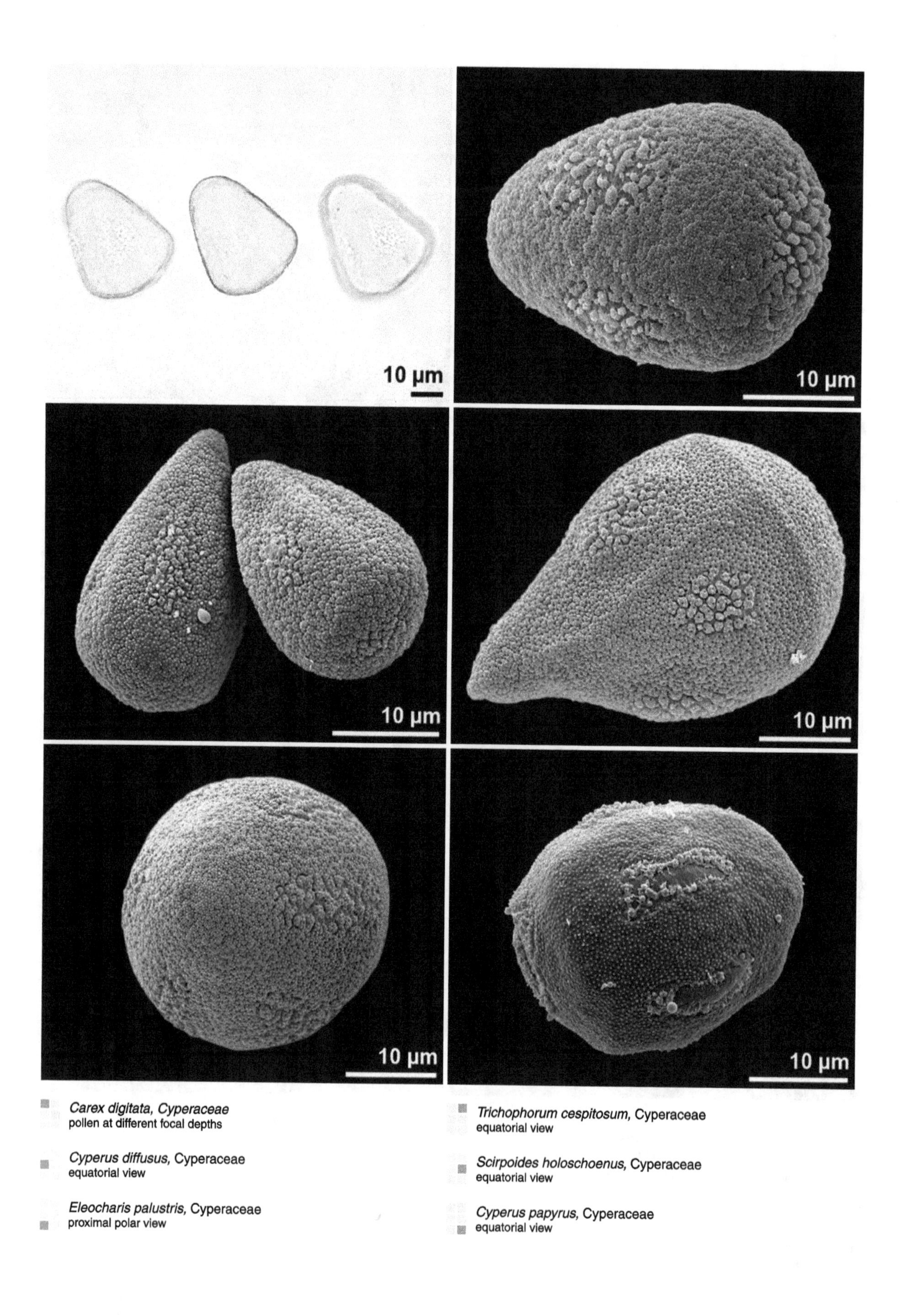

Carex digitata, Cyperaceae
pollen at different focal depths

Trichophorum cespitosum, Cyperaceae
equatorial view

Cyperus diffusus, Cyperaceae
equatorial view

Scirpoides holoschoenus, Cyperaceae
equatorial view

Eleocharis palustris, Cyperaceae
proximal polar view

Cyperus papyrus, Cyperaceae
equatorial view

tetrad

unit of four pollen grains

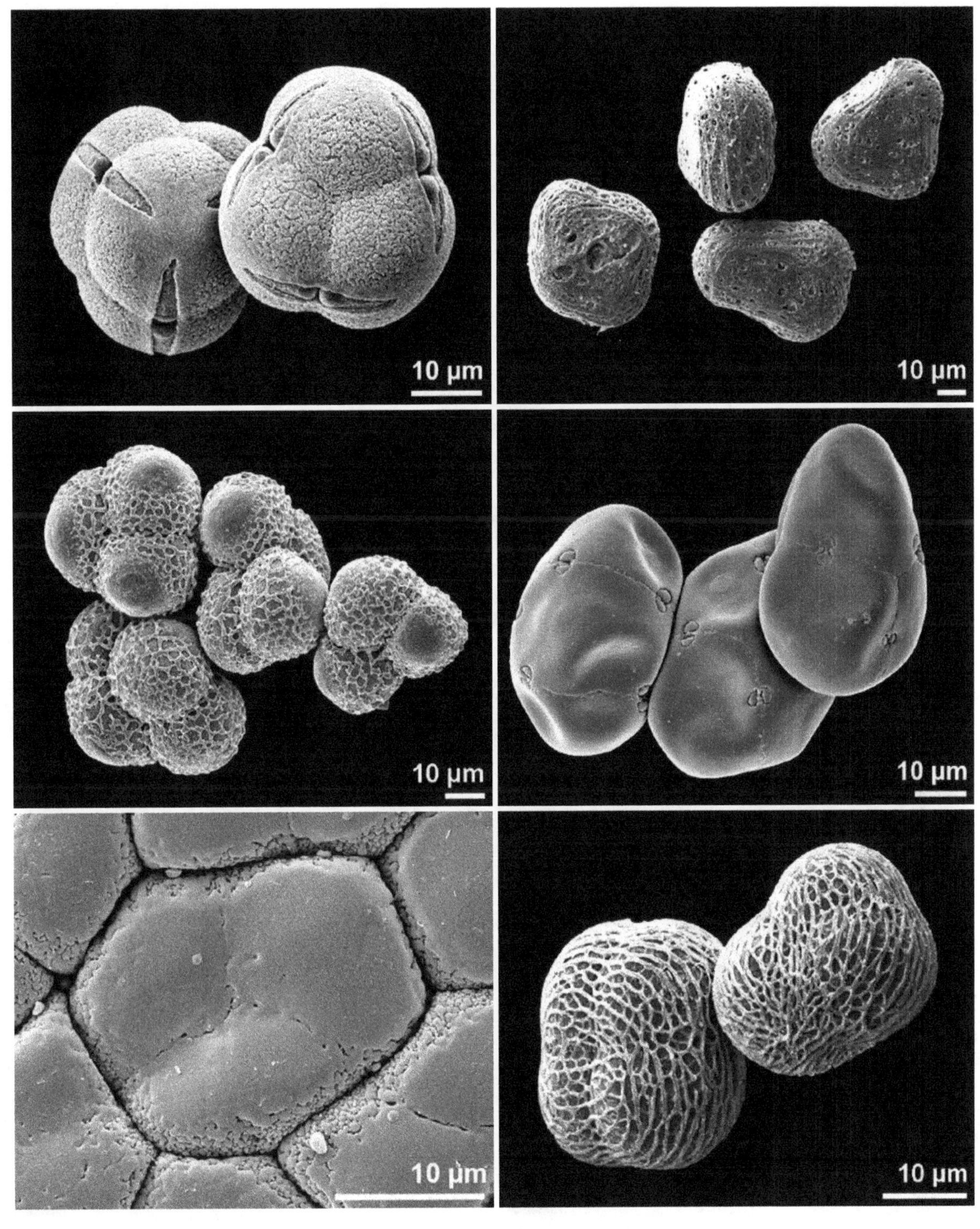

- *Erica herbacea,* Ericaceae
  tetrahedral tetrads
- *Epipactis helleborine,* Orchidaceae
  decussate tetrads
- *Epidendrum centropetalum,* Orchidaceae
  tetrad, part of pollinium
- *Chlorospatha kolbii,* Araceae
  differnt planar tetrads
- *Cyprinia gracilis,* Apocynaceae
  planar tetrads, dry pollen
- *Xanthosoma ceronii,* Araceae
  tetrad planar (left) and decussate (right)

tetrad tetrahedral

unit of four pollen grains in which the centers of the grains define a tetrahedron

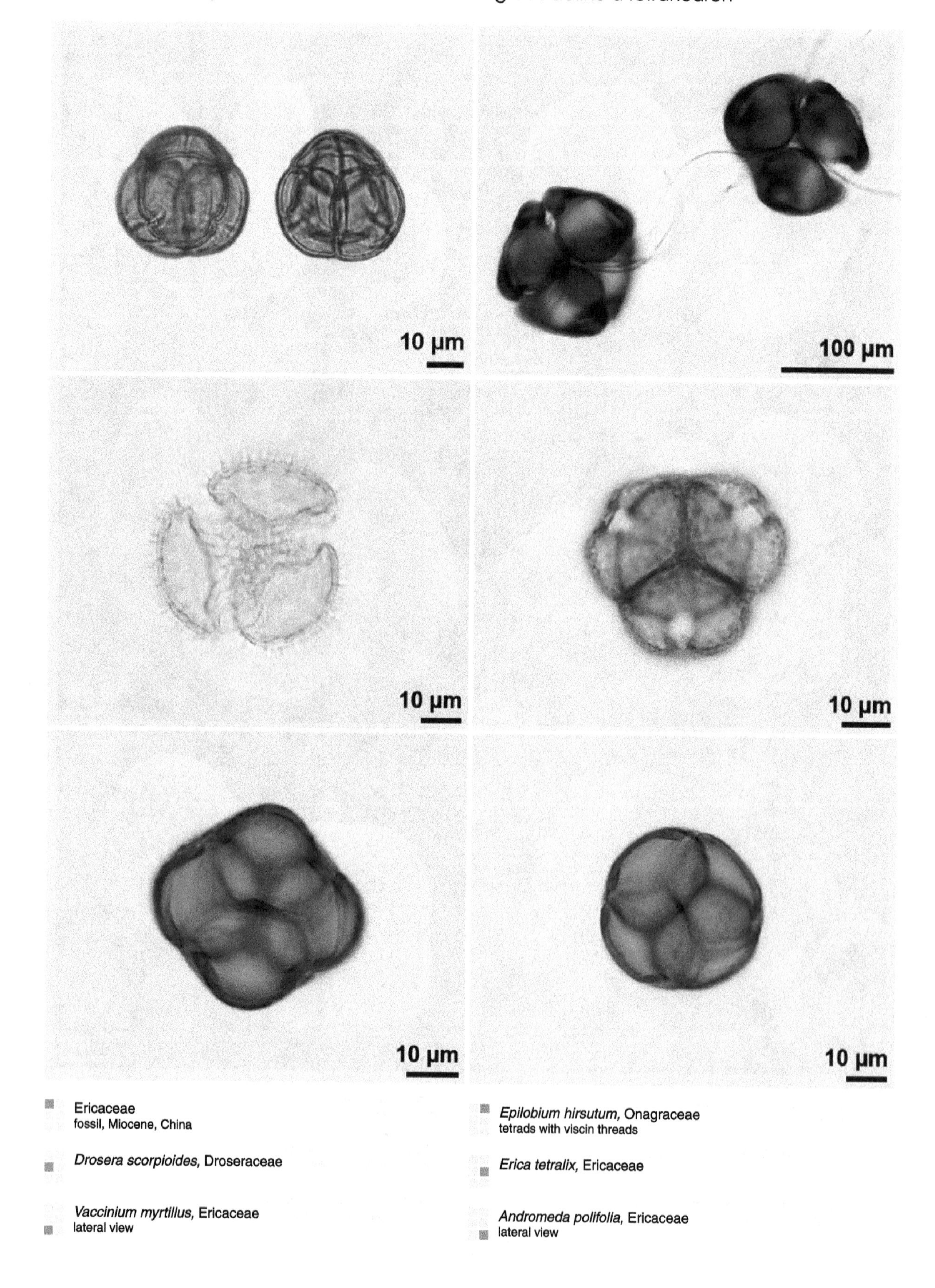

Ericaceae
fossil, Miocene, China

Epilobium hirsutum, Onagraceae
tetrads with viscin threads

Drosera scorpioides, Droseraceae

Erica tetralix, Ericaceae

Vaccinium myrtillus, Ericaceae
lateral view

Andromeda polifolia, Ericaceae
lateral view

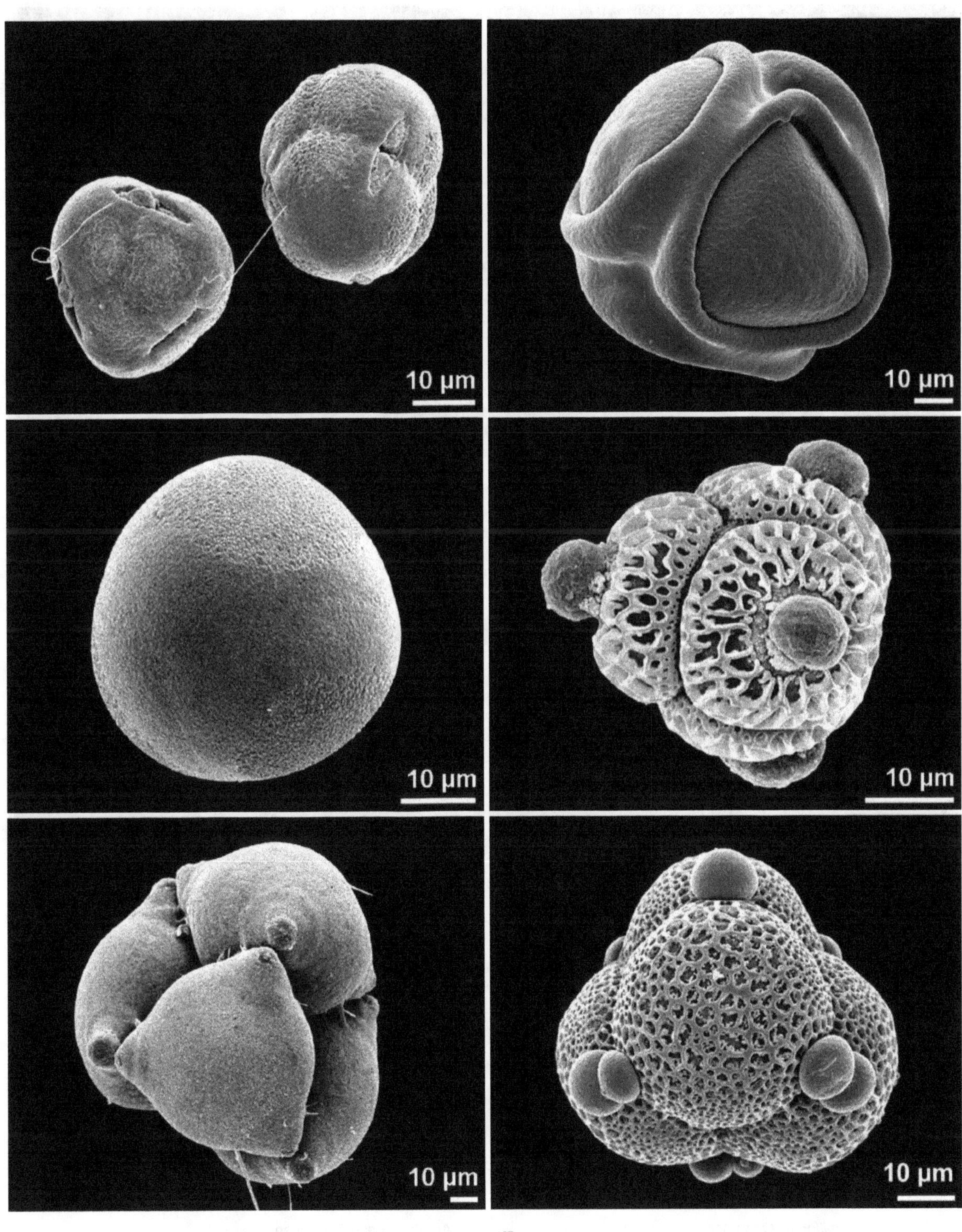

Rhododendron hirsutum, Ericaceae
tetrads with viscin threads

Luzula campestris, Juncaceae
ulcerate pollen

Epilobium montanum, Onagraceae
viscin threads

Victoria regia, Nymphaeaceae
dry pollen

Drimys granatensis, Winteraceae
ulcerate pollen

Oxyanthus subpunctatus, Rubiaceae
apical view

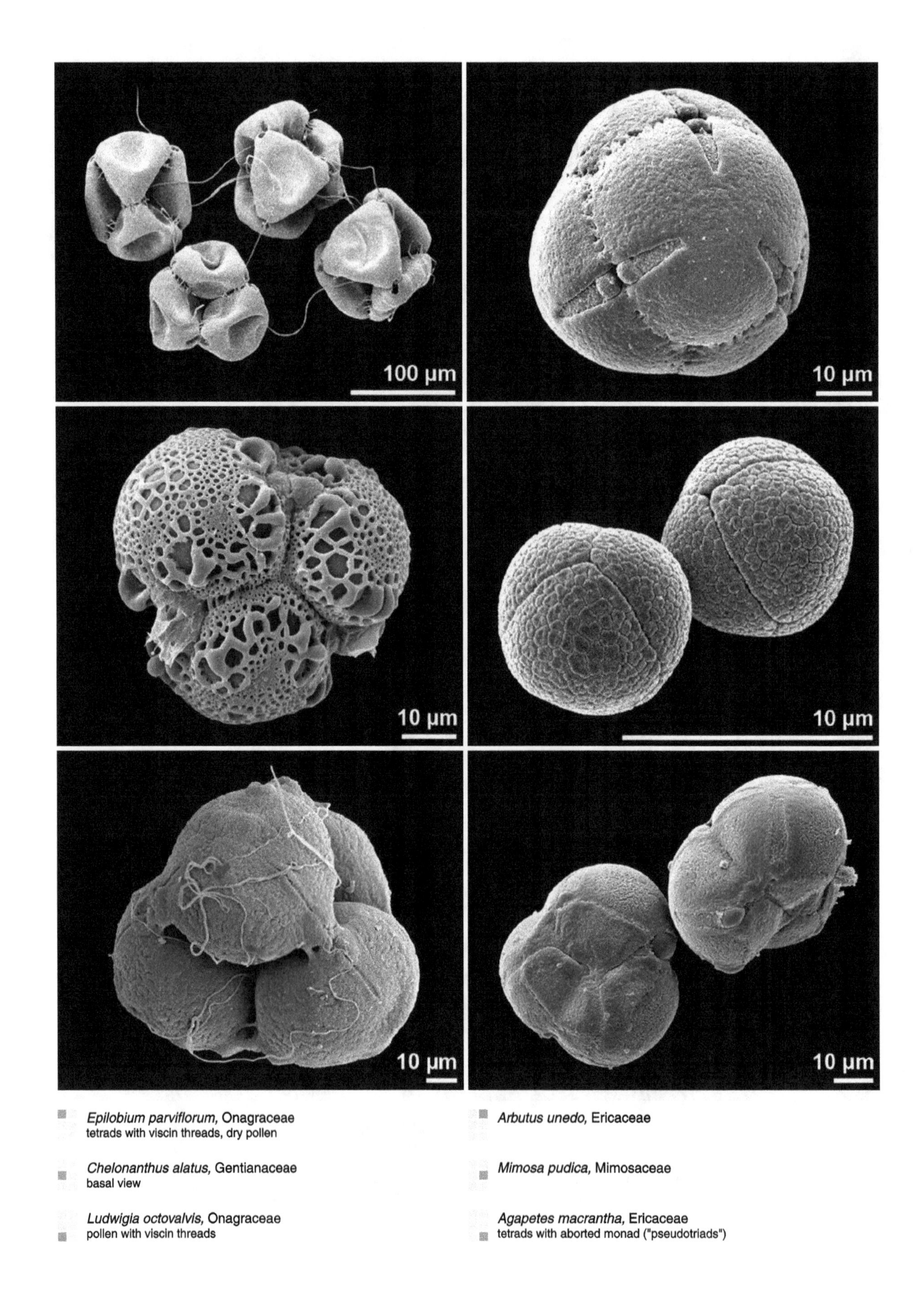

Epilobium parviflorum, Onagraceae
tetrads with viscin threads, dry pollen

Arbutus unedo, Ericaceae

Chelonanthus alatus, Gentianaceae
basal view

Mimosa pudica, Mimosaceae

Ludwigia octovalvis, Onagraceae
pollen with viscin threads

Agapetes macrantha, Ericaceae
tetrads with aborted monad ("pseudotriads")

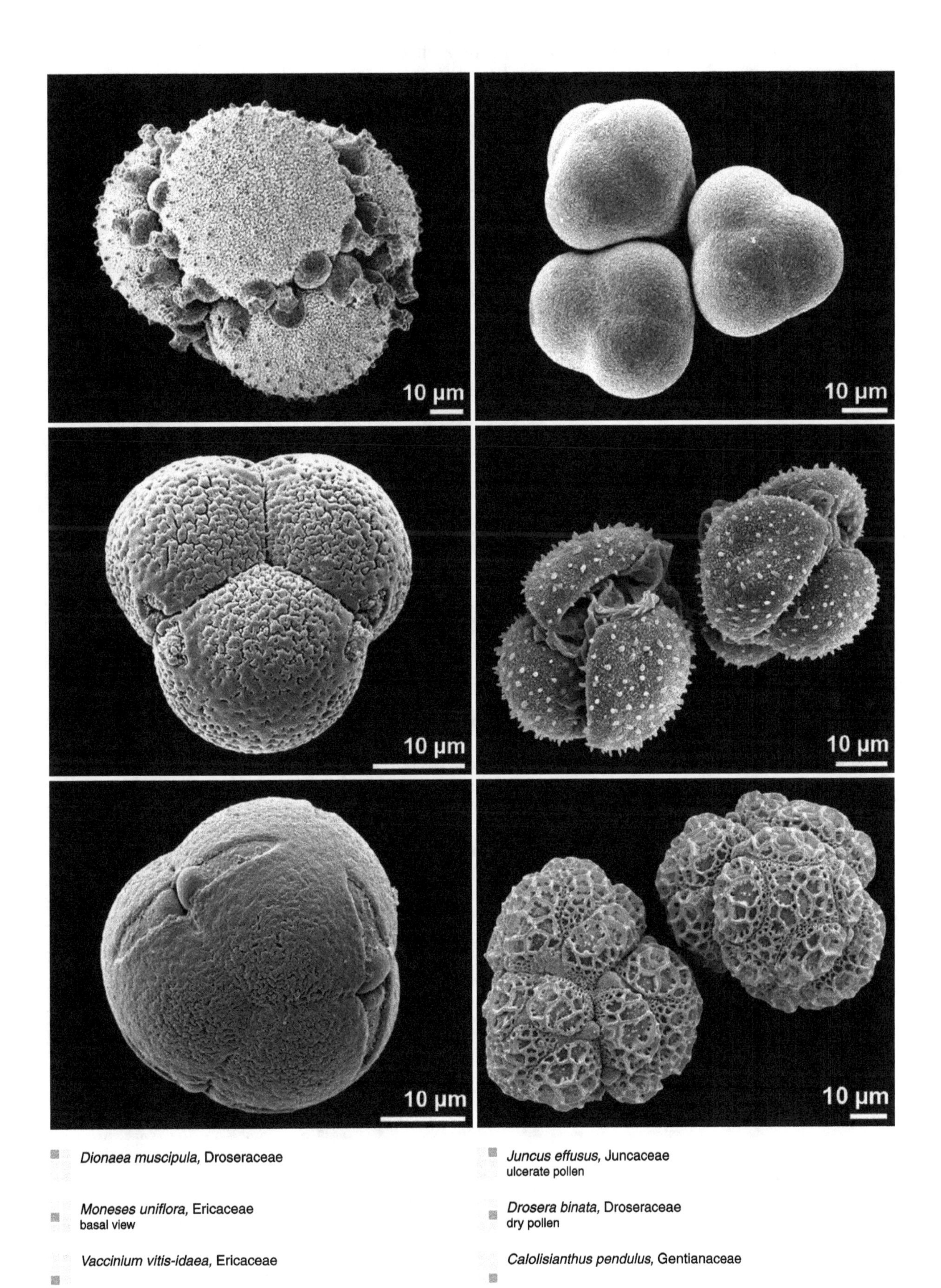

Dionaea muscipula, Droseraceae

Juncus effusus, Juncaceae
ulcerate pollen

Moneses uniflora, Ericaceae
basal view

Drosera binata, Droseraceae
dry pollen

Vaccinium vitis-idaea, Ericaceae

Calolisianthus pendulus, Gentianaceae

tetrad decussate

unit of four pollen grains arranged in two pairs in two different plains

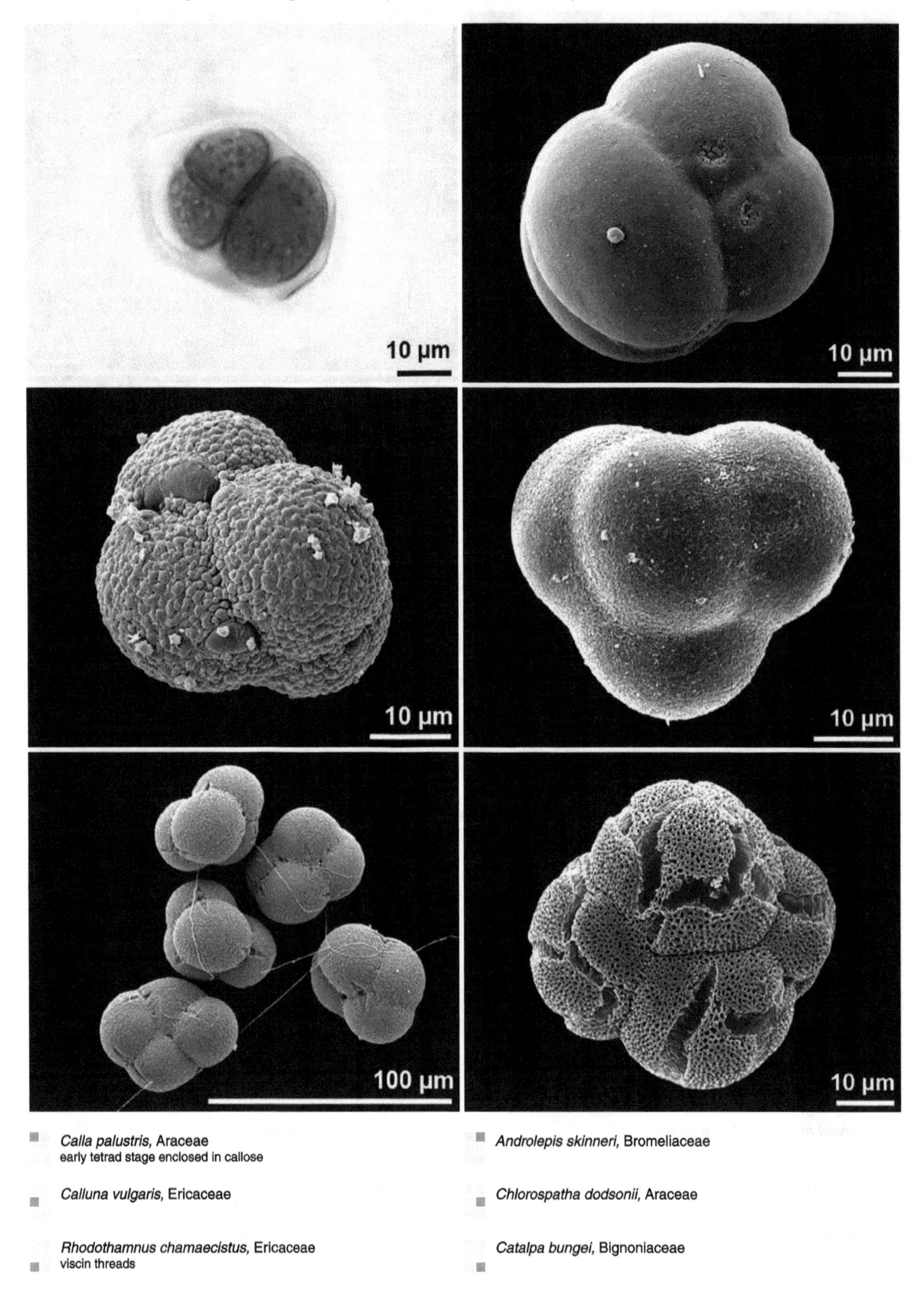

- *Calla palustris*, Araceae
  early tetrad stage enclosed in callose
- *Androlepis skinneri*, Bromeliaceae
- *Calluna vulgaris*, Ericaceae
- *Chlorospatha dodsonii*, Araceae
- *Rhodothamnus chamaecistus*, Ericaceae
  viscin threads
- *Catalpa bungei*, Bignoniaceae

tetrad planar

unit of four pollen grains arranged in one plane: tetragonal, T-shaped, linear

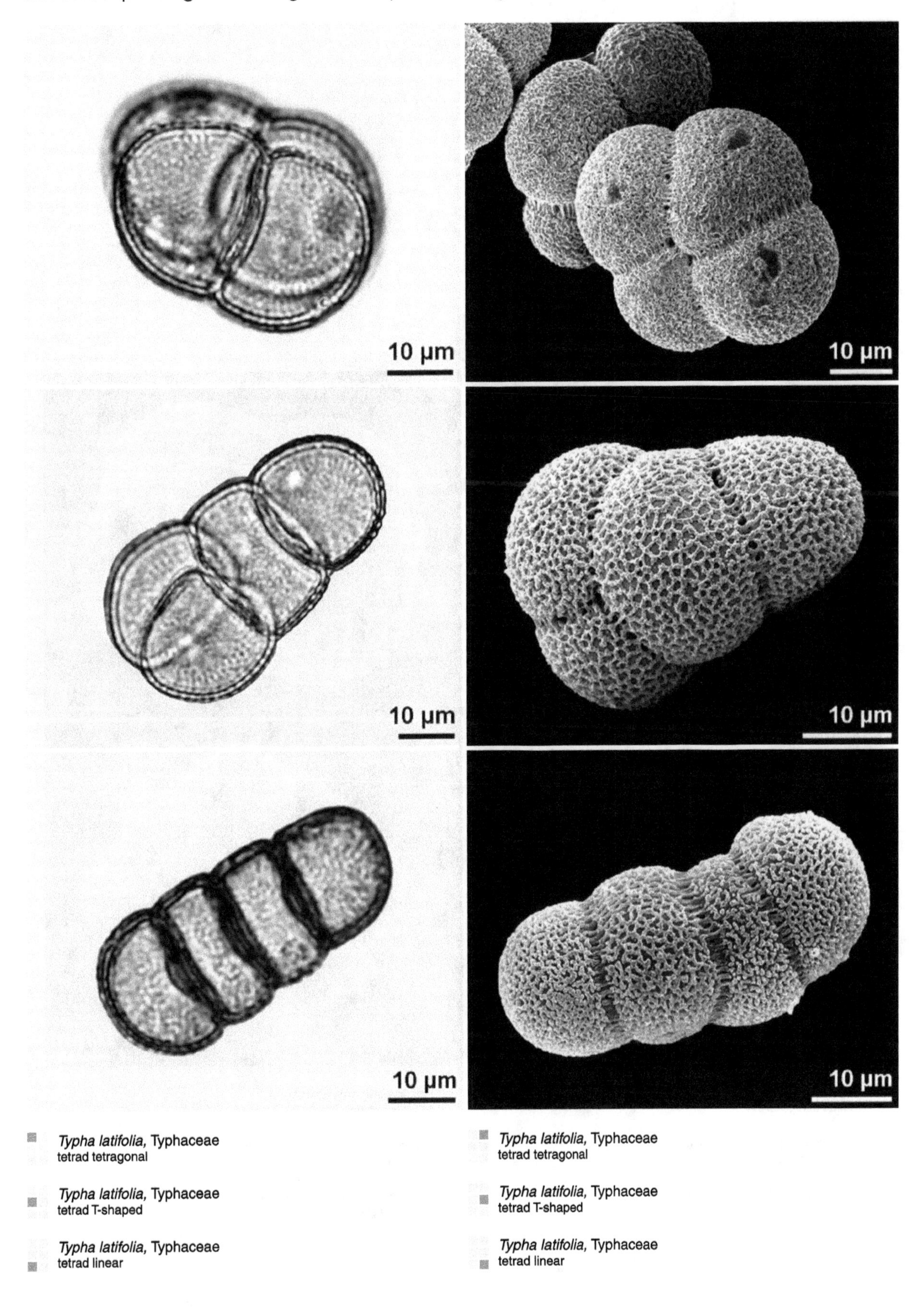

- *Typha latifolia*, Typhaceae
  tetrad tetragonal
- *Typha latifolia*, Typhaceae
  tetrad T-shaped
- *Typha latifolia*, Typhaceae
  tetrad linear

- *Typha latifolia*, Typhaceae
  tetrad tetragonal
- *Typha latifolia*, Typhaceae
  tetrad T-shaped
- *Typha latifolia*, Typhaceae
  tetrad linear

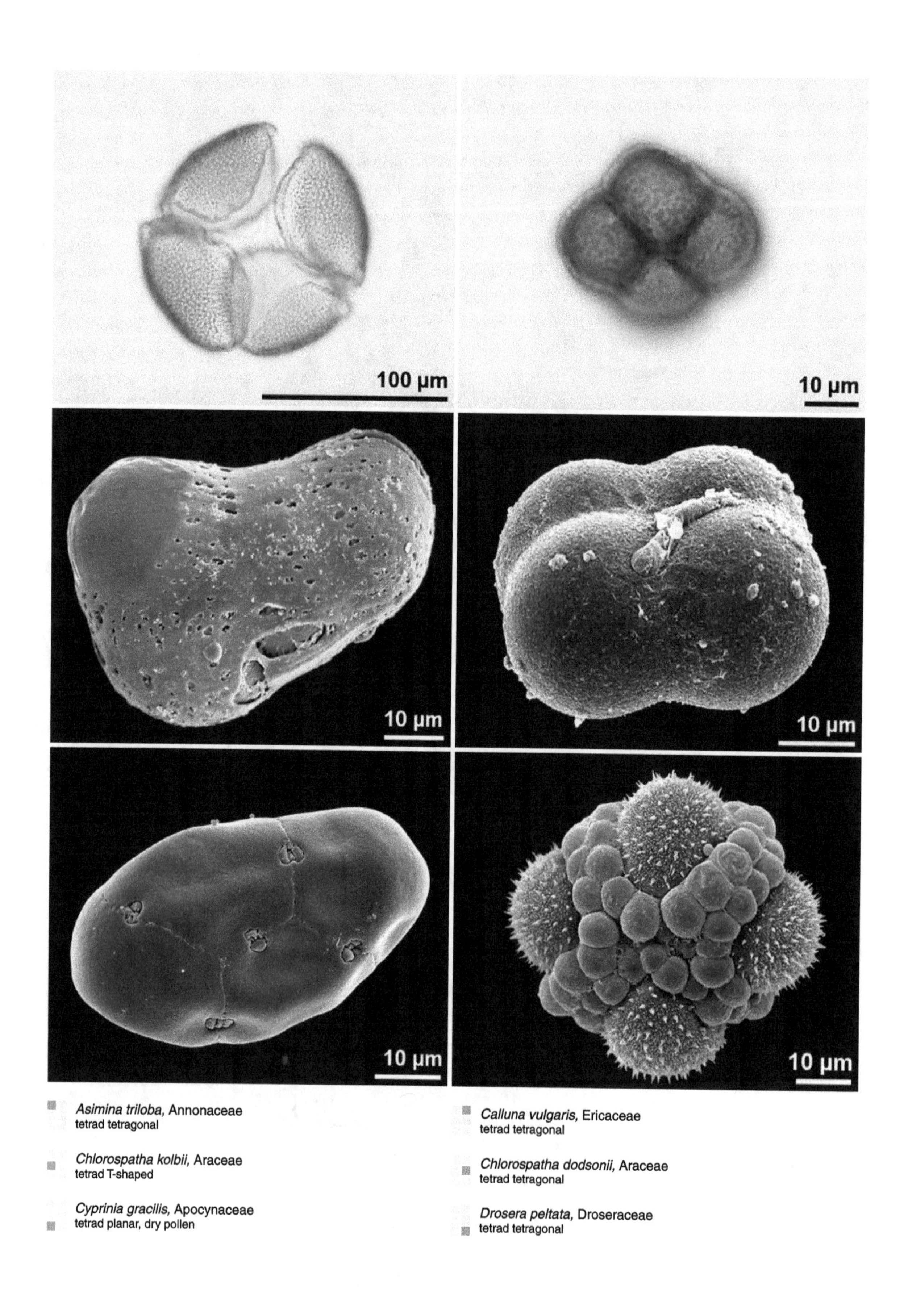

Asimina triloba, Annonaceae
tetrad tetragonal

Chlorospatha kolbii, Araceae
tetrad T-shaped

Cyprinia gracilis, Apocynaceae
tetrad planar, dry pollen

Calluna vulgaris, Ericaceae
tetrad tetragonal

Chlorospatha dodsonii, Araceae
tetrad tetragonal

Drosera peltata, Droseraceae
tetrad tetragonal

polyad

unit of more than four pollen grains (multiple of 4)

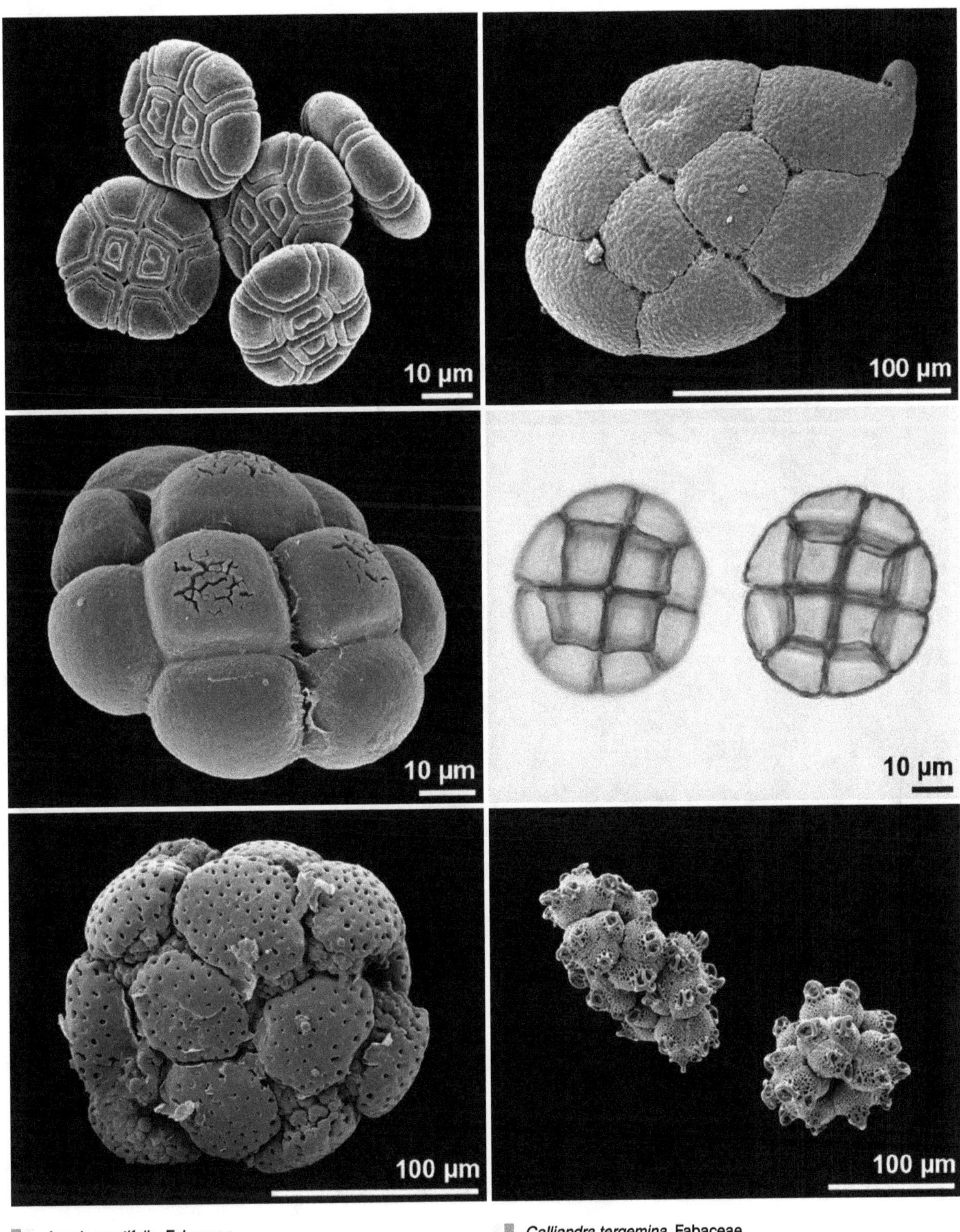

- *Acacia myrtifolia,* Fabaceae
  polyad with 8 monads, dry pollen
- *Pithecellobium dulce,* Fabaceae
  polyad with 16 monads
- *Cymbopetalum aequale,* Annonaceae
  polyad with 16 monads
- *Calliandra tergemina,* Fabaceae
  polyad with 8 monads
- *Acacia* sp., Fabaceae
  polyad with 16 monads
- *Chelonanthus purpurascens,* Gentianaceae
  polyad with 32 monads (8 united tetrads)

Acacia sp., Fabaceae
polyad of 8 monads, irregularly arranged, dry pollen

Acacia dealbata, Fabaceae
polyad with 16 monads

Acacia karroo, Fabaceae
polyad with 16 monads

Acacia karroo, Fabaceae
polyad with 16 monads, dry pollen

Albizia julibrissin, Fabaceae
polyad with 16 monads

Albizia saman, Fabaceae
polyad with 32 monads

massula

unit of more than four pollen grains but less than the locular content of a theca
Comment: In angiosperms only used for Orchidaceae with sectile pollinia

- *Traunsteinera globosa*, Orchidaceae
- *Epipogium aphyllum*, Orchidaceae
- *Herminium monorchis*, Orchidaceae
  massulae forming pollinium
- *Gennaria diphylla*, Orchidaceae
  massulae forming pollinium
- *Orchis italica*, Orchidaceae
- *Orchis purpurea*, Orchidaceae

pollinium

unit of a more or less interconnected loculiform pollen mass
Comment: loculi may be subdivided by septae, thus resulting in more than two pollinia

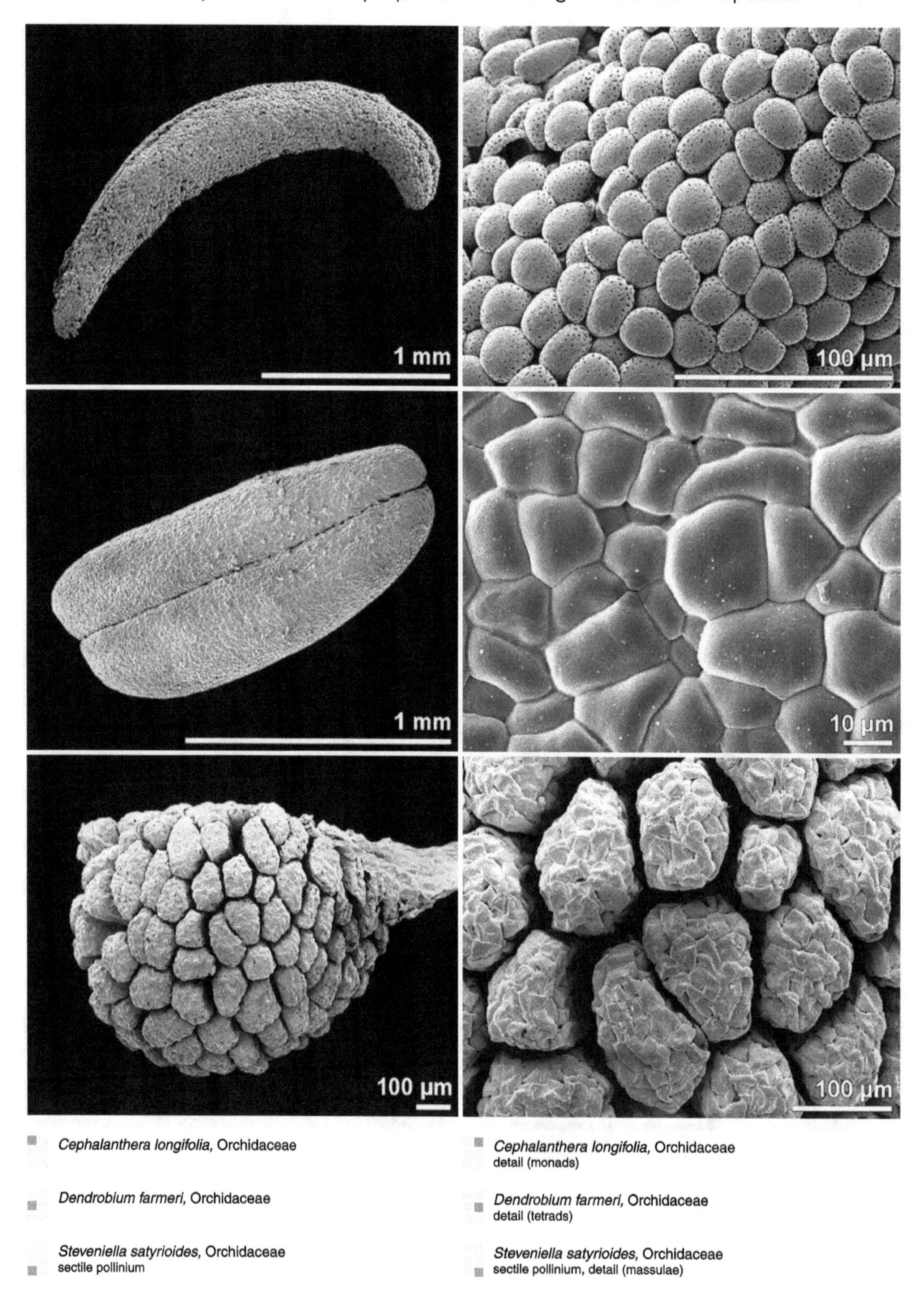

Cephalanthera longifolia, Orchidaceae

Cephalanthera longifolia, Orchidaceae
detail (monads)

Dendrobium farmeri, Orchidaceae

Dendrobium farmeri, Orchidaceae
detail (tetrads)

Steveniella satyrioides, Orchidaceae
sectile pollinium

Steveniella satyrioides, Orchidaceae
sectile pollinium, detail (massulae)

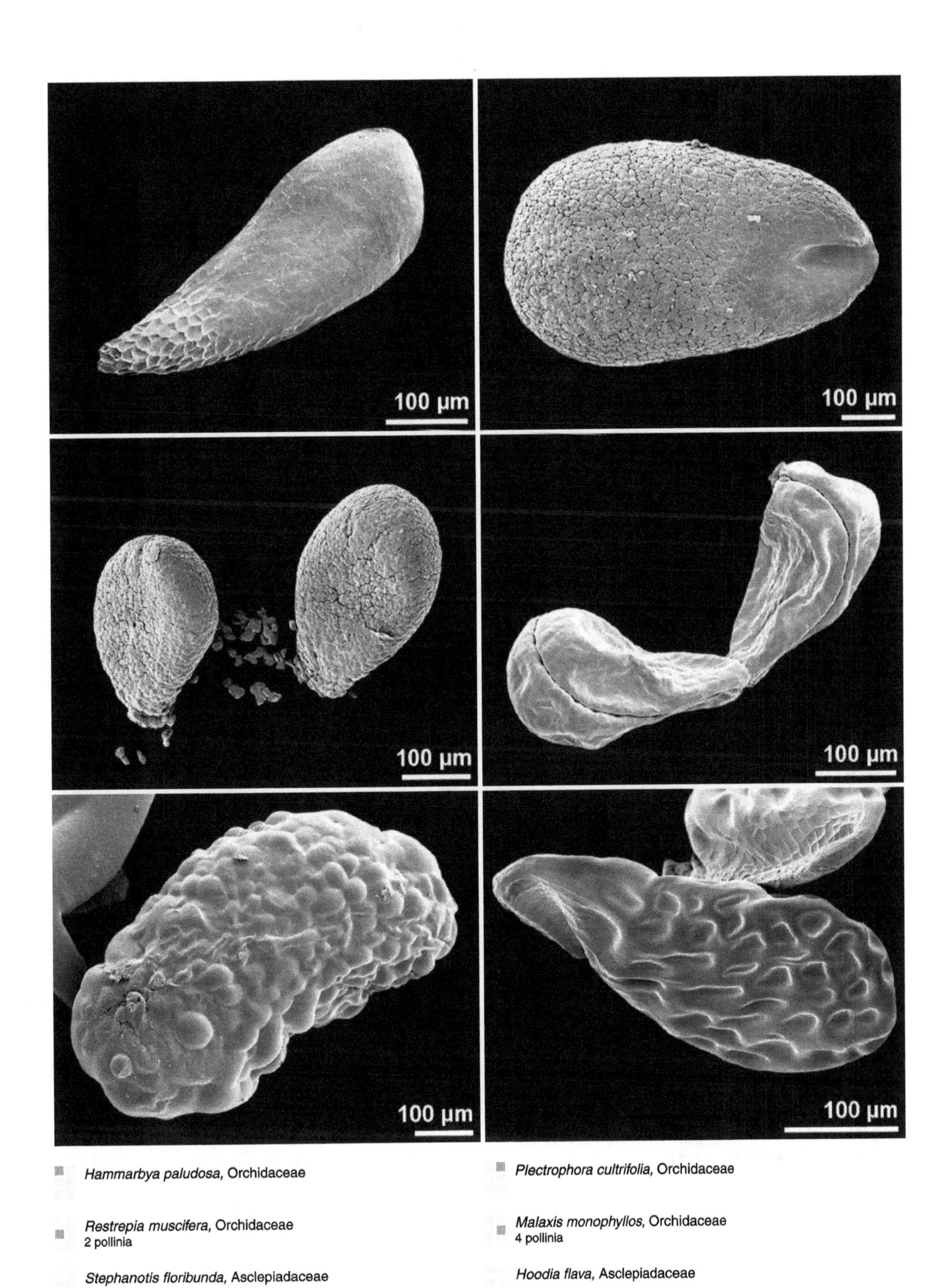

Hammarbya paludosa, Orchidaceae

Plectrophora cultrifolia, Orchidaceae

Restrepia muscifera, Orchidaceae
2 pollinia

Malaxis monophyllos, Orchidaceae
4 pollinia

Stephanotis floribunda, Asclepiadaceae

Hoodia flava, Asclepiadaceae

pollinarium

dispersal unit of pollinium (or pollinia) plus secretions and/or tissues that aid in the removal of the structure from the flower

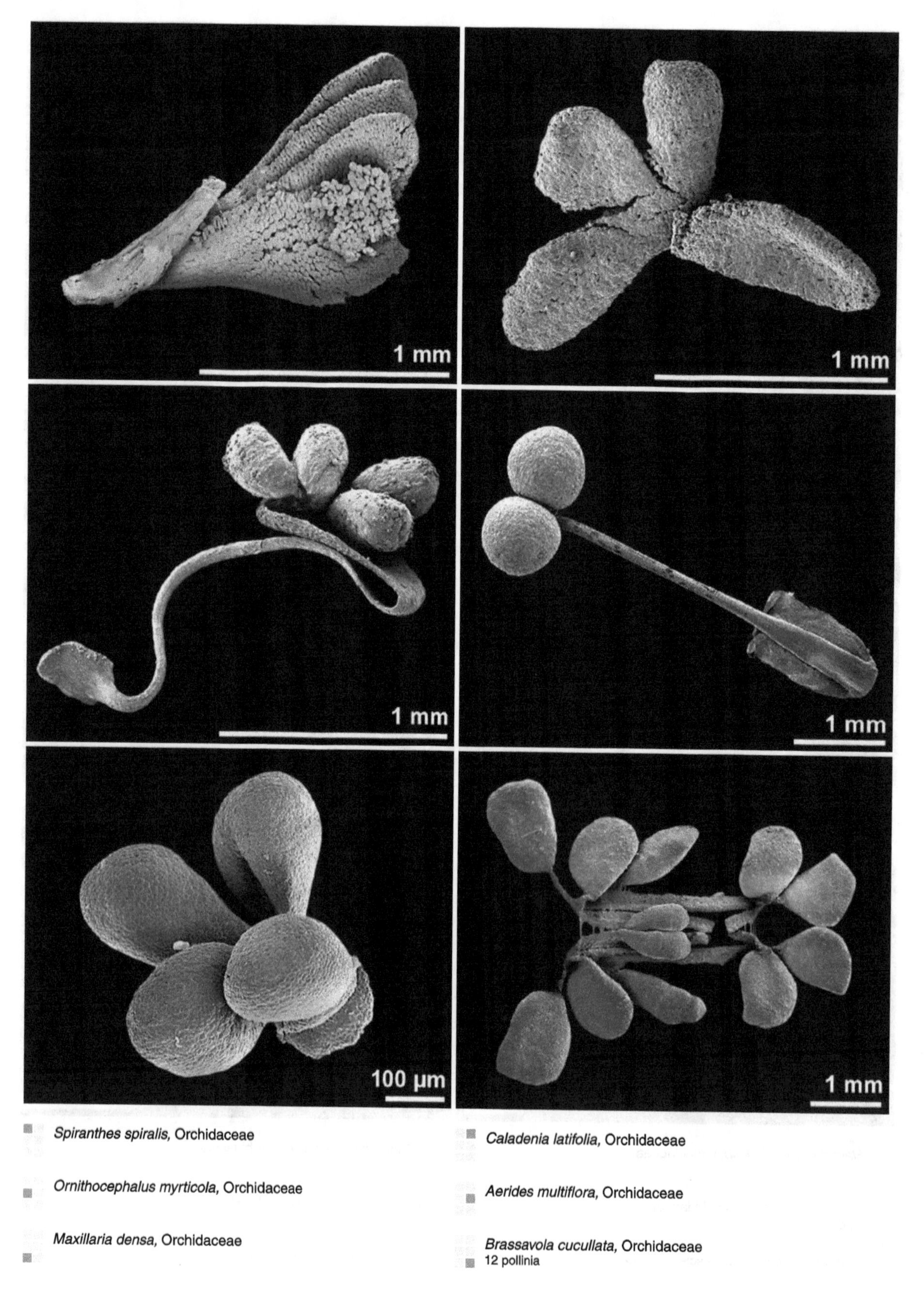

- *Spiranthes spiralis*, Orchidaceae
- *Caladenia latifolia*, Orchidaceae
- *Ornithocephalus myrticola*, Orchidaceae
- *Aerides multiflora*, Orchidaceae
- *Maxillaria densa*, Orchidaceae
- *Brassavola cucullata*, Orchidaceae 12 pollinia

- *Coelogyne fimbriata*, Orchidaceae
- *Pleurothallis loranthophylla*, Orchidaceae
- *Oncidium maizaefolium*, Orchidaceae
- *Schoenorchis fragrans*, Orchidaceae
- *Haraella odorata*, Orchidaceae
- *Stanhopea oculata*, Orchidaceae

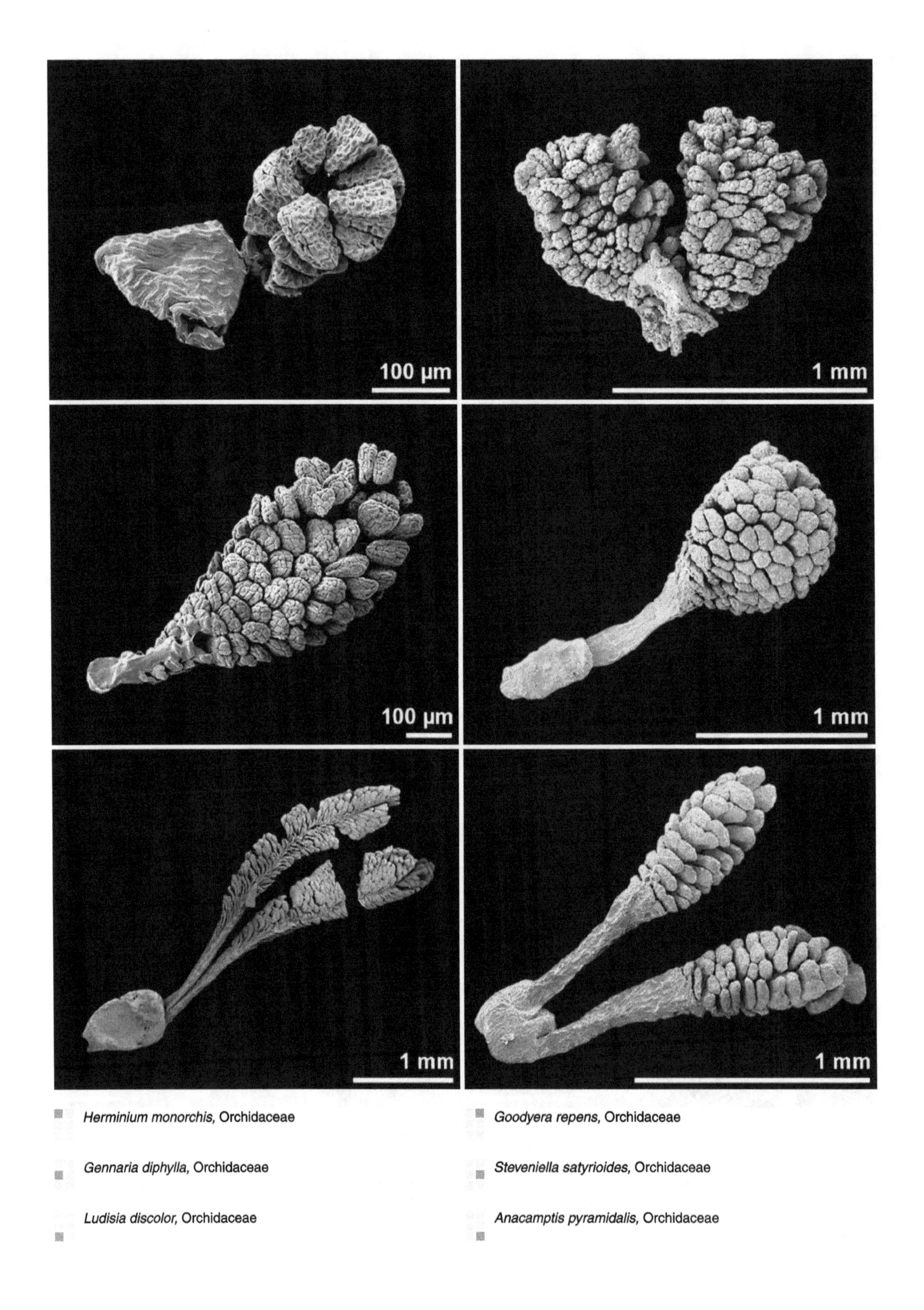

Herminium monorchis, Orchidaceae

Goodyera repens, Orchidaceae

Gennaria diphylla, Orchidaceae

Steveniella satyrioides, Orchidaceae

Ludisia discolor, Orchidaceae

Anacamptis pyramidalis, Orchidaceae

Stephanotis floribunda, Asclepiadaceae

Ceropegia sandersonii, Asclepiadaceae

Hoya carnosa, Asclepiadaceae

Hoya multiflora, Asclepiadaceae

Frerea indica, Asclepiadaceae

Orbeanthus hardyi, Asclepiadaceae

8

Structure of Pollen Grain

outline circular

outline elliptic

outline lobate

outline triangular

outline quadrangular

outline polygonal

P/E-ratio, oblate

P/E-ratio, isodiametric

P/E-ratio, prolate

isopolar

heteropolar

shape

saccus/saccate, corpus

saccus, monosaccate

saccus, bisaccate

saccus, trisaccate

infoldings, apertures sunken

infoldings, boat-shaped

infoldings, cup-shaped

infoldings, interapertural area sunken

infoldings, irregular

outline circular

outline describes the contour of pollen grains in polar and/or equatorial view

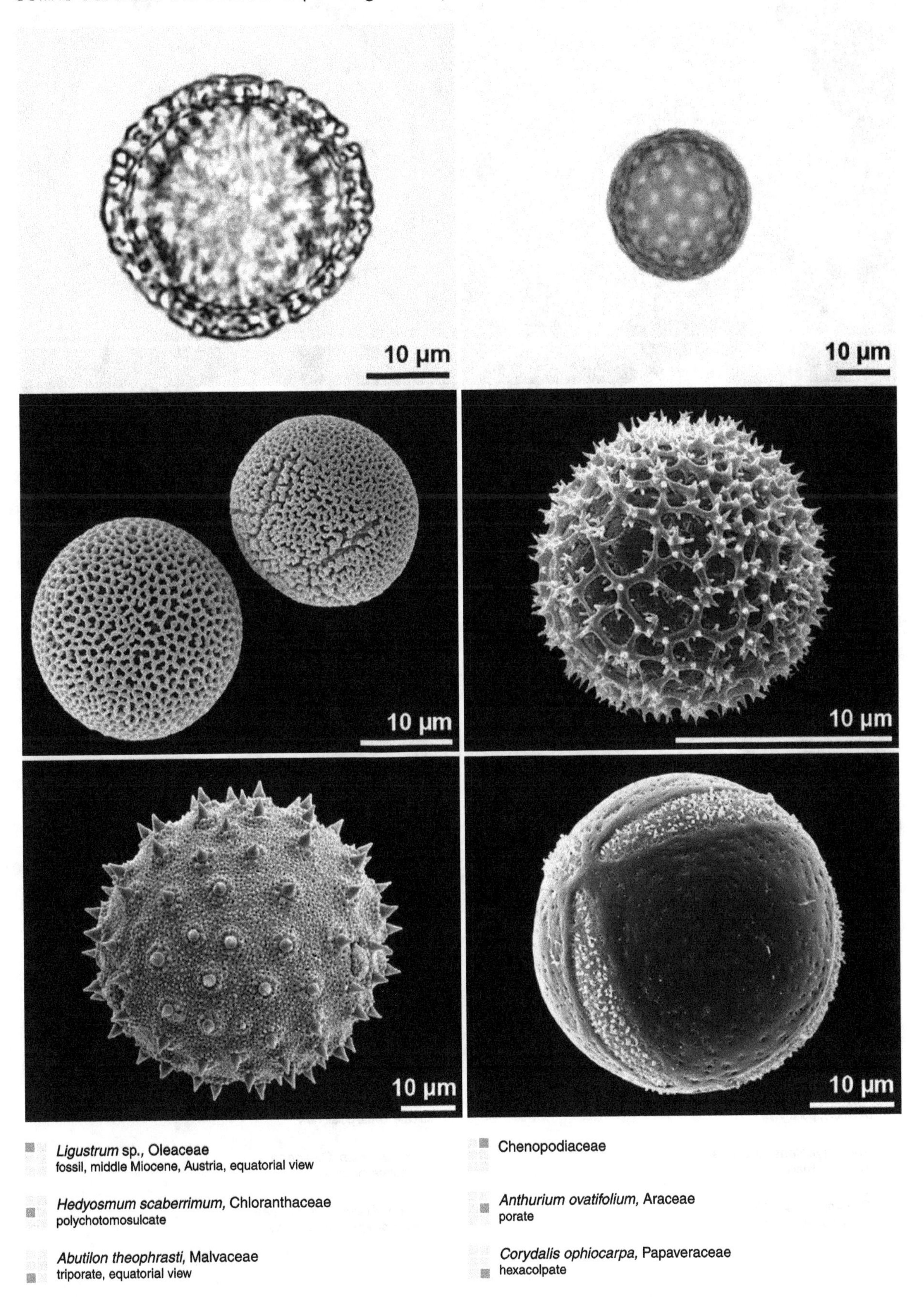

- *Ligustrum* sp., Oleaceae
  fossil, middle Miocene, Austria, equatorial view
- Chenopodiaceae
- *Hedyosmum scaberrimum*, Chloranthaceae
  polychotomosulcate
- *Anthurium ovatifolium*, Araceae
  porate
- *Abutilon theophrasti*, Malvaceae
  triporate, equatorial view
- *Corydalis ophiocarpa*, Papaveraceae
  hexacolpate

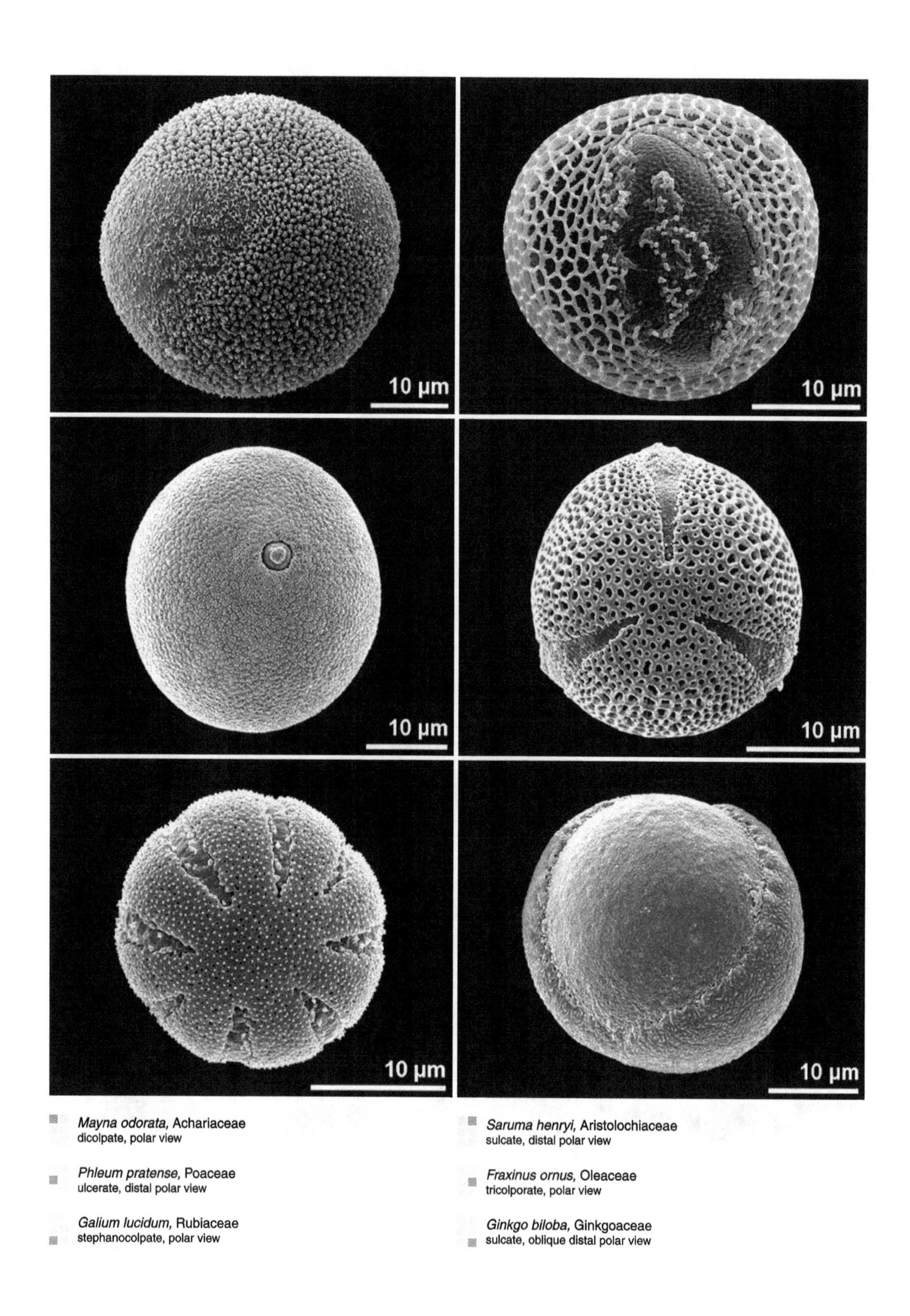

Mayna odorata, Achariaceae
dicolpate, polar view

Saruma henryi, Aristolochiaceae
sulcate, distal polar view

Phleum pratense, Poaceae
ulcerate, distal polar view

Fraxinus ornus, Oleaceae
tricolporate, polar view

Galium lucidum, Rubiaceae
stephanocolpate, polar view

Ginkgo biloba, Ginkgoaceae
sulcate, oblique distal polar view

outline elliptic

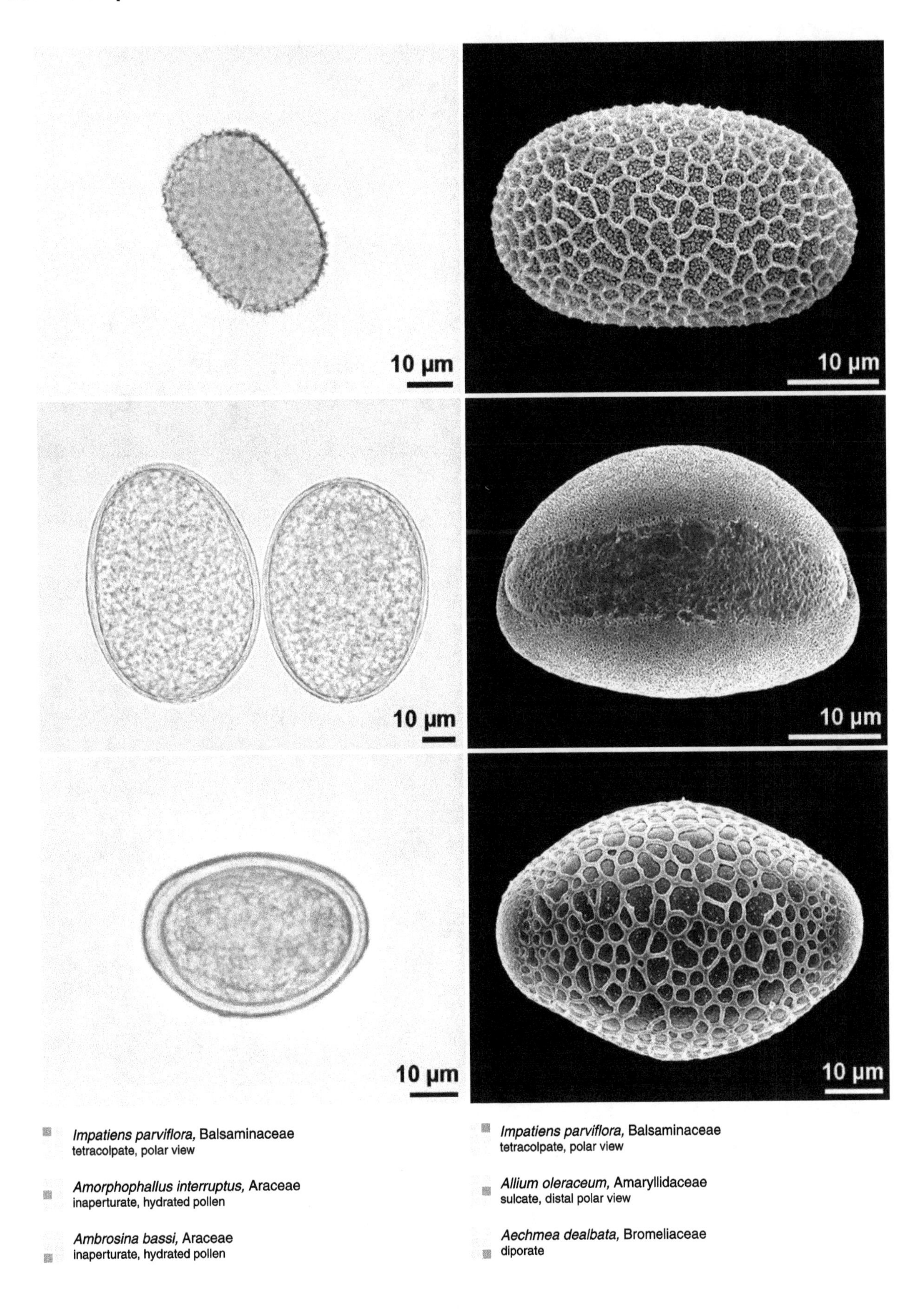

Impatiens parviflora, Balsaminaceae
tetracolpate, polar view

Amorphophallus interruptus, Araceae
inaperturate, hydrated pollen

Ambrosina bassi, Araceae
inaperturate, hydrated pollen

Impatiens parviflora, Balsaminaceae
tetracolpate, polar view

Allium oleraceum, Amaryllidaceae
sulcate, distal polar view

Aechmea dealbata, Bromeliaceae
diporate

- *Salvia coccinea,* Lamiaceae
  hexacolpate, polar view
- *Billbergia porteana,* Bromeliaceae
  sulcate, distal polar view
- *Galeopsis tetrahit,* Lamiaceae
  tricolpate, dry pollen
- *Commelina erecta,* Commelinaceae
  sulcate, proximal polar view
- *Zamia loddigesii,* Zamiaceae
  sulcate, dry pollen
- *Physostegia virginiana,* Lamiaceae
  tricolpate, dry pollen

outline lobate

outline in polar view of a pollen grain with bulged interapertural areas (mainly in dry pollen grains)

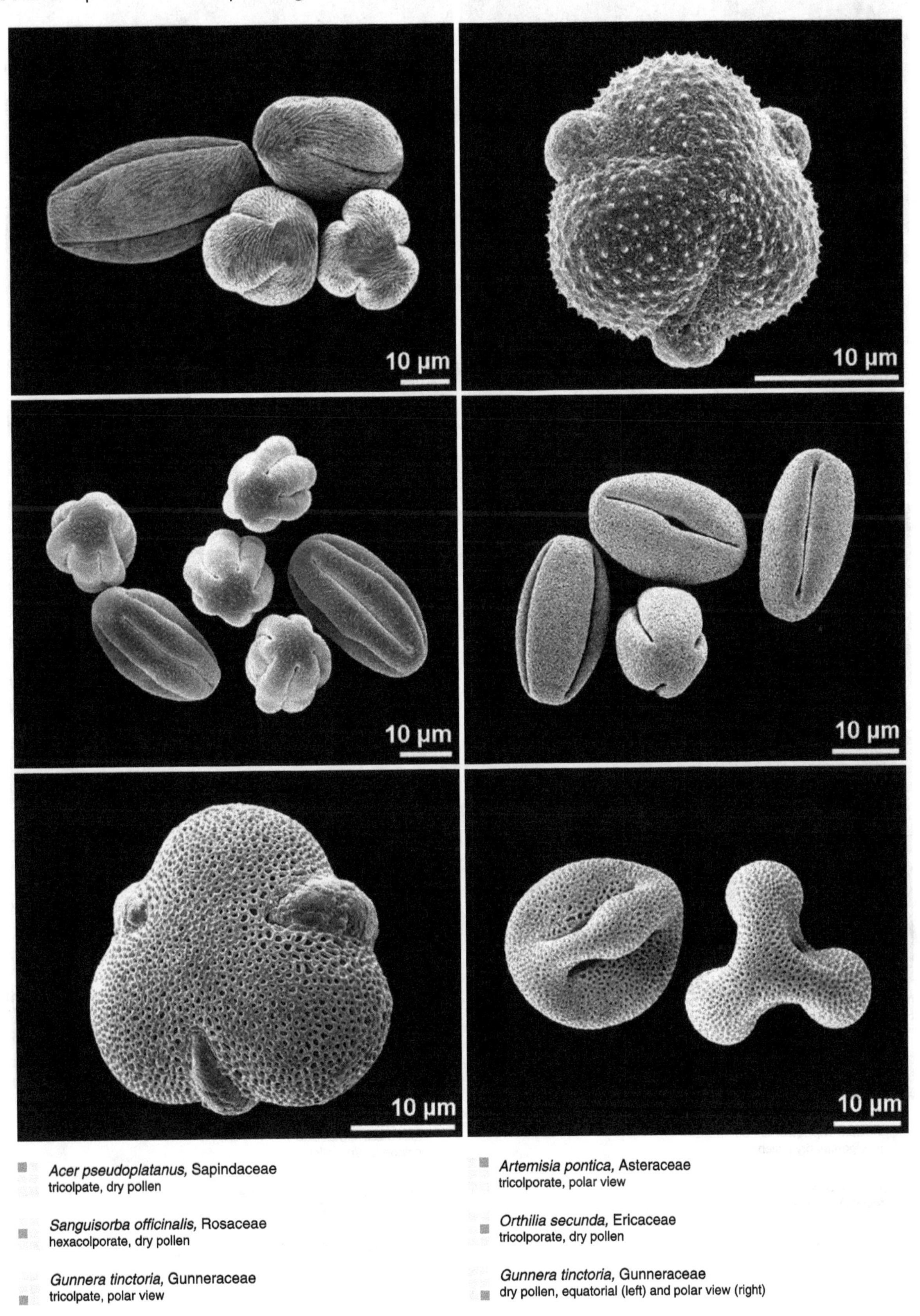

- *Acer pseudoplatanus,* Sapindaceae
  tricolpate, dry pollen
- *Artemisia pontica,* Asteraceae
  tricolporate, polar view
- *Sanguisorba officinalis,* Rosaceae
  hexacolporate, dry pollen
- *Orthilia secunda,* Ericaceae
  tricolporate, dry pollen
- *Gunnera tinctoria,* Gunneraceae
  tricolpate, polar view
- *Gunnera tinctoria,* Gunneraceae
  dry pollen, equatorial (left) and polar view (right)

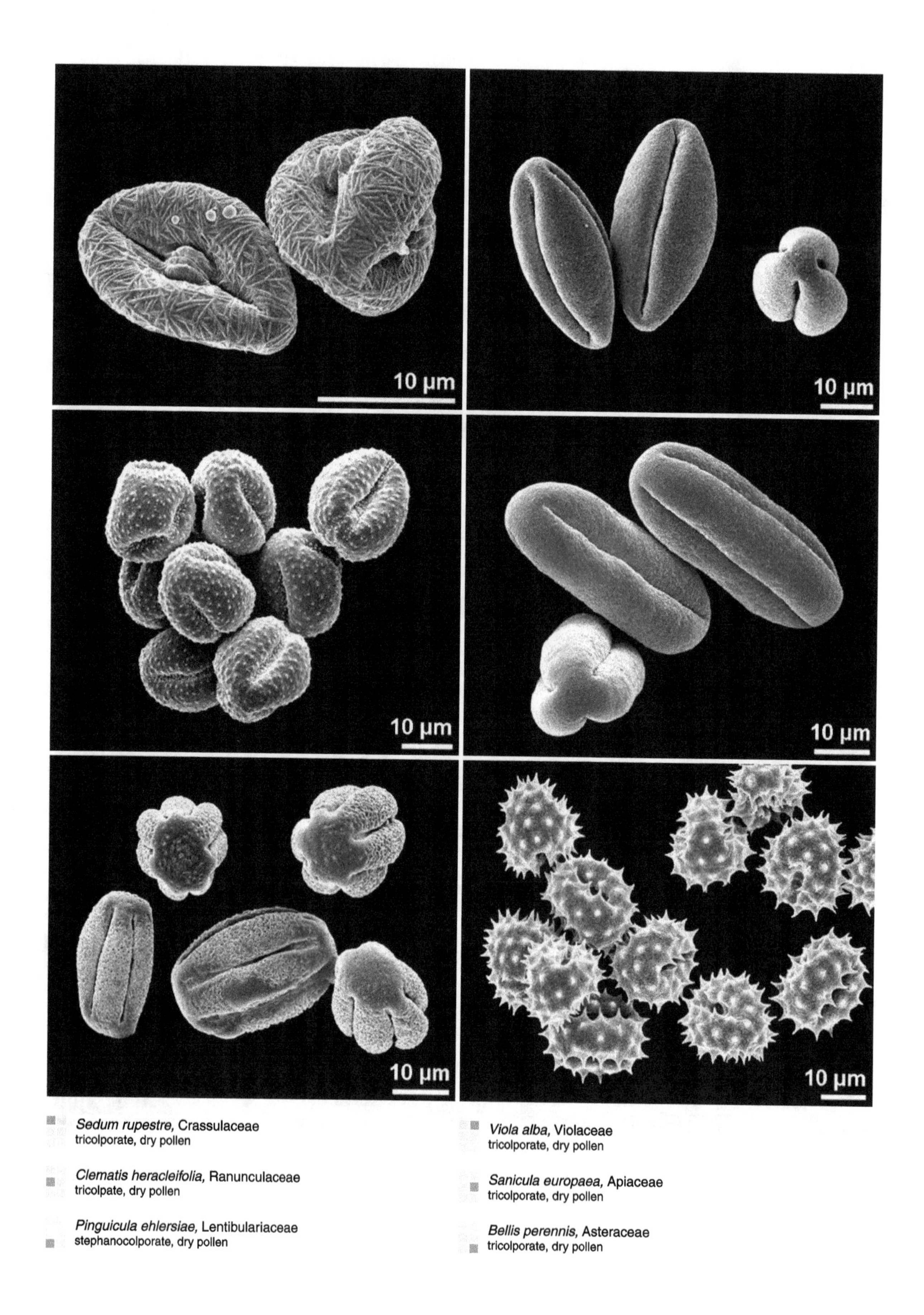

- *Sedum rupestre,* Crassulaceae
  tricolporate, dry pollen
- *Viola alba,* Violaceae
  tricolporate, dry pollen
- *Clematis heracleifolia,* Ranunculaceae
  tricolpate, dry pollen
- *Sanicula europaea,* Apiaceae
  tricolporate, dry pollen
- *Pinguicula ehlersiae,* Lentibulariaceae
  stephanocolporate, dry pollen
- *Bellis perennis,* Asteraceae
  tricolporate, dry pollen

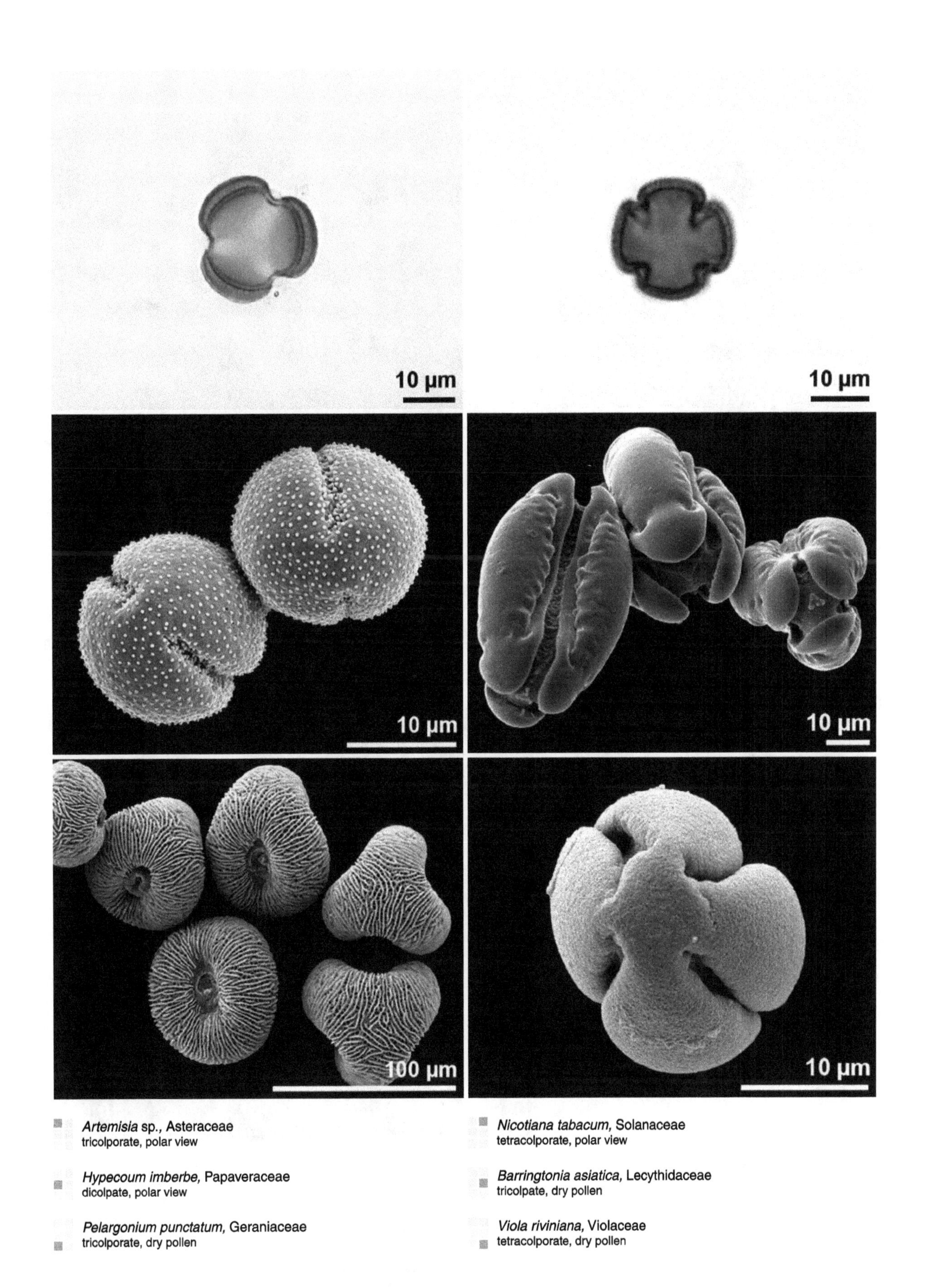

Artemisia sp., Asteraceae
tricolporate, polar view

Hypecoum imberbe, Papaveraceae
dicolpate, polar view

Pelargonium punctatum, Geraniaceae
tricolporate, dry pollen

Nicotiana tabacum, Solanaceae
tetracolporate, polar view

Barringtonia asiatica, Lecythidaceae
tricolpate, dry pollen

Viola riviniana, Violaceae
tetracolporate, dry pollen

outline triangular

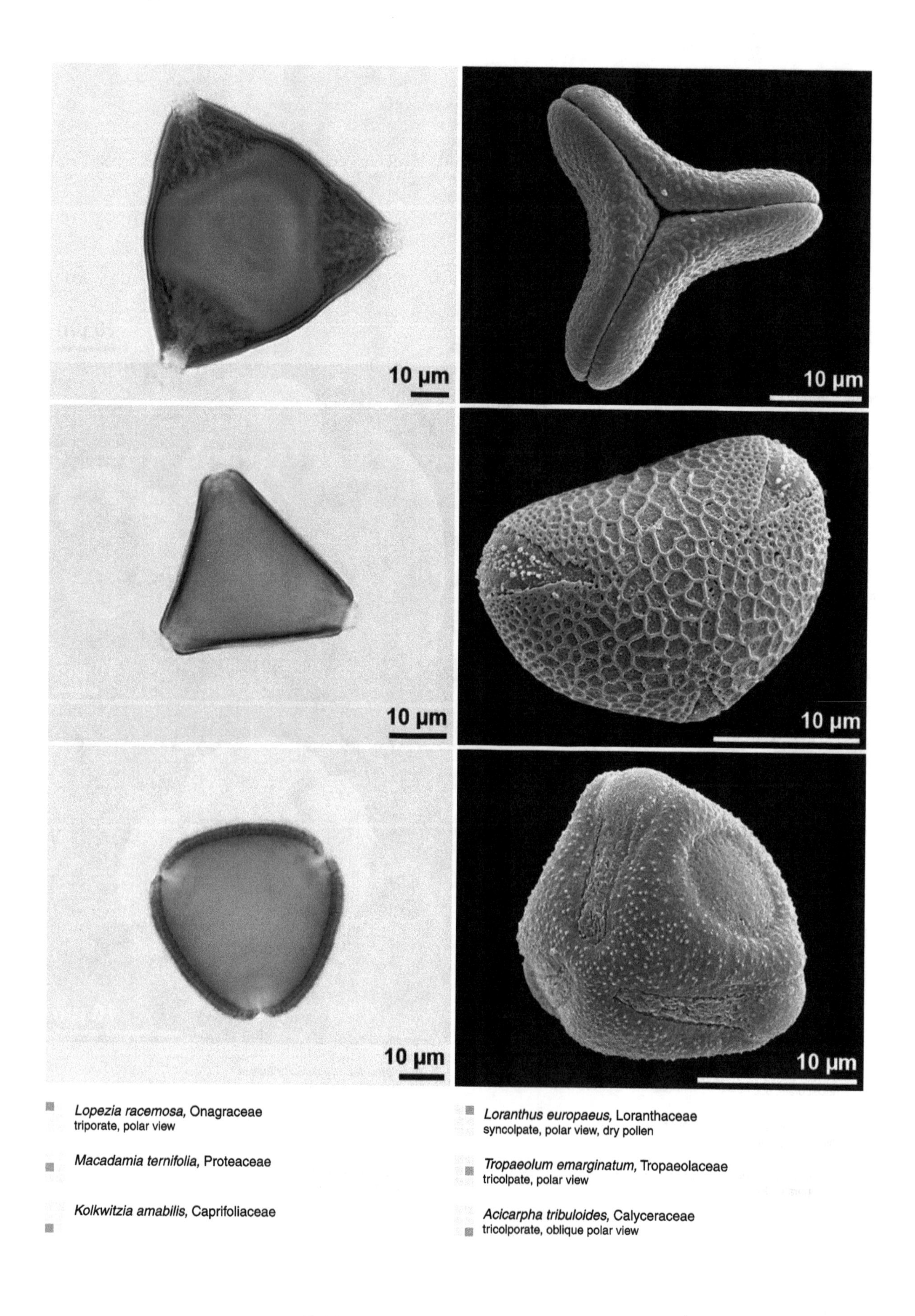

Lopezia racemosa, Onagraceae
triporate, polar view

Macadamia ternifolia, Proteaceae

Kolkwitzia amabilis, Caprifoliaceae

Loranthus europaeus, Loranthaceae
syncolpate, polar view, dry pollen

Tropaeolum emarginatum, Tropaeolaceae
tricolpate, polar view

Acicarpha tribuloides, Calyceraceae
tricolporate, oblique polar view

Callistemon coccineus, Myrtaceae
tricolporate, synaperturate, polar view

Echinops ritro, Asteraceae
tricolporate, polar view

Paullinia tomentosa, Sapindaceae
triporate, polar view

Hypoestes phyllostachya, Acanthaceae
tricolporate, dry pollen

Bupleurum rotundifolium, Apiaceae
tricolporate, polar view

Primula denticulata, Primulaceae
tricolporate, synaperturate, dry pollen

Orlaya grandiflora, Apiaceae
tricolporate, dry pollen

Circaea lutetiana, Onagraceae
triporate, pollen with viscin threads, polar view

Sempervivum globiferum, Crassulaceae
tricolporate, polar view

Cunonia capensis, Cunoniaceae
hexaporate, polar view

Dipsacus fullonum, Caprifoliaceae
triporate, polar view

Potentilla inclinata, Rosaceae
tricolporate, polar view

outline quadrangular

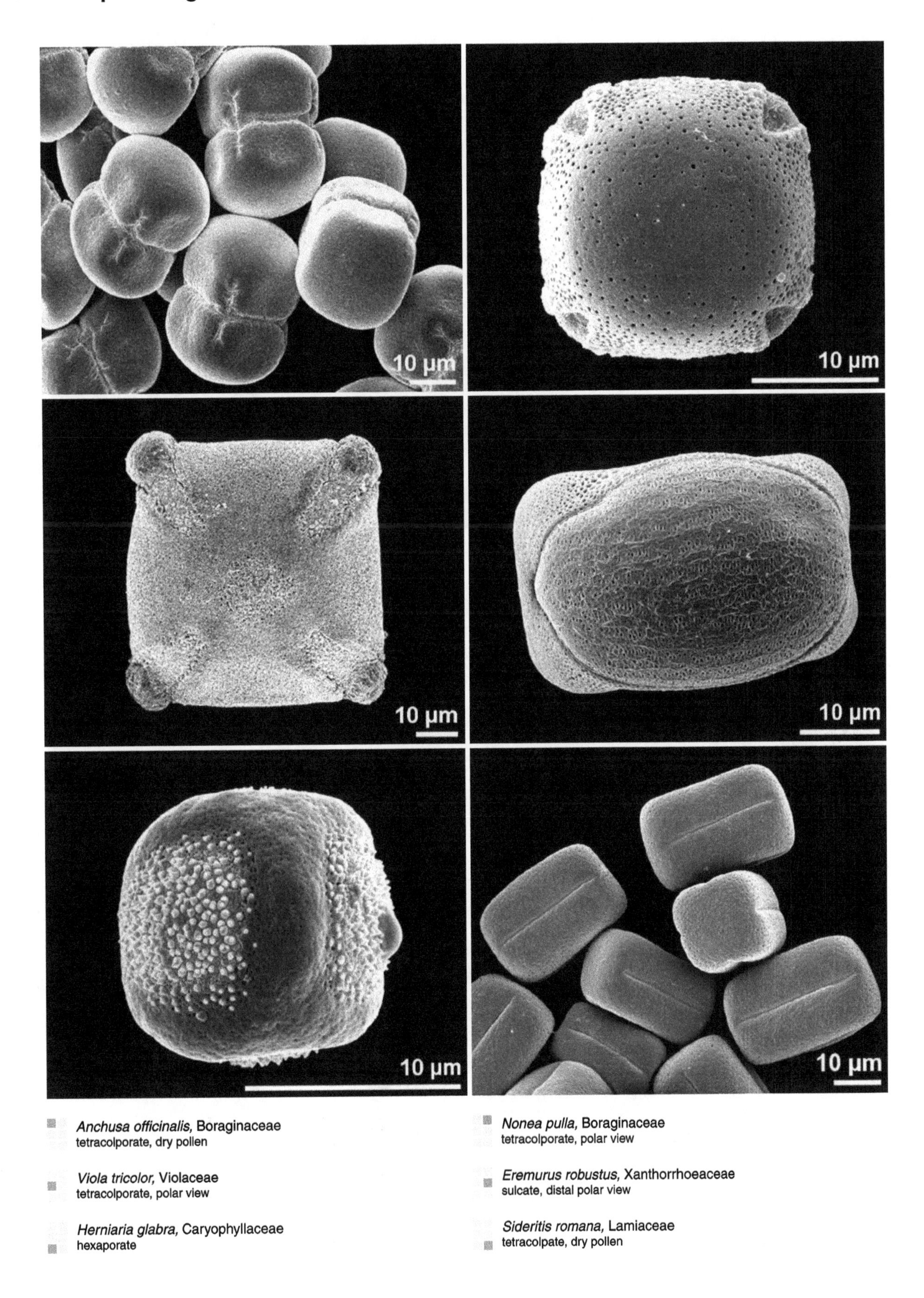

Anchusa officinalis, Boraginaceae
tetracolporate, dry pollen

Nonea pulla, Boraginaceae
tetracolporate, polar view

Viola tricolor, Violaceae
tetracolporate, polar view

Eremurus robustus, Xanthorrhoeaceae
sulcate, distal polar view

Herniaria glabra, Caryophyllaceae
hexaporate

Sideritis romana, Lamiaceae
tetracolpate, dry pollen

outline polygonal

Viola arvensis, Violaceae
pentacolporate, polar view

Viola arvensis, Violaceae
pentacolporate, polar view

Opuntia basilaris, Cactaceae
pantocolpate, dry pollen

Talinum paniculatum, Talinaceae
pantocolpate, dry pollen

Sarracenia alata, Sarraceniaceae
stephanocolporate, polar view

Stellaria holostea, Caryophyllaceae
pantoporate, dry pollen

P/E-ratio, oblate

P/E-ratio refers to the length of the polar axis between the two poles compared to the equatorial diameter
oblate: pollen grain with a polar axis shorter than the equatorial diameter

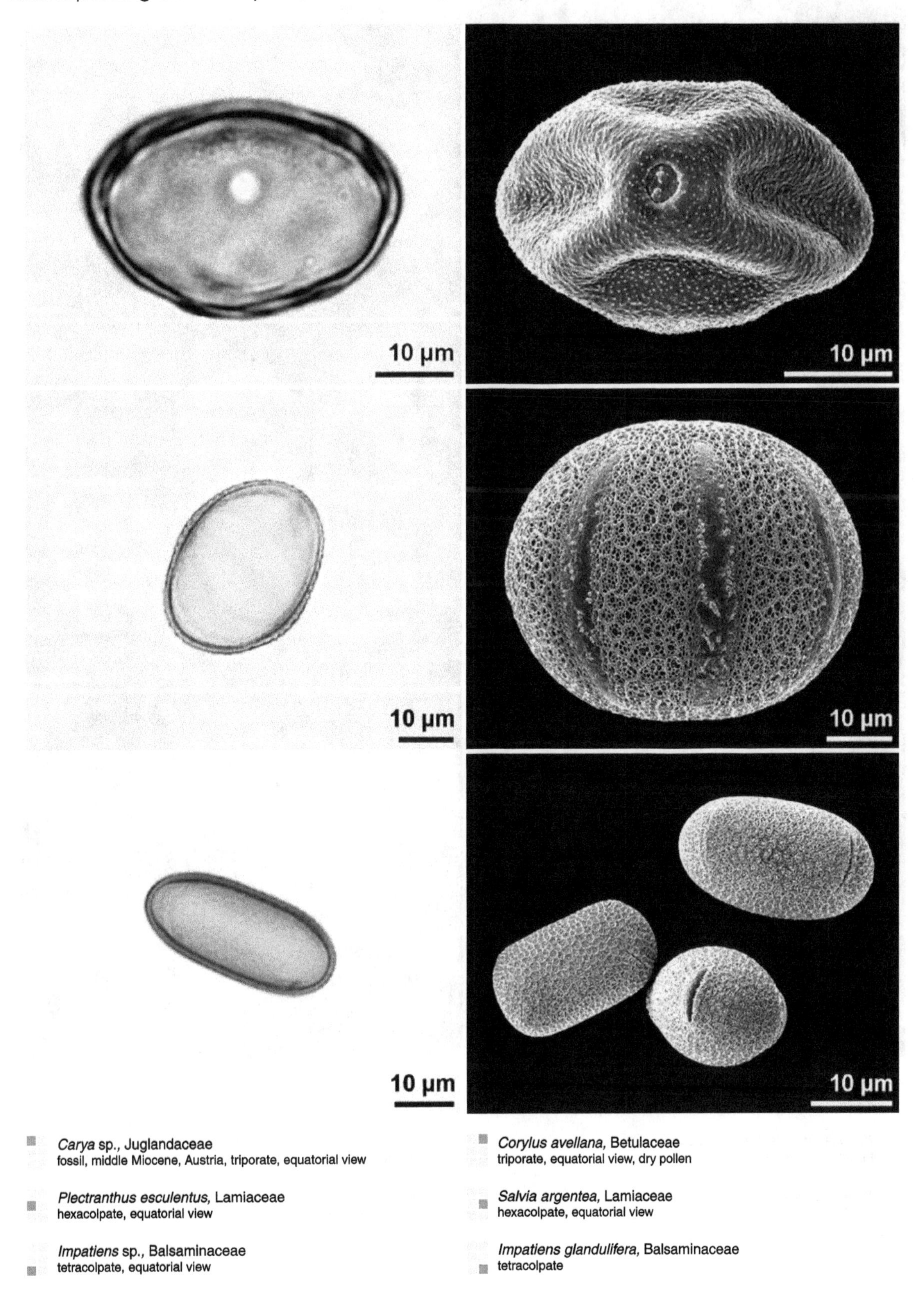

- *Carya* sp., Juglandaceae
  fossil, middle Miocene, Austria, triporate, equatorial view
- *Corylus avellana*, Betulaceae
  triporate, equatorial view, dry pollen
- *Plectranthus esculentus*, Lamiaceae
  hexacolpate, equatorial view
- *Salvia argentea*, Lamiaceae
  hexacolpate, equatorial view
- *Impatiens* sp., Balsaminaceae
  tetracolpate, equatorial view
- *Impatiens glandulifera*, Balsaminaceae
  tetracolpate

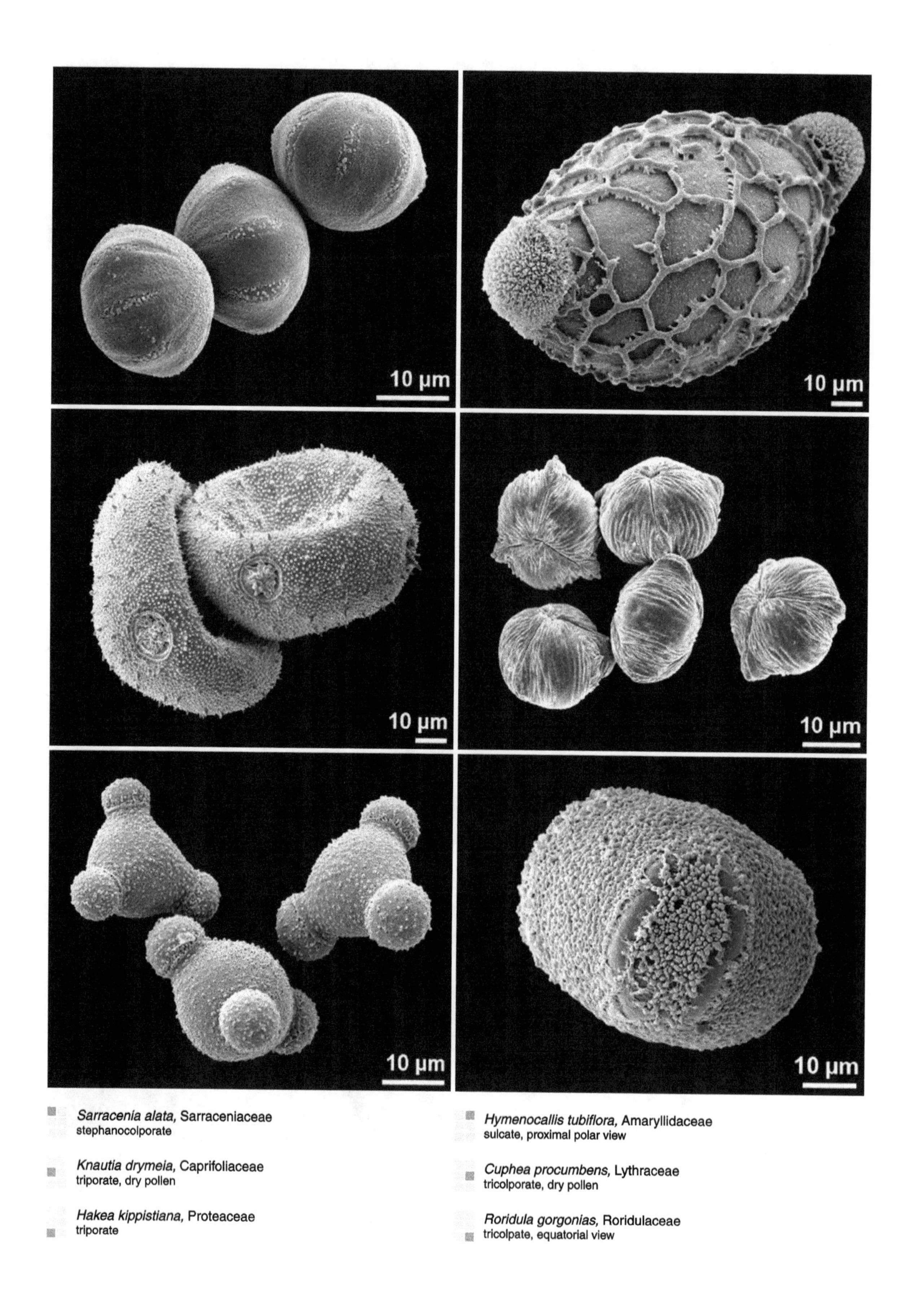

- *Sarracenia alata*, Sarraceniaceae
  stephanocolporate
- *Knautia drymeia*, Caprifoliaceae
  triporate, dry pollen
- *Hakea kippistiana*, Proteaceae
  triporate
- *Hymenocallis tubiflora*, Amaryllidaceae
  sulcate, proximal polar view
- *Cuphea procumbens*, Lythraceae
  tricolporate, dry pollen
- *Roridula gorgonias*, Roridulaceae
  tricolpate, equatorial view

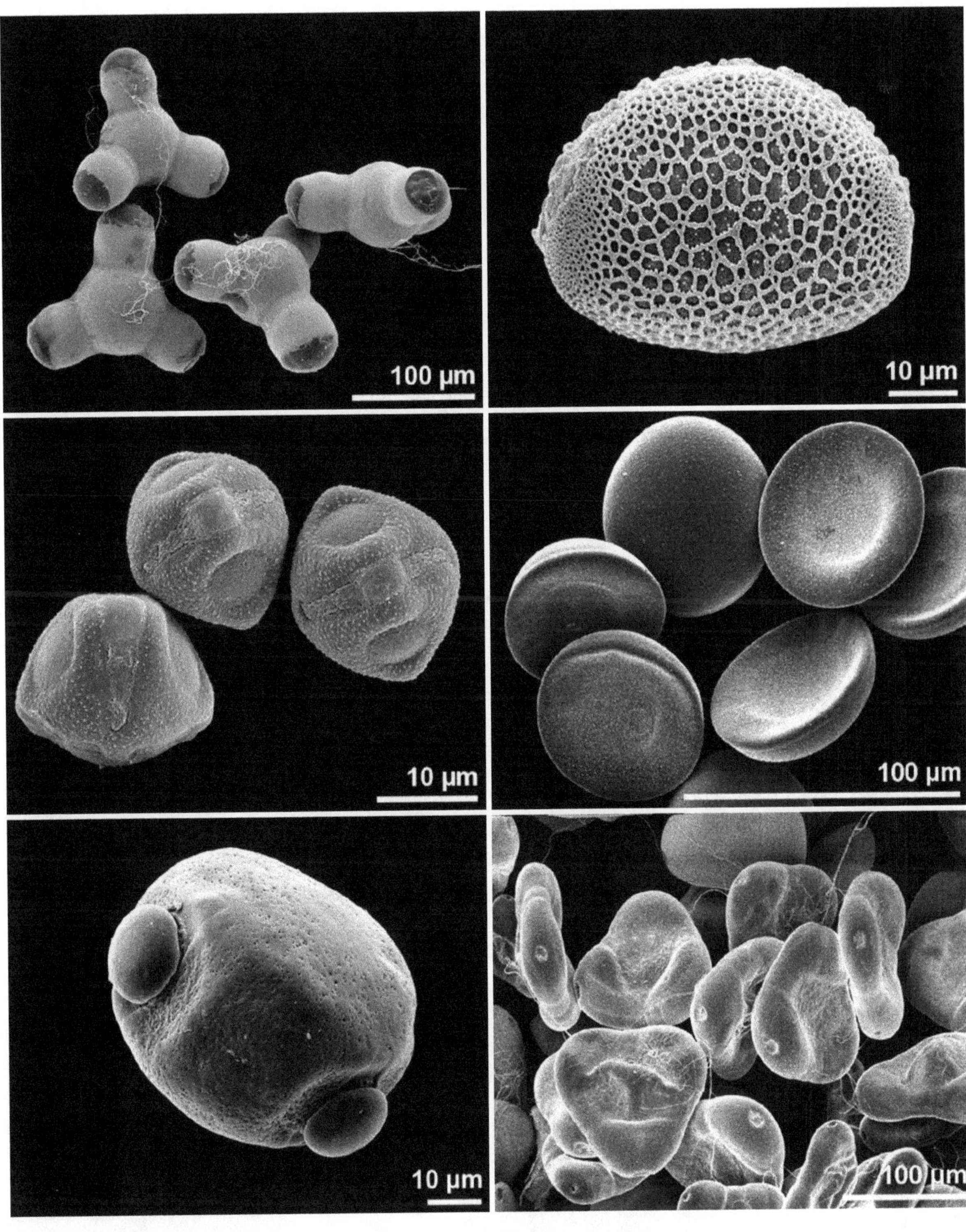

Clarkia unguiculata, Onagraceae
triporate, pollen with viscin threads

Acicarpha tribuloides, Calyceraceae
tricolporate

Amsonia ciliata, Apocynaceae
tricolporate, oblique equatorial view

Vriesea neoglutinosa, Bromeliaceae
sulcate, equatorial view

Heliconia sp., Heliconiaceae
ulcerate, dry pollen

Clarkia purpurea, Onagraceae
triporate, dry pollen

P/E-ratio, isodiametric

isodiametric: pollen grain with a polar axis equal to the equatorial diameter

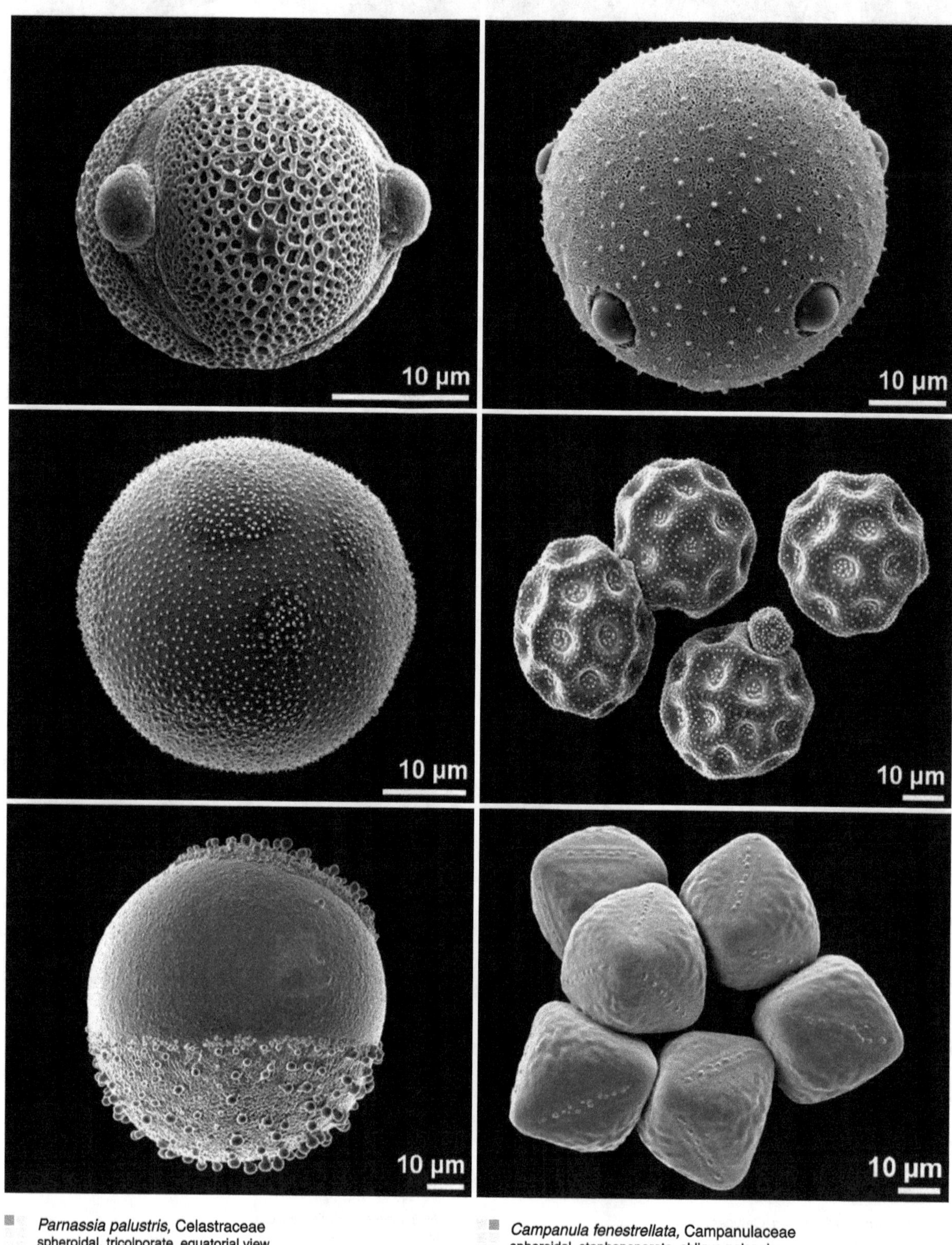

Parnassia palustris, Celastraceae
spheroidal, tricolporate, equatorial view

Campanula fenestrellata, Campanulaceae
spheroidal, stephanoporate, oblique polar view

Roemeria hybrida, Papaveraceae
spheroidal, pantoporate

Silene nutans, Caryophyllaceae
polygonal, pantoporate, dry pollen

Iris pumila, Iridaceae
spheroidal, sulcate, distal polar view

Sarcocapnos enneaphylla, Papaveraceae
hexacolpate

Whitfieldia lateritia, Acanthaceae
diporate

Whitfieldia lateritia, Acanthaceae
dry pollen

Schoepfia schreberi, Schoepfiaceae
tetraaperturate

Thesium arvense, Santalaceae
triradiate colpi, dry pollen

Pedicularis gyroflexa, Orobanchaceae
ring-like aperture, dry pollen

Basella alba, Basellaceae
hexacolpate

P/E-ratio, prolate

prolate: pollen grain with a polar axis longer than the equatorial diameter

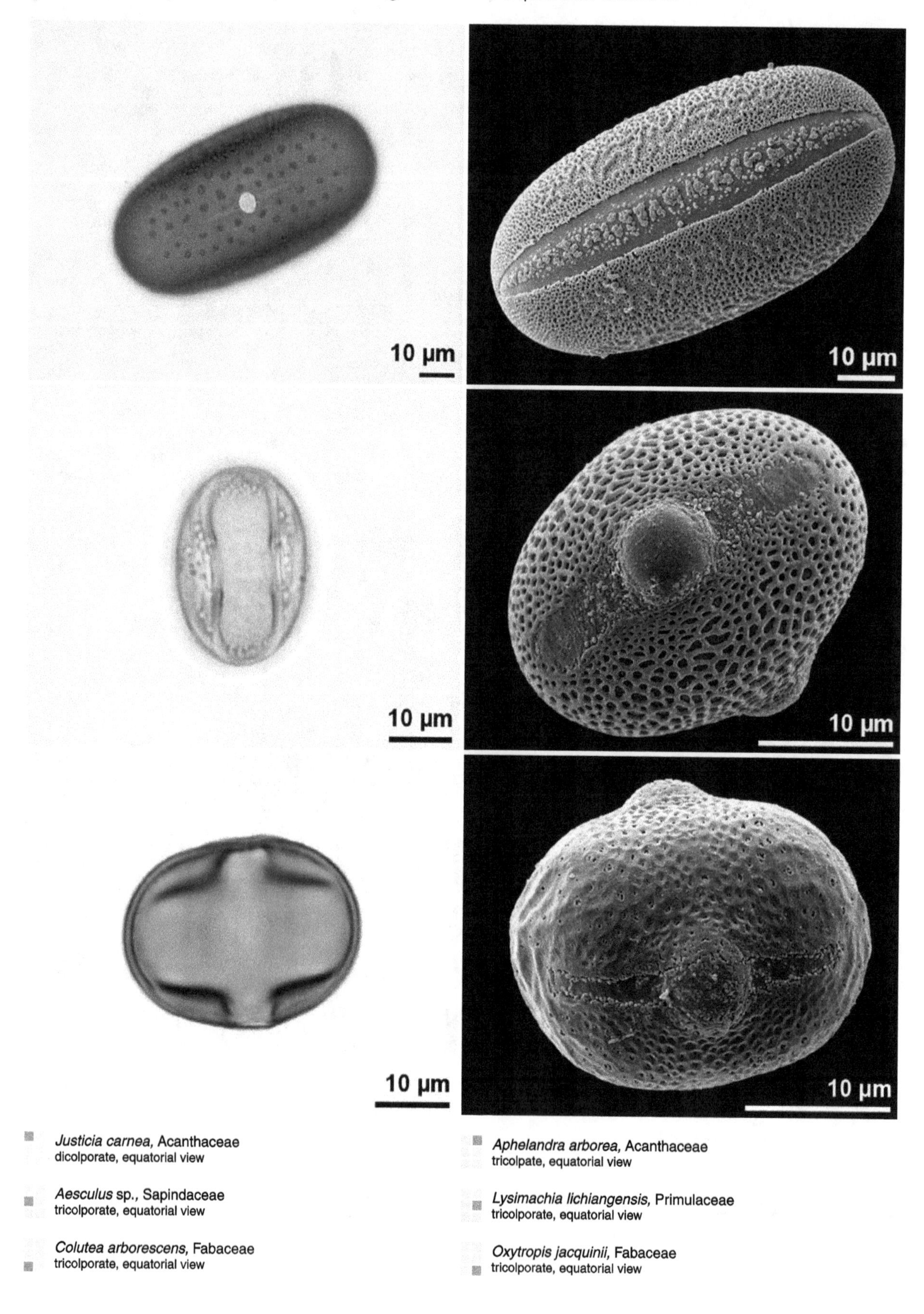

Justicia carnea, Acanthaceae
dicolporate, equatorial view

Aesculus sp., Sapindaceae
tricolporate, equatorial view

Colutea arborescens, Fabaceae
tricolporate, equatorial view

Aphelandra arborea, Acanthaceae
tricolpate, equatorial view

Lysimachia lichiangensis, Primulaceae
tricolporate, equatorial view

Oxytropis jacquinii, Fabaceae
tricolporate, equatorial view

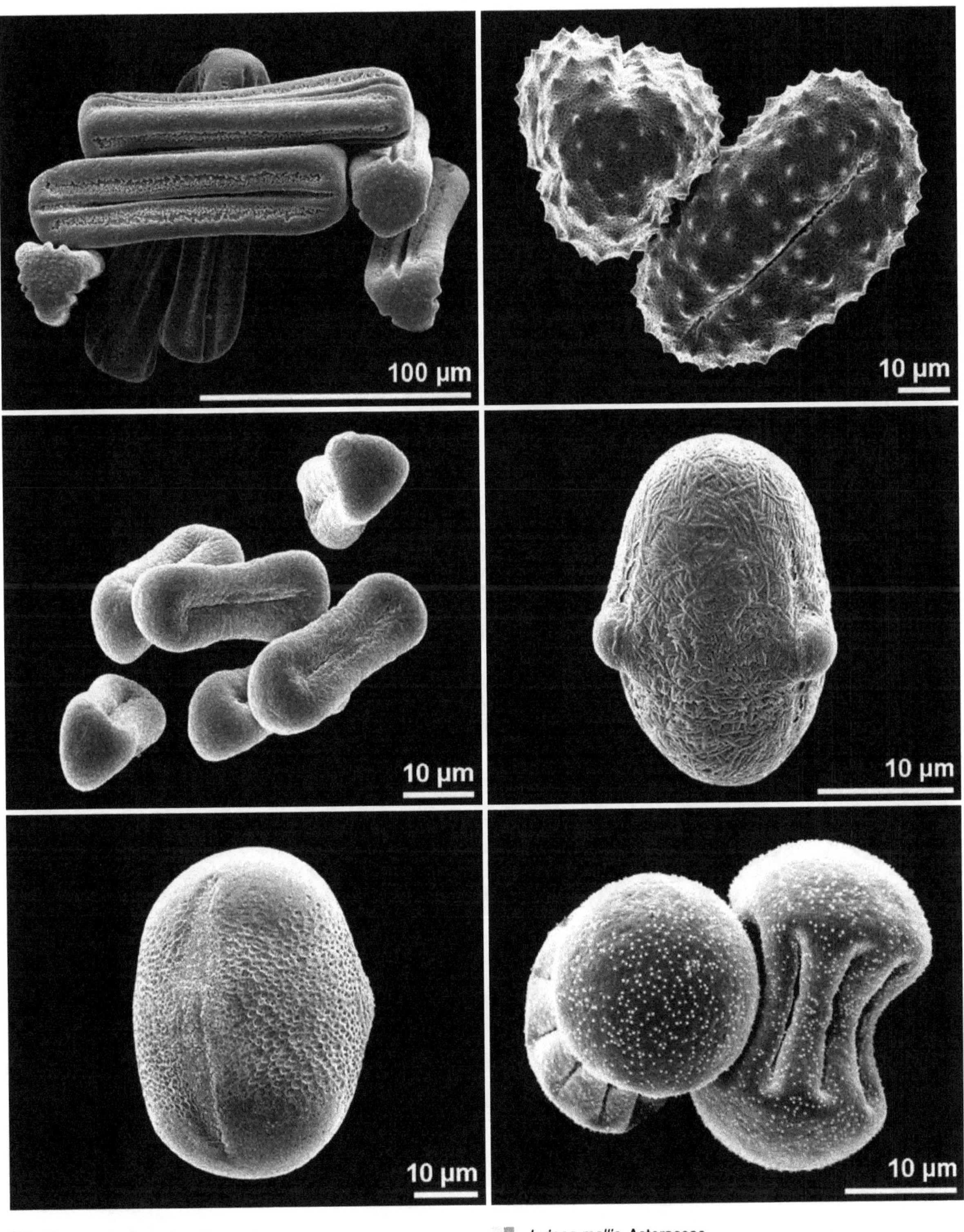

Crossandra flava, Acanthaceae
tricolpate, dry pollen

Jurinea mollis, Asteraceae
tricolporate, dry pollen

Torilis arvensis, Apiaceae
tricolporate, dry pollen

Peucedanum cervaria, Apiaceae
tricolporate, equatorial view

Astragalus onobrychis, Fabaceae
tricolporate, equatorial view

Symphytum officinale, Boraginaceae
stephanocolporate, dry pollen

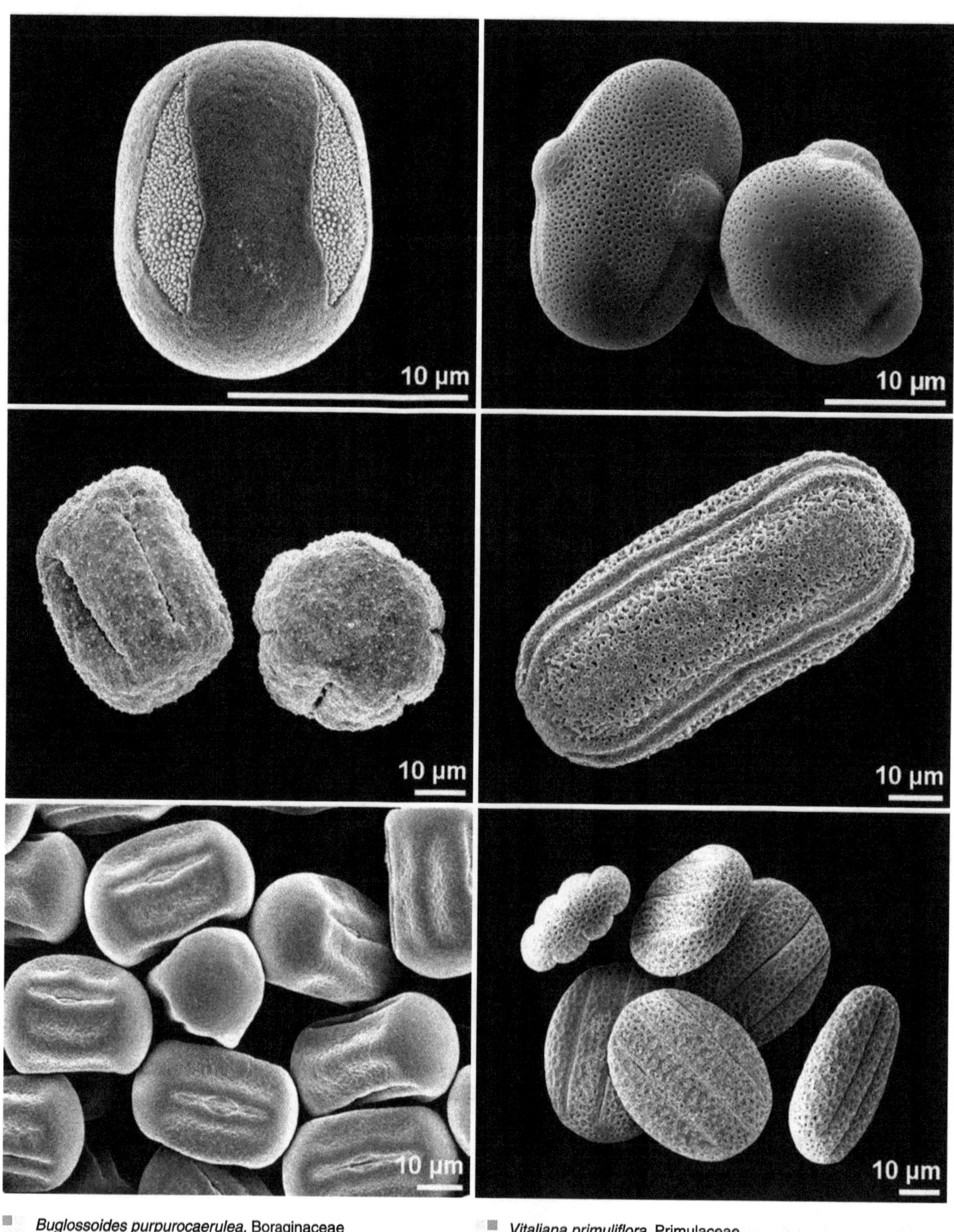

Buglossoides purpurocaerulea, Boraginaceae
tetracolporate, equatorial view

Vitaliana primuliflora, Primulaceae
tricolporate

Platycodon grandiflorus, Campanulaceae
stephanocolpate, dry pollen

Stenandrium guineense, Acanthaceae
tricolpate, equatorial view

Lathyrus tuberosus, Fabaceae
tricolporate, dry pollen

Salvia sclarea, Lamiaceae
hexacolpate, dry pollen

isopolar

pollen grain with identical proximal and distal faces

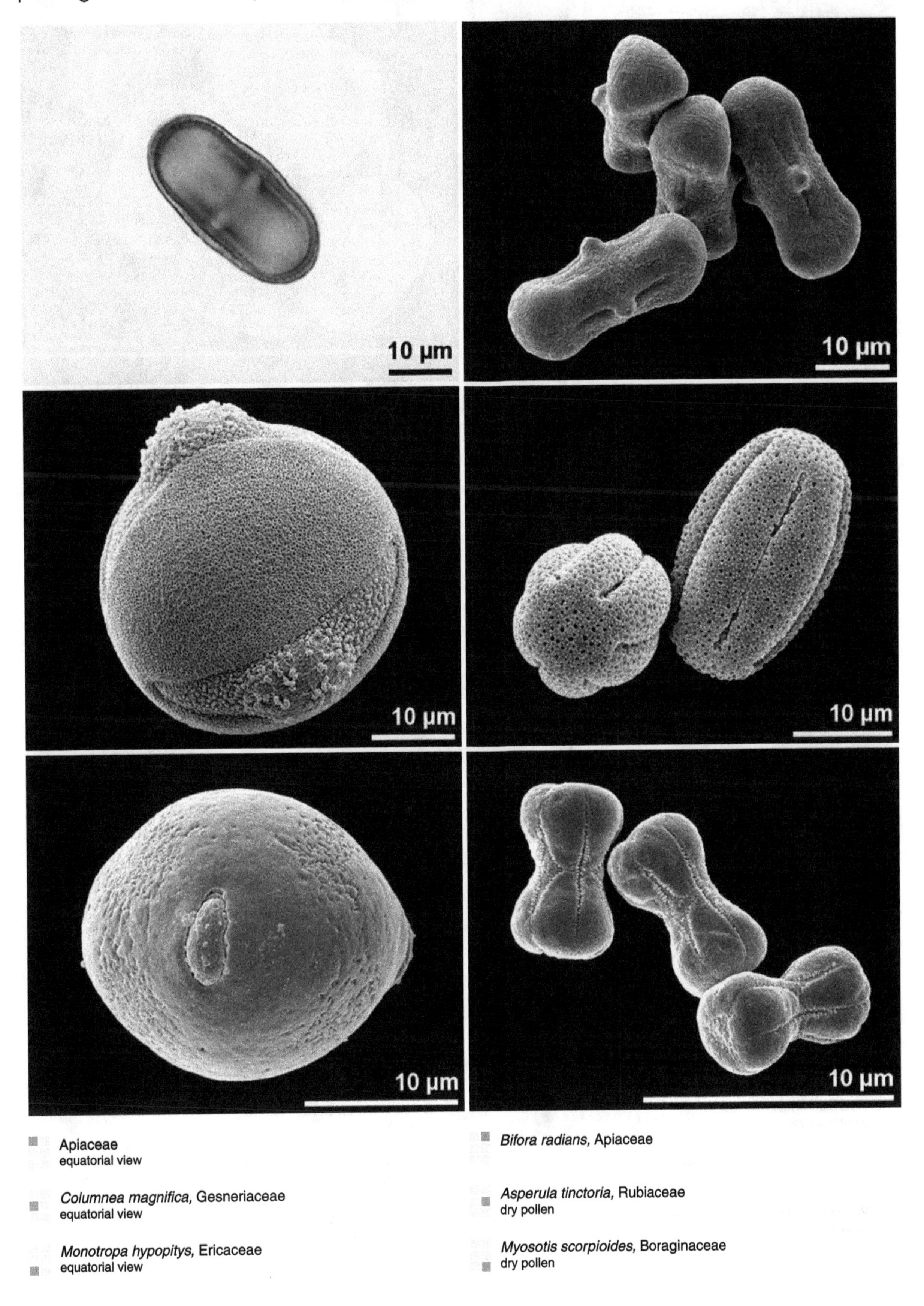

- Apiaceae
  equatorial view
- *Bifora radians*, Apiaceae
- *Columnea magnifica*, Gesneriaceae
  equatorial view
- *Asperula tinctoria*, Rubiaceae
  dry pollen
- *Monotropa hypopitys*, Ericaceae
  equatorial view
- *Myosotis scorpioides*, Boraginaceae
  dry pollen

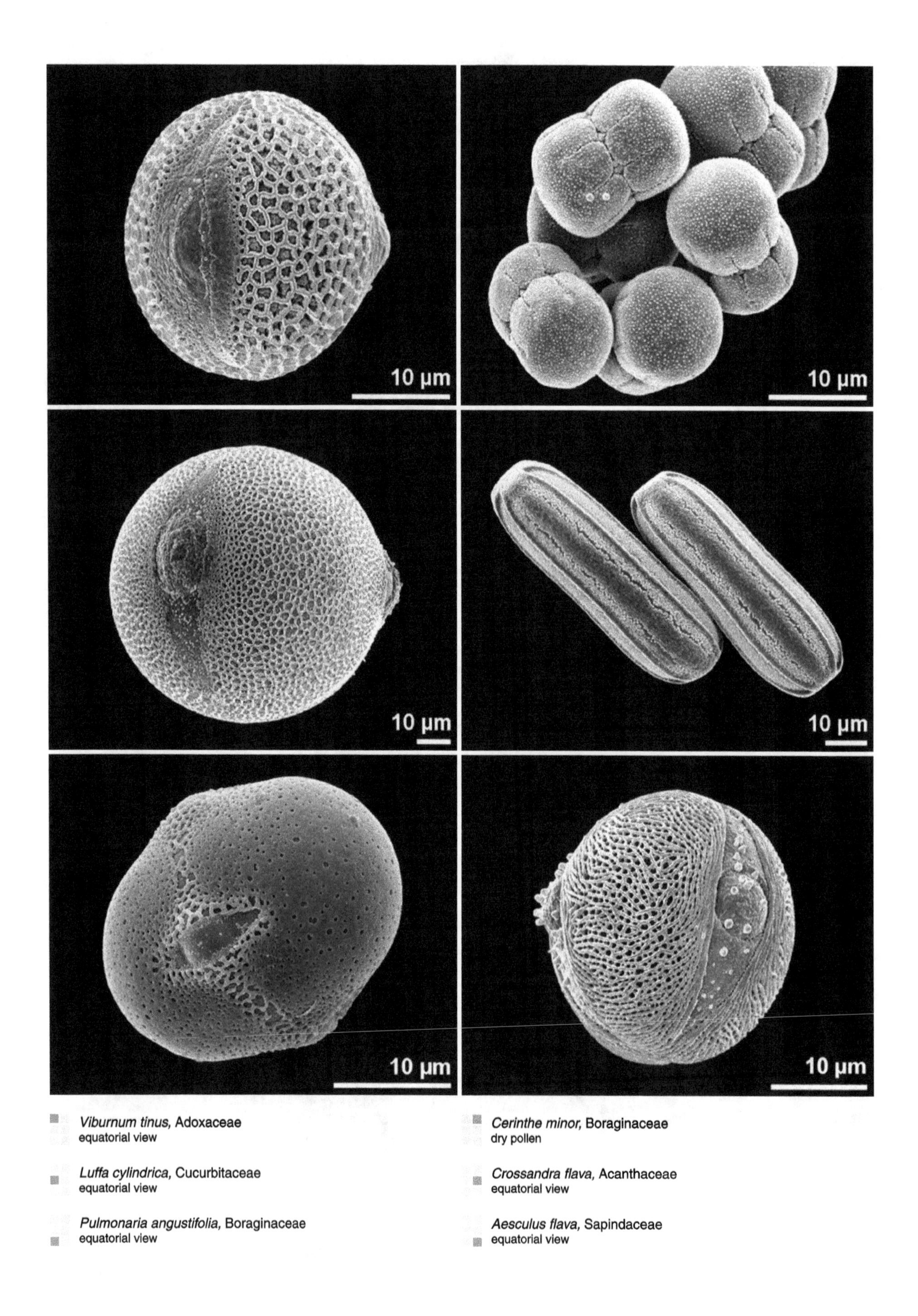

Viburnum tinus, Adoxaceae
equatorial view

Cerinthe minor, Boraginaceae
dry pollen

Luffa cylindrica, Cucurbitaceae
equatorial view

Crossandra flava, Acanthaceae
equatorial view

Pulmonaria angustifolia, Boraginaceae
equatorial view

Aesculus flava, Sapindaceae
equatorial view

heteropolar

pollen grain with different proximal and distal faces

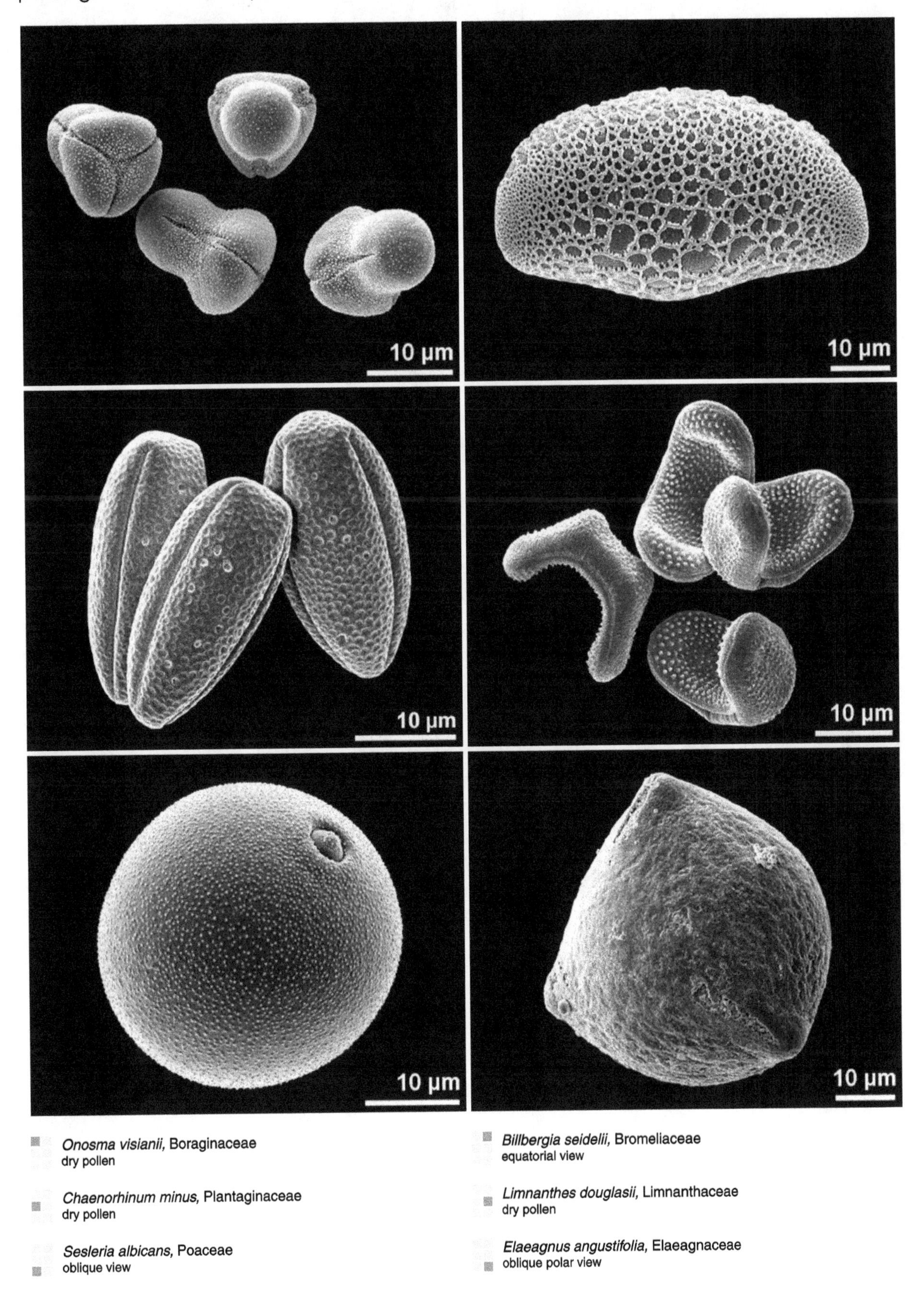

Onosma visianii, Boraginaceae
dry pollen

Chaenorhinum minus, Plantaginaceae
dry pollen

Sesleria albicans, Poaceae
oblique view

Billbergia seidelii, Bromeliaceae
equatorial view

Limnanthes douglasii, Limnanthaceae
dry pollen

Elaeagnus angustifolia, Elaeagnaceae
oblique polar view

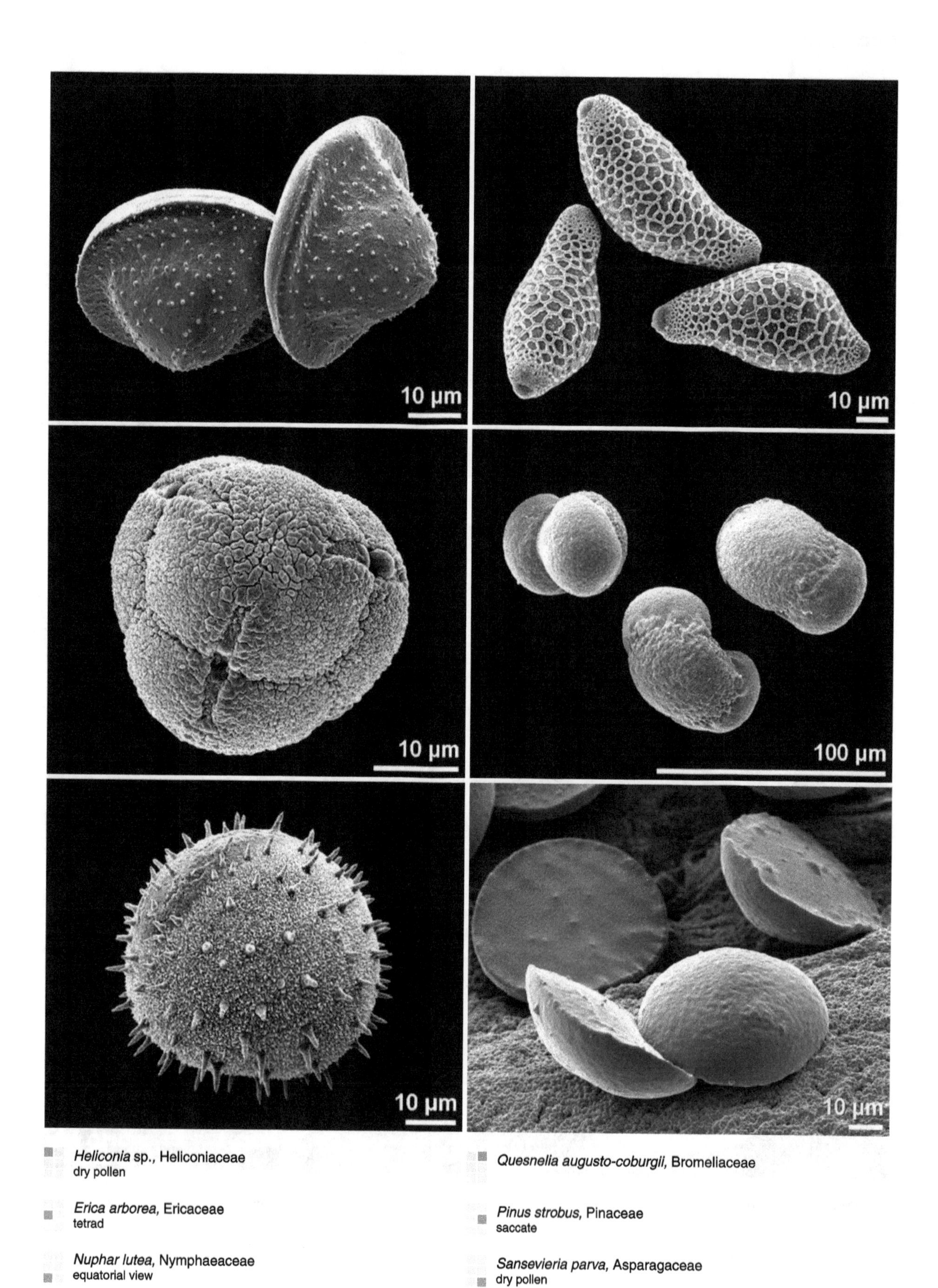

Heliconia sp., Heliconiaceae
dry pollen

Quesnelia augusto-coburgii, Bromeliaceae

Erica arborea, Ericaceae
tetrad

Pinus strobus, Pinaceae
saccate

Nuphar lutea, Nymphaeaceae
equatorial view

Sansevieria parva, Asparagaceae
dry pollen

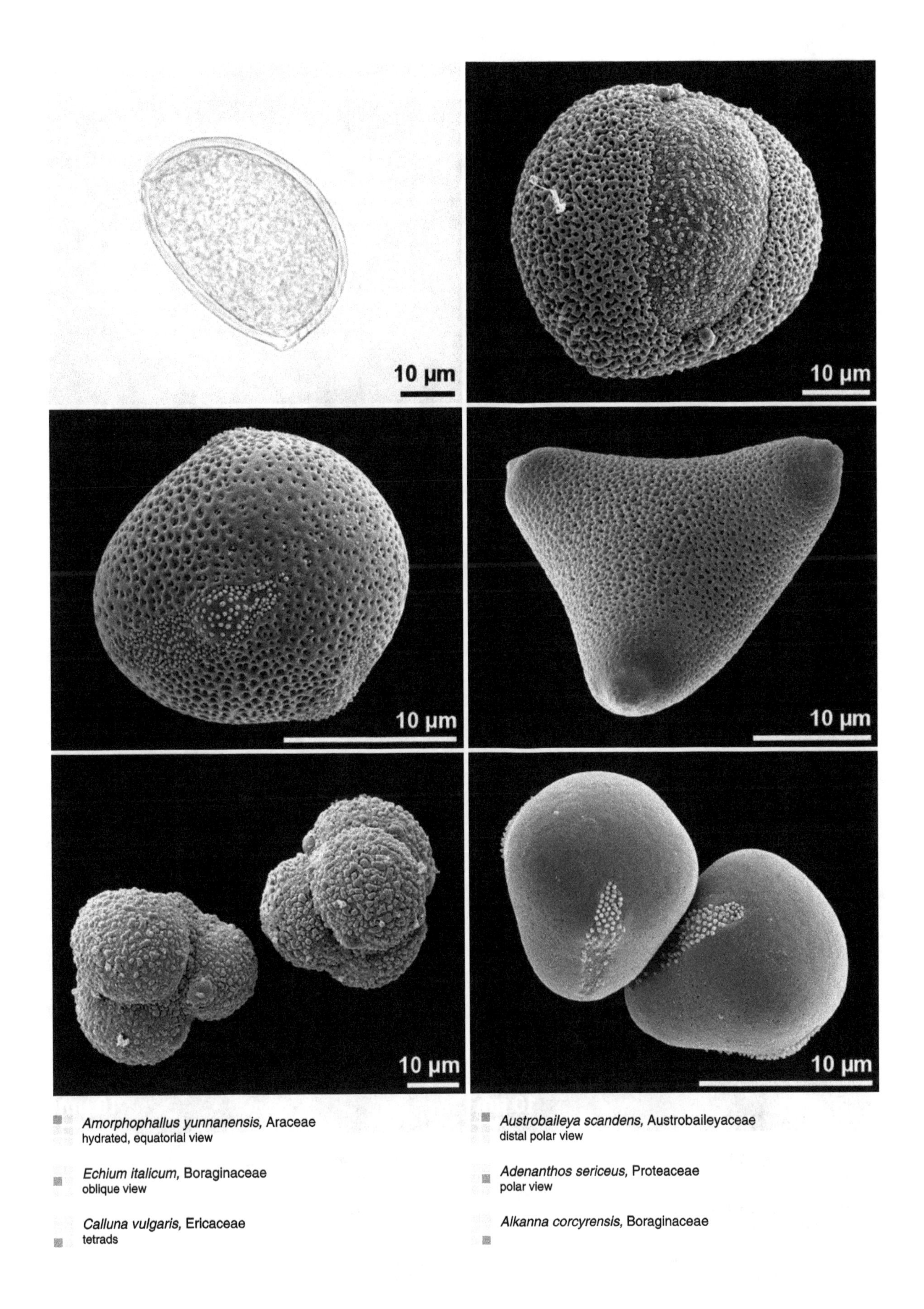

Amorphophallus yunnanensis, Araceae
hydrated, equatorial view

Austrobaileya scandens, Austrobaileyaceae
distal polar view

Echium italicum, Boraginaceae
oblique view

Adenanthos sericeus, Proteaceae
polar view

Calluna vulgaris, Ericaceae
tetrads

Alkanna corcyrensis, Boraginaceae

shape

3-dimensional form of a pollen grain in relation to the P/E-ratio

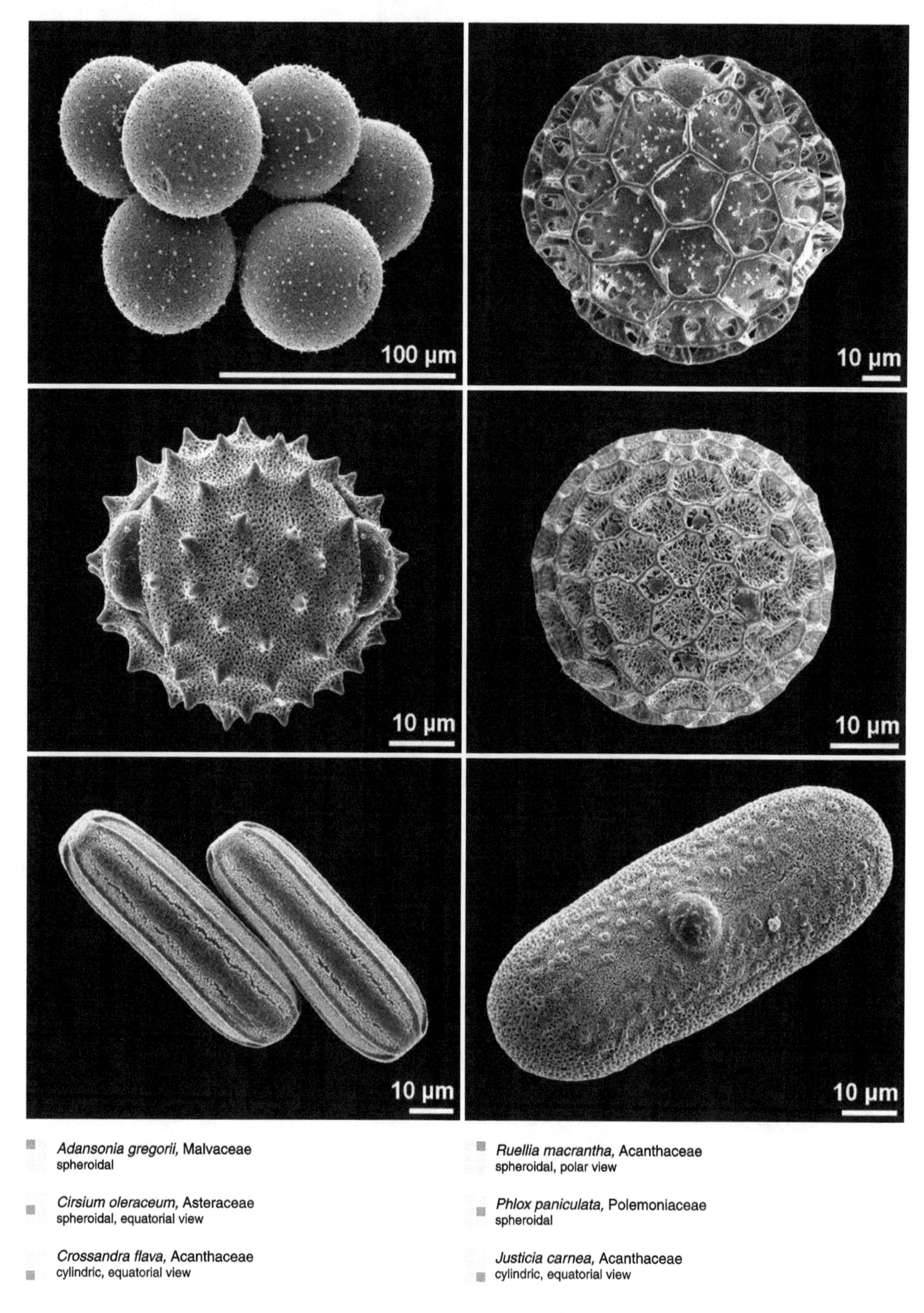

Adansonia gregorii, Malvaceae
spheroidal

Ruellia macrantha, Acanthaceae
spheroidal, polar view

Cirsium oleraceum, Asteraceae
spheroidal, equatorial view

Phlox paniculata, Polemoniaceae
spheroidal

Crossandra flava, Acanthaceae
cylindric, equatorial view

Justicia carnea, Acanthaceae
cylindric, equatorial view

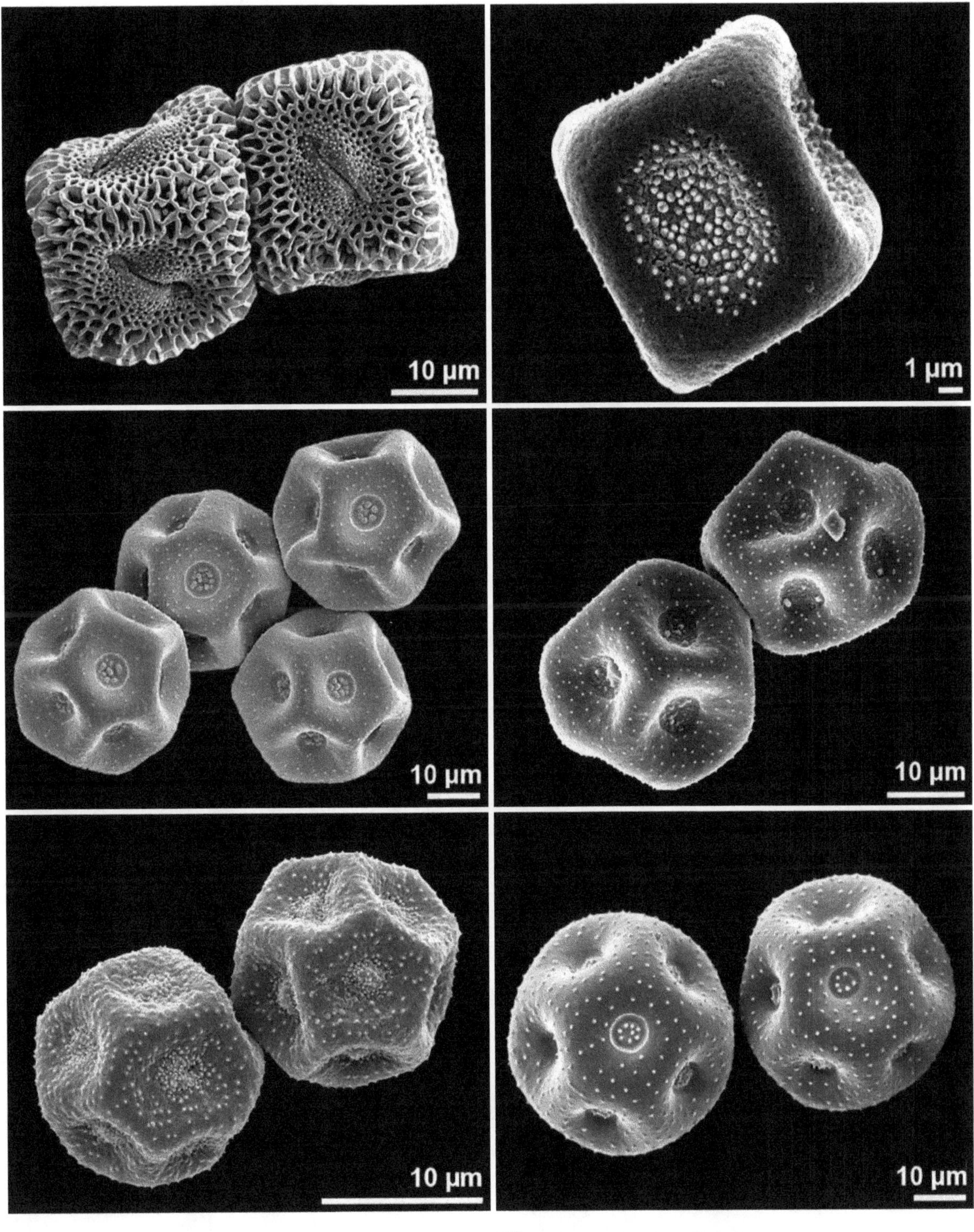

Basella alba, Basellaceae
cubical, dry pollen

Herniaria glabra, Caryophyllaceae
cubical, dry pollen

Cerastium dubium, Caryophyllaceae
polygonal, dry pollen

Eremogone procera, Caryophyllaceae
polygonal, dry pollen

Paronychia polygonifolia, Caryophyllaceae
polygonal, dry pollen

Paronychia polygonifolia, Caryophyllaceae
polygonal, dry pollen

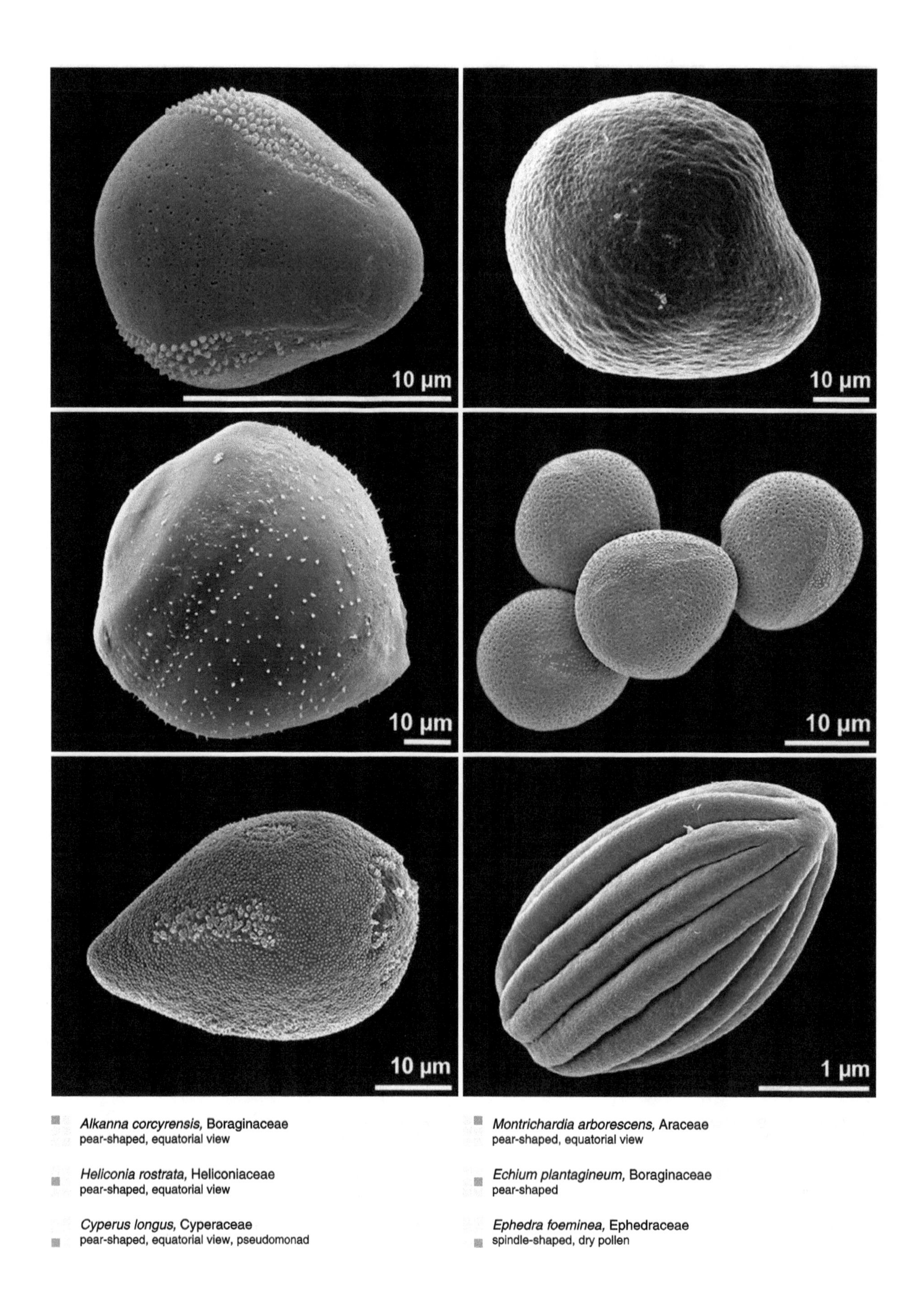

Alkanna corcyrensis, Boraginaceae
pear-shaped, equatorial view

Heliconia rostrata, Heliconiaceae
pear-shaped, equatorial view

Cyperus longus, Cyperaceae
pear-shaped, equatorial view, pseudomonad

Montrichardia arborescens, Araceae
pear-shaped, equatorial view

Echium plantagineum, Boraginaceae
pear-shaped

Ephedra foeminea, Ephedraceae
spindle-shaped, dry pollen

Brugmansia suaveolens, Solanaceae
barrel-shaped, dry pollen

Anthyllis vulneraria, Fabaceae
barrel-shaped, dry pollen

Corydalis cheilanthifolia, Papaveraceae
triangular pyramid, dry pollen

Schoepfia schreberi, Schoepfiaceae
triangular pyramid

Aechmea drakeana, Bromeliaceae
wedge-shaped

Cardiospermum halicacabum, Sapindaceae
convex triangular

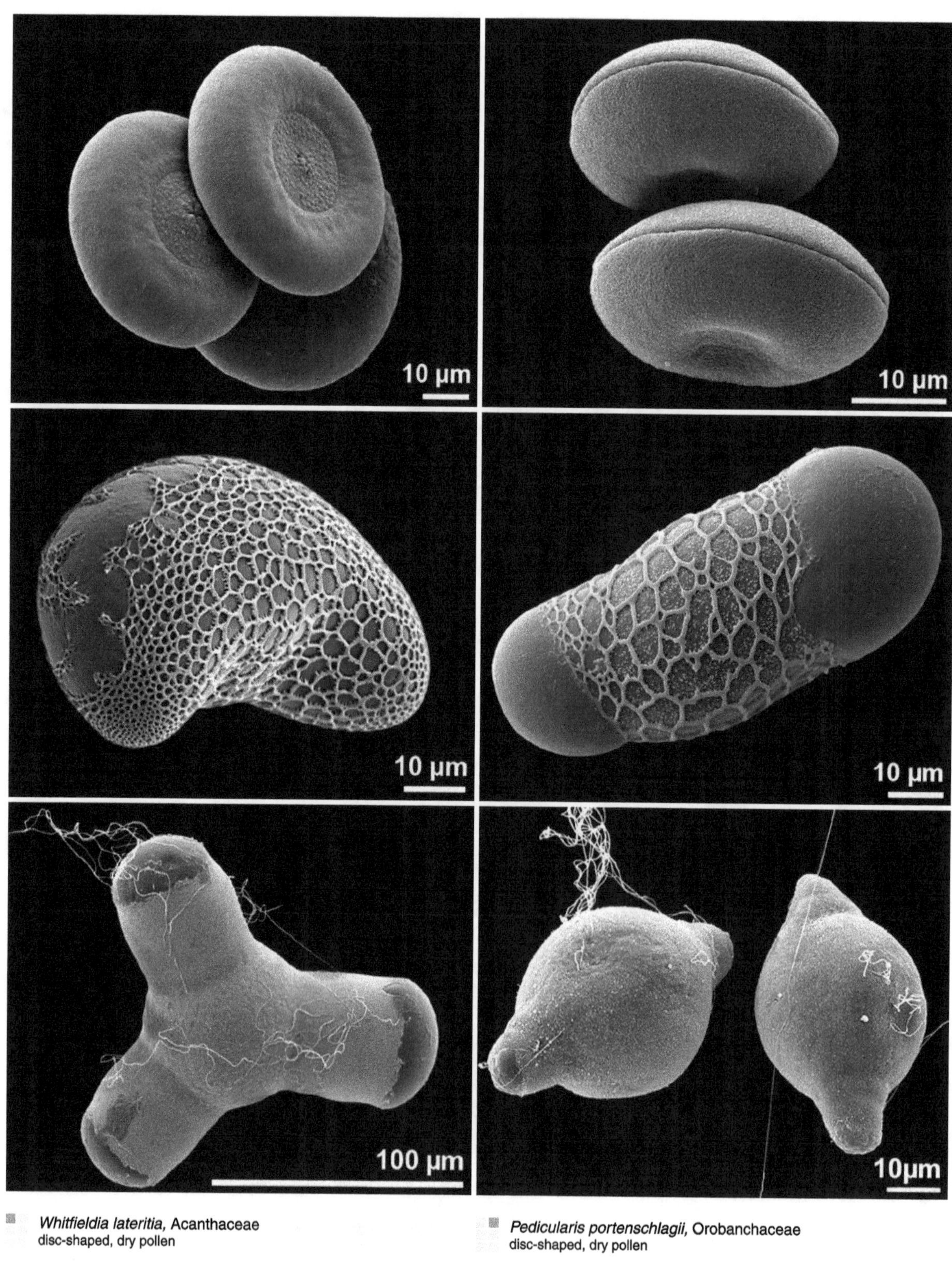

Whitfieldia lateritia, Acanthaceae
disc-shaped, dry pollen

Pedicularis portenschlagii, Orobanchaceae
disc-shaped, dry pollen

Billbergia pyramidalis, Bromeliaceae
bean-shaped

Quesnelia imbricata, Bromeliaceae
cylindric

Clarkia unguiculata, Onagraceae
triangular star

Fuchsia paniculata, Onagraceae
lemon-shaped

Acicarpha tribuloides, Calyceraceae
triangular dipyramid, dry pollen

Myosotis alpestris, Boraginaceae
dumbbell-shaped, dry pollen

Sansevieria suffruticosa, Asparagaceae
cup-shaped, dry pollen

Nicandra physalodes, Solanaceae
dry pollen

Loranthus europaeus, Loranthaceae
dry pollen

Juncus jacquinii, Juncaceae
tetrad, monads cup-shaped, dry pollen

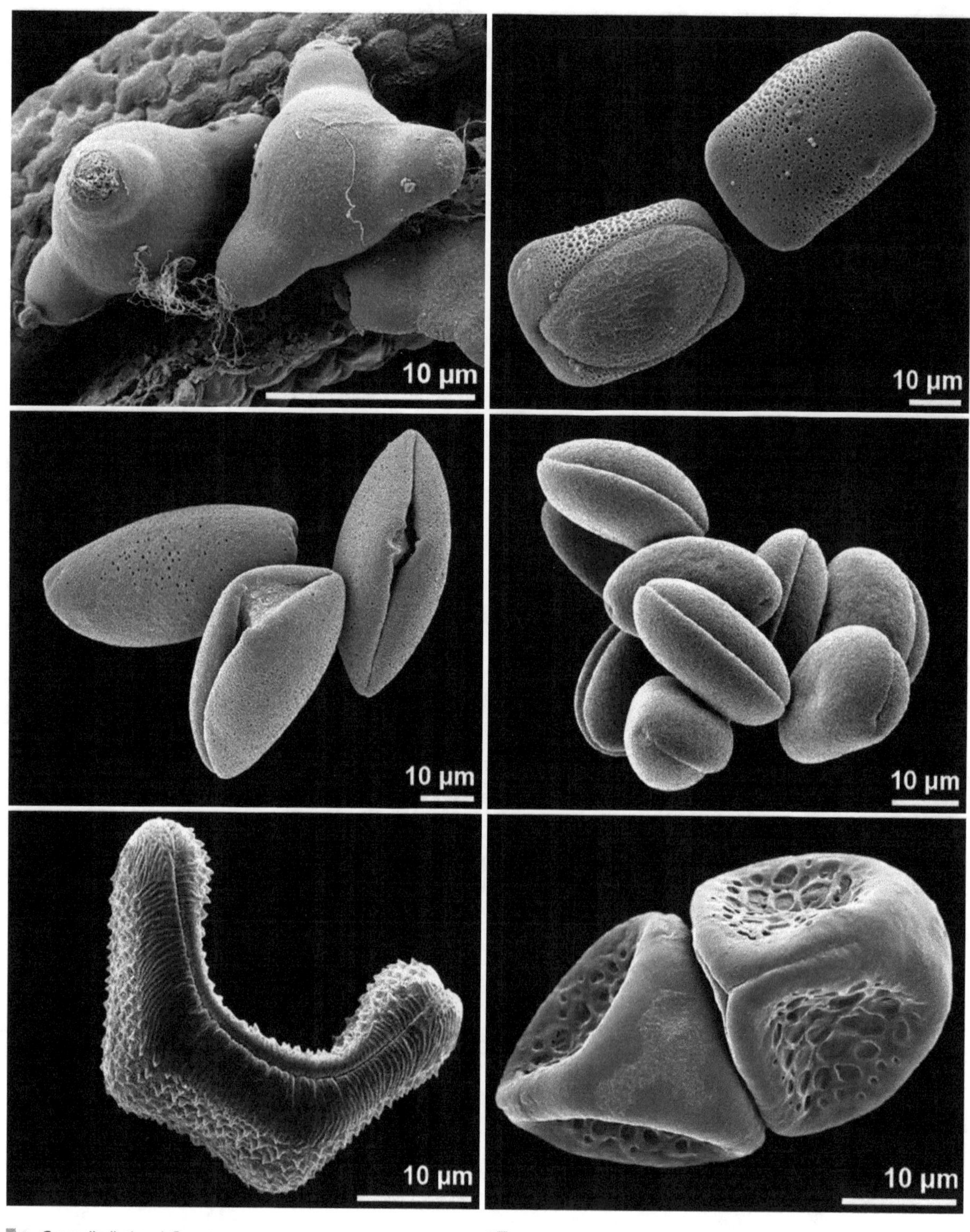

Gaura lindheimeri, Onagraceae

Eremurus robustus, Xanthorrhoeaceae
rounded cuboid

Hyacinthoides italica, Asparagaceae
boat-shaped, dry pollen

Galanthus nivalis, Amaryllidaceae
boat-shaped, dry pollen

Limnanthes douglasii, Limnanthaceae
U-shaped, dry pollen

Thesium dollineri, Santalaceae
convex triangular pyramid, dry pollen

saccus/saccate, corpus

saccus: exinous expansion forming an air sac
corpus: body of a saccate pollen grain

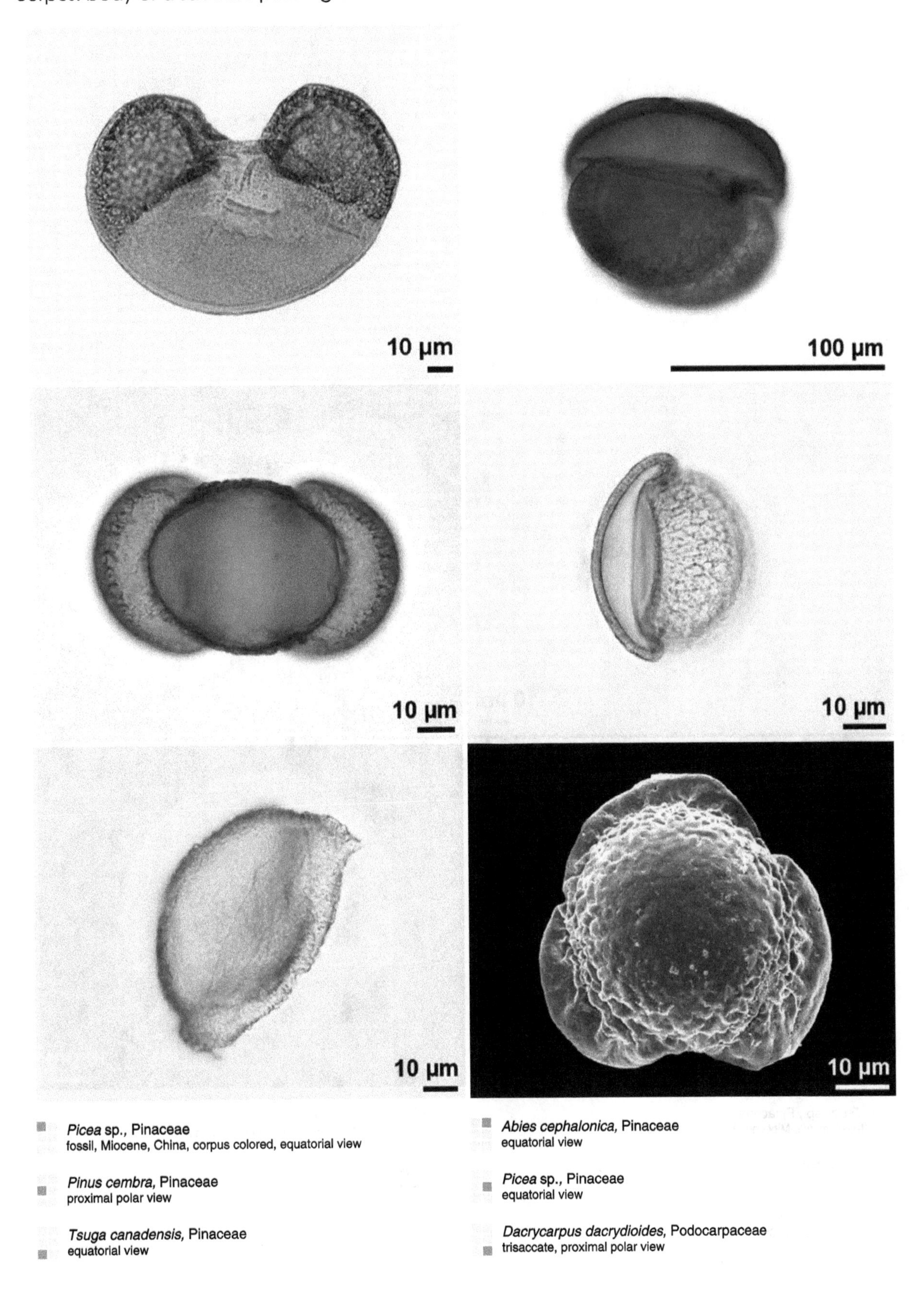

- *Picea* sp., Pinaceae
  fossil, Miocene, China, corpus colored, equatorial view
- *Abies cephalonica*, Pinaceae
  equatorial view
- *Pinus cembra*, Pinaceae
  proximal polar view
- *Picea* sp., Pinaceae
  equatorial view
- *Tsuga canadensis*, Pinaceae
  equatorial view
- *Dacrycarpus dacrydioides*, Podocarpaceae
  trisaccate, proximal polar view

saccus, monosaccate

monosaccate: pollen grain with a single saccus

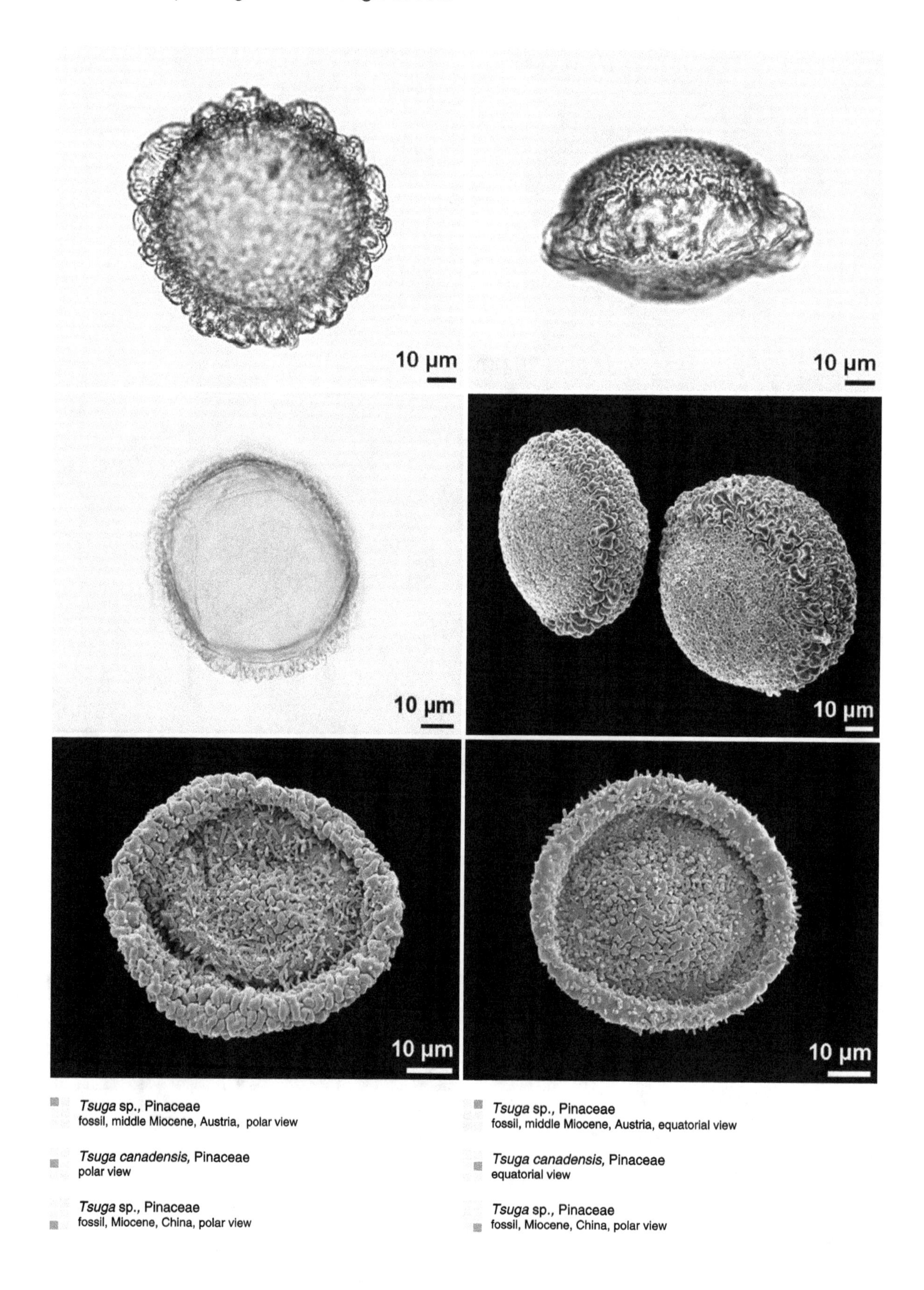

- *Tsuga* sp., Pinaceae
  fossil, middle Miocene, Austria, polar view
- *Tsuga* sp., Pinaceae
  fossil, middle Miocene, Austria, equatorial view
- *Tsuga canadensis*, Pinaceae
  polar view
- *Tsuga canadensis*, Pinaceae
  equatorial view
- *Tsuga* sp., Pinaceae
  fossil, Miocene, China, polar view
- *Tsuga* sp., Pinaceae
  fossil, Miocene, China, polar view

saccus, bisaccate

bisaccate: pollen grain with two sacci

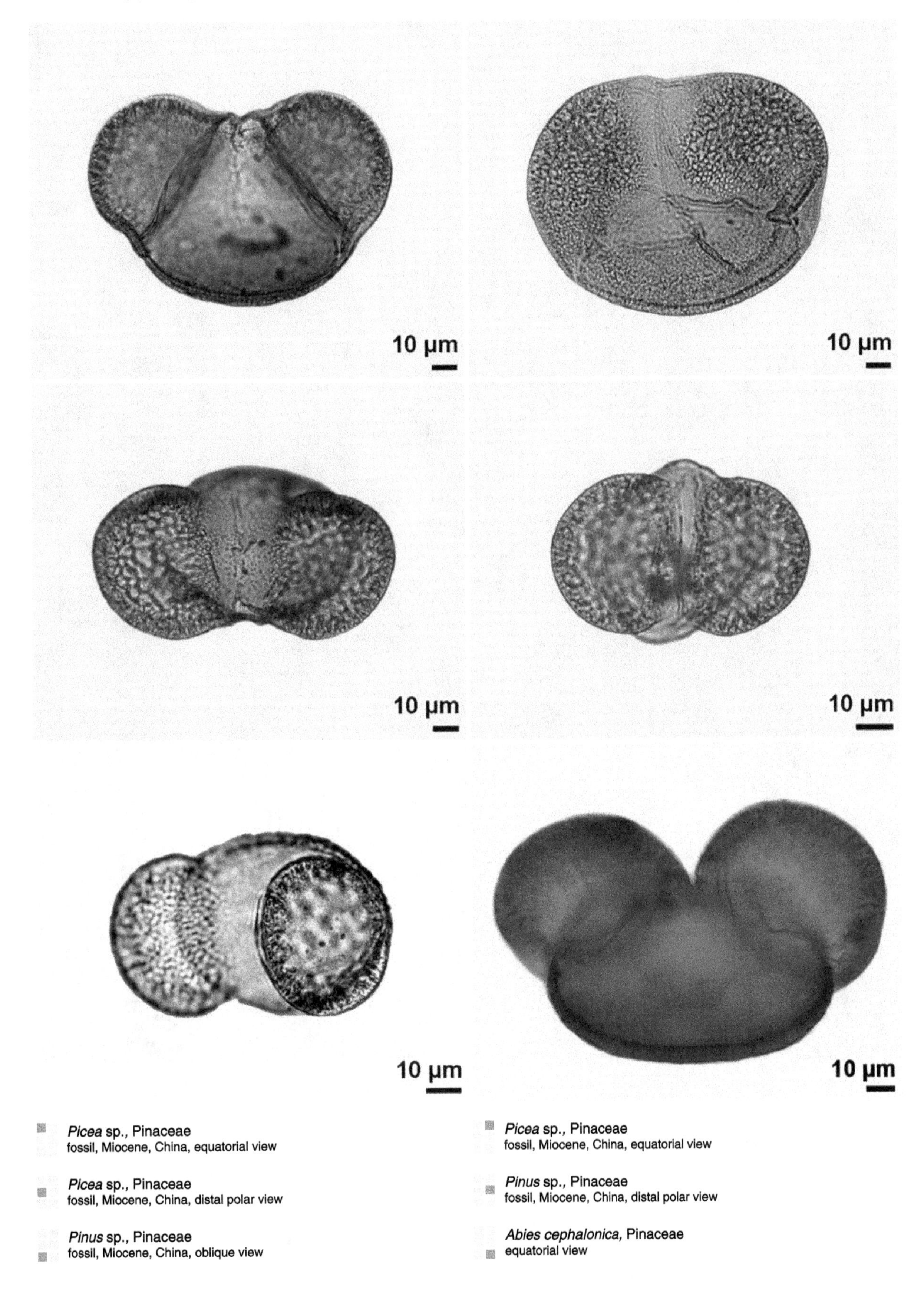

Picea sp., Pinaceae
fossil, Miocene, China, equatorial view

Picea sp., Pinaceae
fossil, Miocene, China, equatorial view

Picea sp., Pinaceae
fossil, Miocene, China, distal polar view

Pinus sp., Pinaceae
fossil, Miocene, China, distal polar view

Pinus sp., Pinaceae
fossil, Miocene, China, oblique view

Abies cephalonica, Pinaceae
equatorial view

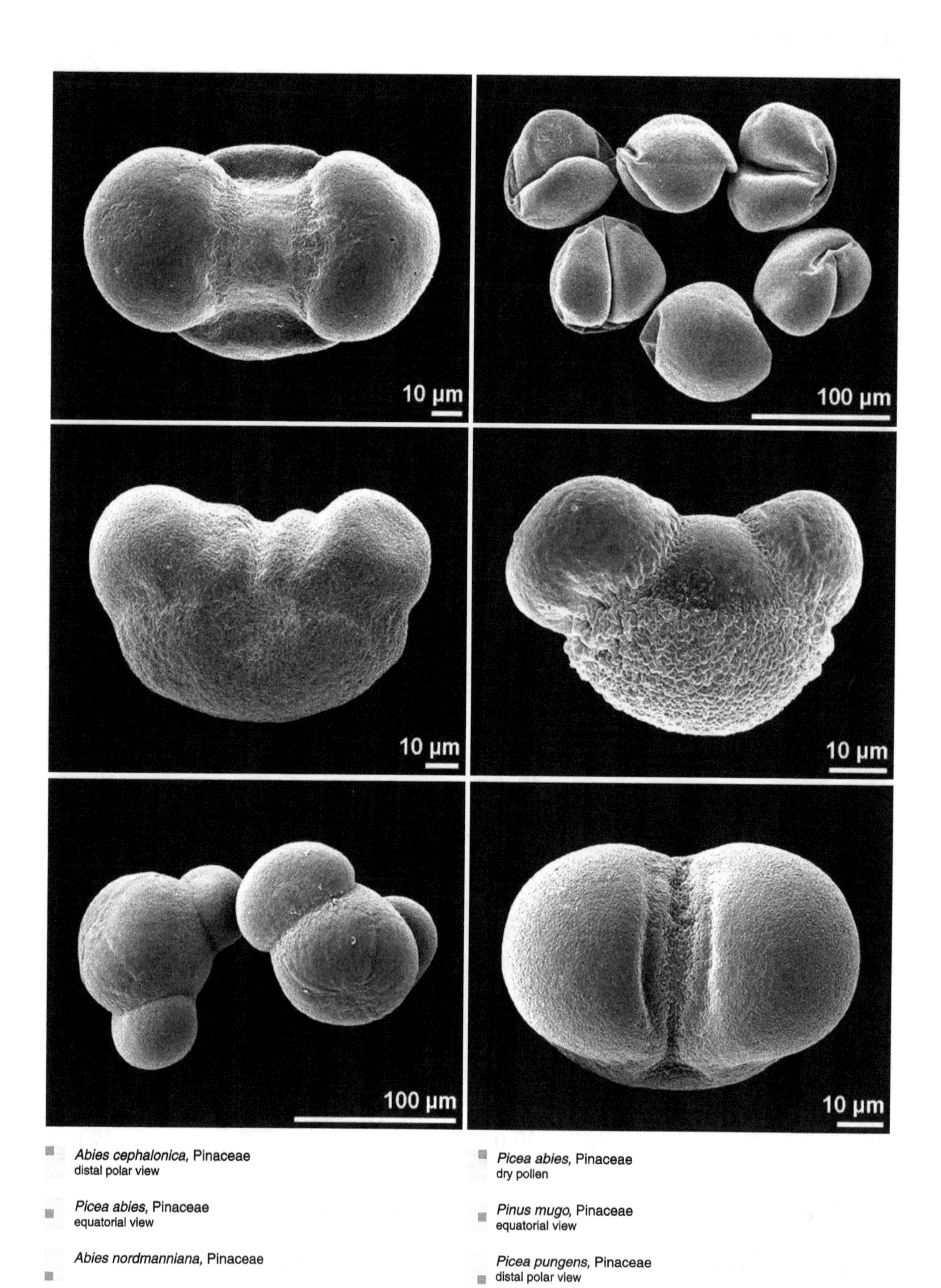

Abies cephalonica, Pinaceae
distal polar view

Picea abies, Pinaceae
dry pollen

Picea abies, Pinaceae
equatorial view

Pinus mugo, Pinaceae
equatorial view

Abies nordmanniana, Pinaceae

Picea pungens, Pinaceae
distal polar view

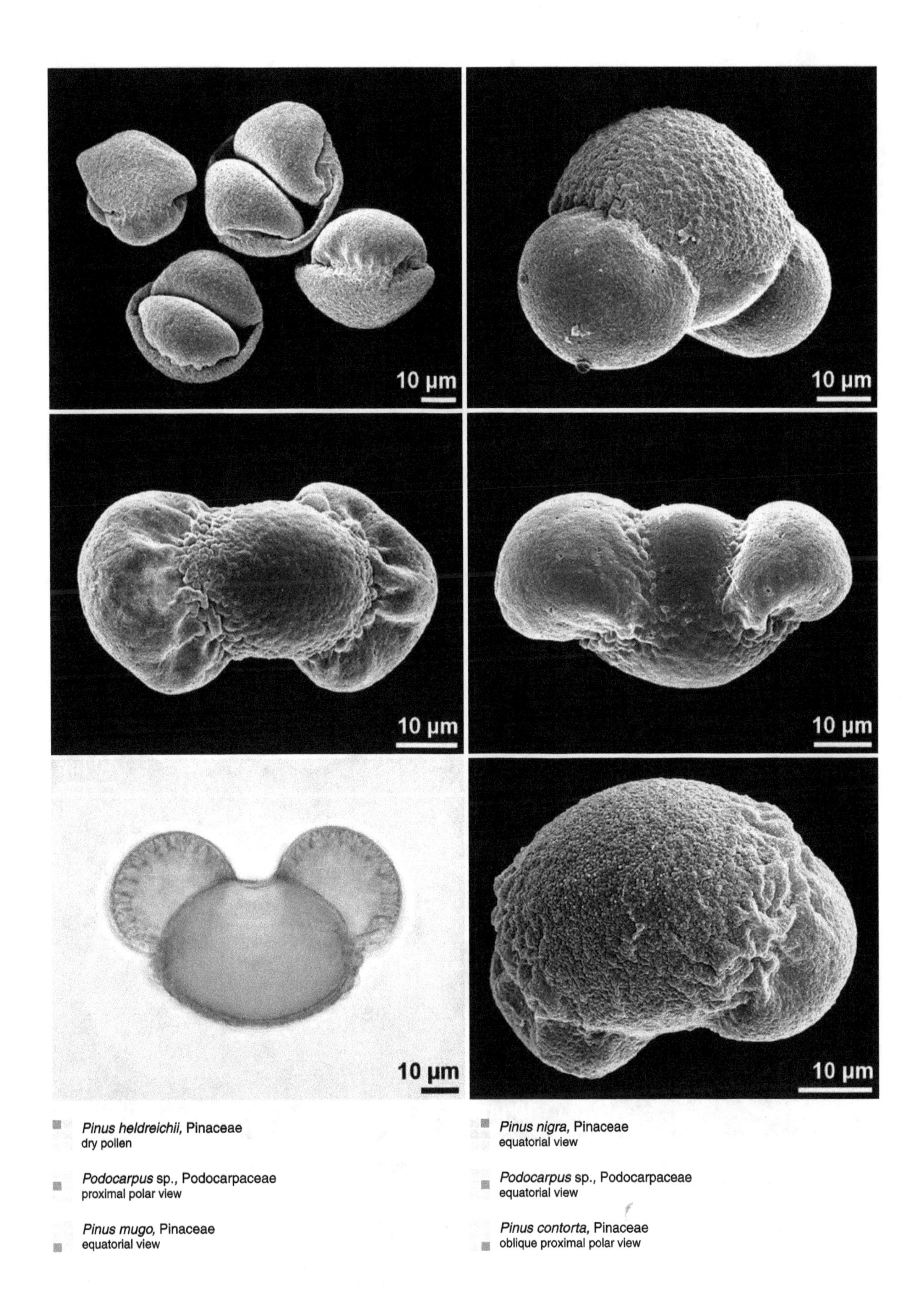

Pinus heldreichii, Pinaceae
dry pollen

Pinus nigra, Pinaceae
equatorial view

Podocarpus sp., Podocarpaceae
proximal polar view

Podocarpus sp., Podocarpaceae
equatorial view

Pinus mugo, Pinaceae
equatorial view

Pinus contorta, Pinaceae
oblique proximal polar view

saccus, trisaccate

trisaccate: pollen grain with three sacci

- Podocarpaceae
  fossil, early Miocene, South-Africa, oblique view
- *Pherosphaera hookeriana*, Podocarpaceae
  equatorial view
- *Pherosphaera hookeriana*, Podocarpaceae
  proximal polar view
- *Dacrycarpus dacrydioides*, Podocarpaceae
  distal polar view
- *Abies concolor*, Pinaceae
- *Abies concolor*, Pinaceae
  distal polar view

infoldings, apertures sunken

infoldings (dry pollen): consequence of harmomegathy in dry condition

Fritillaria pontica, Liliaceae
sulcate

Veratrum album, Melanthiaceae
sulcate

Galium odoratum, Rubiaceae
colpate

Sparmannia africana, Tiliaceae
colporate

Roemeria hybrida, Papaveraceae
pantoporate

Bifora radians, Apiaceae
tricolporate

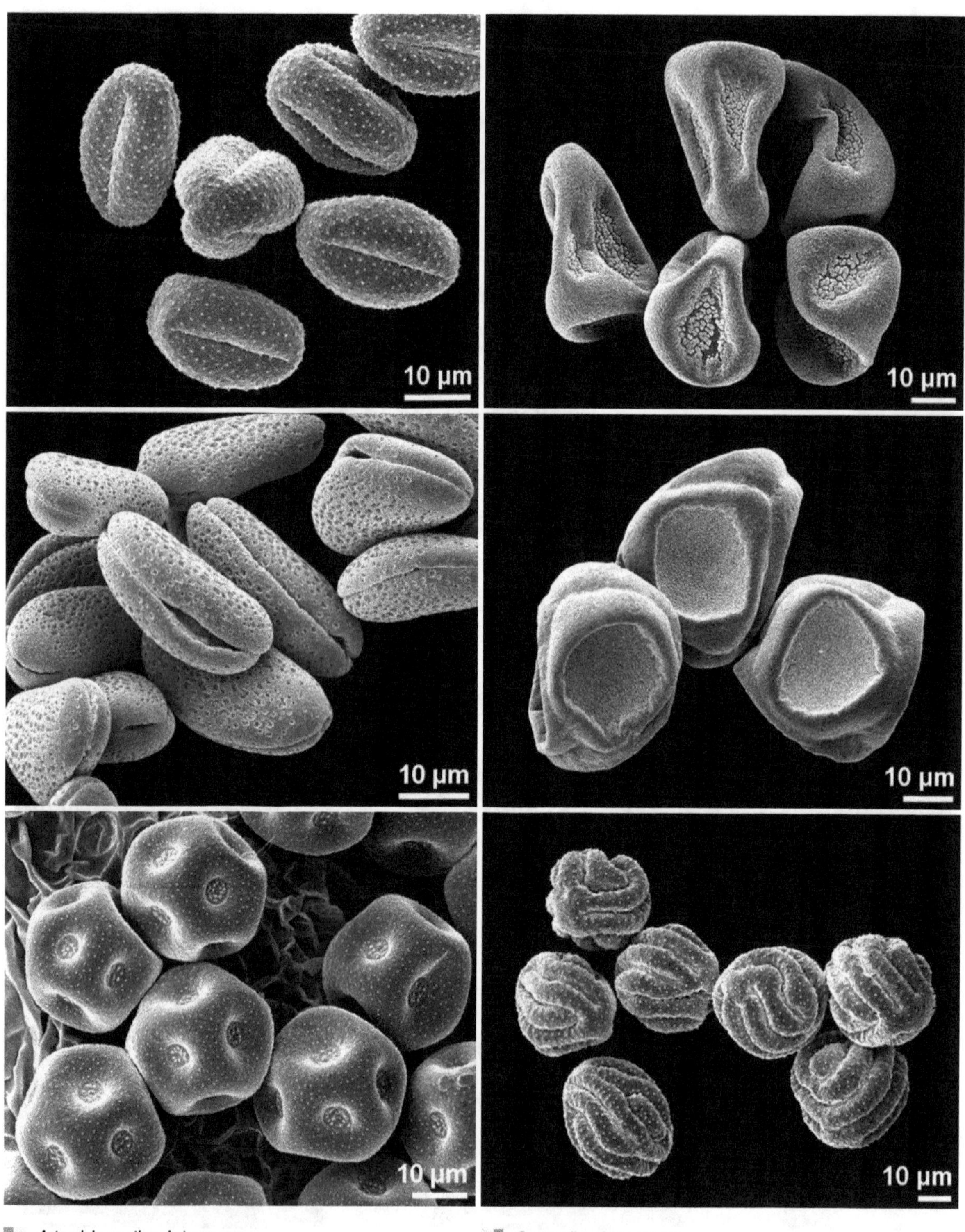

Artemisia pontica, Asteraceae
colporate

Lachenalia aloides, Asparagaceae
sulcate

Moehringia muscosa, Caryophyllaceae
pantoporate

Carex alba, Cyperaceae
pseudomonads, poroidate

Luzula sylvatica, Juncaceae
tetrads, ulcerate

Anemone hortensis, Ranunculaceae
spiraperturate

infoldings, boat-shaped

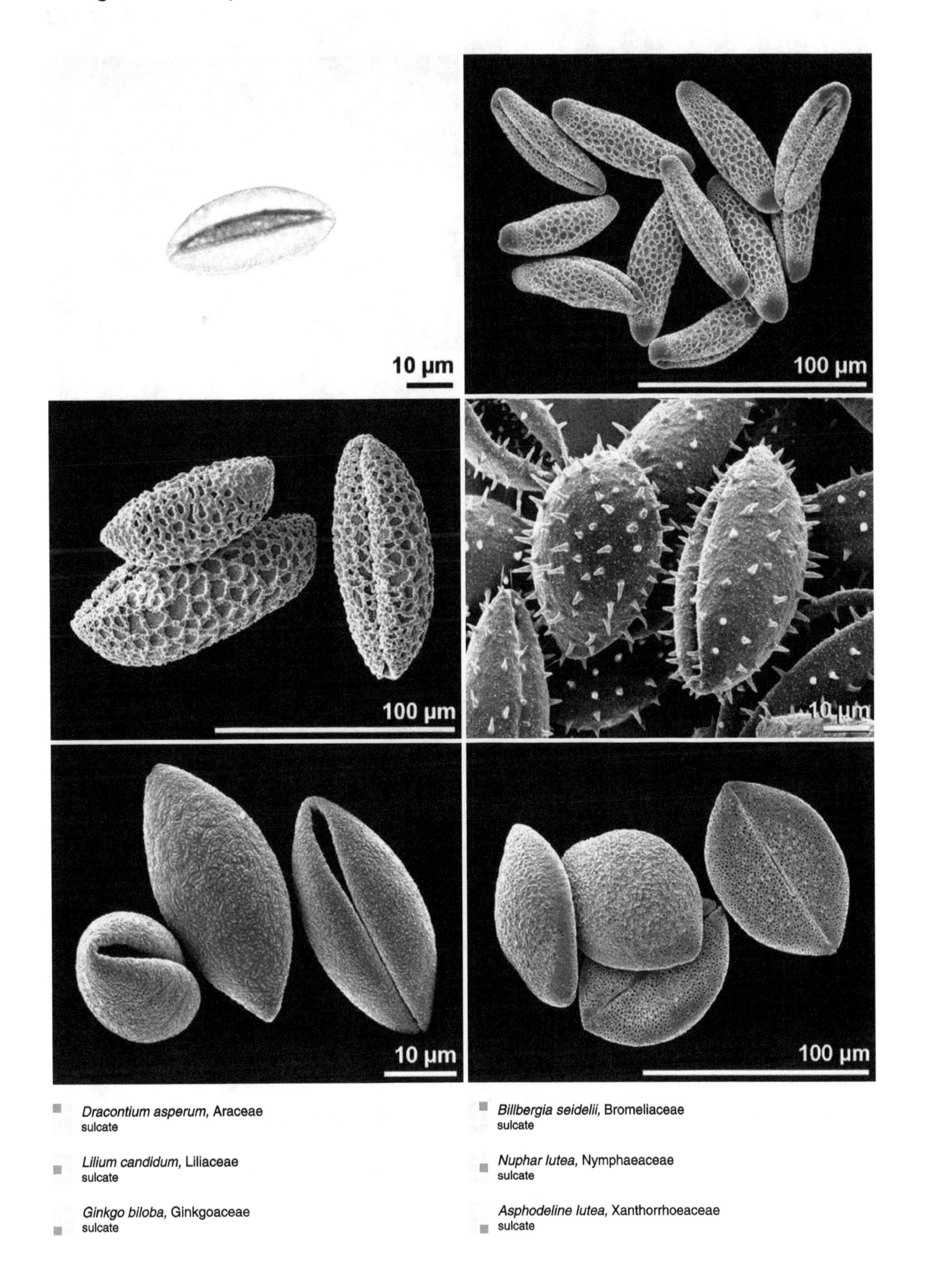

- *Dracontium asperum*, Araceae
  sulcate
- *Lilium candidum*, Liliaceae
  sulcate
- *Ginkgo biloba*, Ginkgoaceae
  sulcate
- *Billbergia seidelii*, Bromeliaceae
  sulcate
- *Nuphar lutea*, Nymphaeaceae
  sulcate
- *Asphodeline lutea*, Xanthorrhoeaceae
  sulcate

Lysichiton americanus, Araceae
sulcate

Gagea lutea, Liliaceae
sulcate

Dioon edule, Zamiaceae
sulcate

Piper nigrum, Piperaceae
sulcate

Sparganium erectum, Typhaceae
ulcerate

Symplocarpus foetidus, Araceae
sulcate

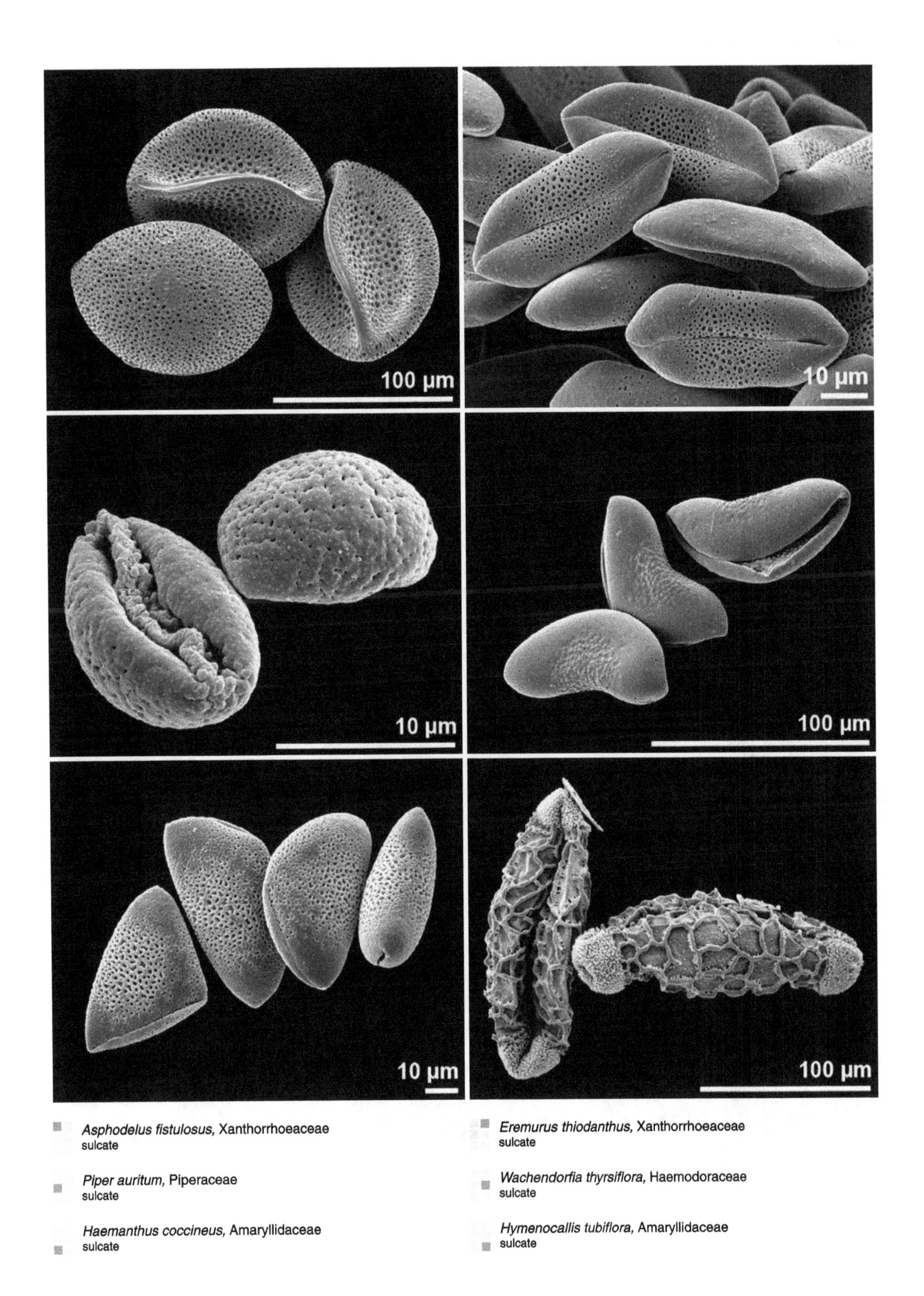

Asphodelus fistulosus, Xanthorrhoeaceae
sulcate

Eremurus thiodanthus, Xanthorrhoeaceae
sulcate

Piper auritum, Piperaceae
sulcate

Wachendorfia thyrsiflora, Haemodoraceae
sulcate

Haemanthus coccineus, Amaryllidaceae
sulcate

Hymenocallis tubiflora, Amaryllidaceae
sulcate

infoldings, cup-shaped

Bougainvillea sp., Nyctaginaceae
colpate

Tilia euchlora, Malvaceae
colporate

Luzula campestris, Juncaceae
tetrads, ulcerate

Heliconia sp., Heliconiaceae
ulcerate

Elaeagnus angustifolia, Elaeagnaceae
colporate

Tsuga canadensis, Pinaceae
leptoma

Adenanthos sericeus, Proteaceae
porate

Petrea volubilis, Verbenaceae
brevicolpate

Cunninghamia lanceolata, Cupressaceae
leptoma

Heliconia stricta, Heliconiaceae
ulcerate

Leucadendron brunioides, Proteaceae
porate

Hibiscus schizopetalus, Malvaceae
porate

infoldings, interapertural area sunken

Alnus glutinosa, Betulaceae
porate

Erica arborea, Ericaceae
tetrads, colporate

Bupleurum rotundifolium, Apiaceae
colporate

Melampyrum arvense, Orobanchaceae
colpate

Leucadendron discolor, Proteaceae
porate

Melastoma sanguineum, Melastomataceae
colporate, heteroaperturate

Verbena officinalis, Verbenaceae
colporate

Ardisia crenata, Primulaceae
syncolporate

Grevillea banksii, Proteaceae
porate

Tsusiophyllum tanakae, Ericaceae
tetrads, colporate

Thesium arvense, Santalaceae
colpate, triradiate colpus

Tropaeolum moritzianum, Tropaeolaceae
colpate

infoldings, irregular

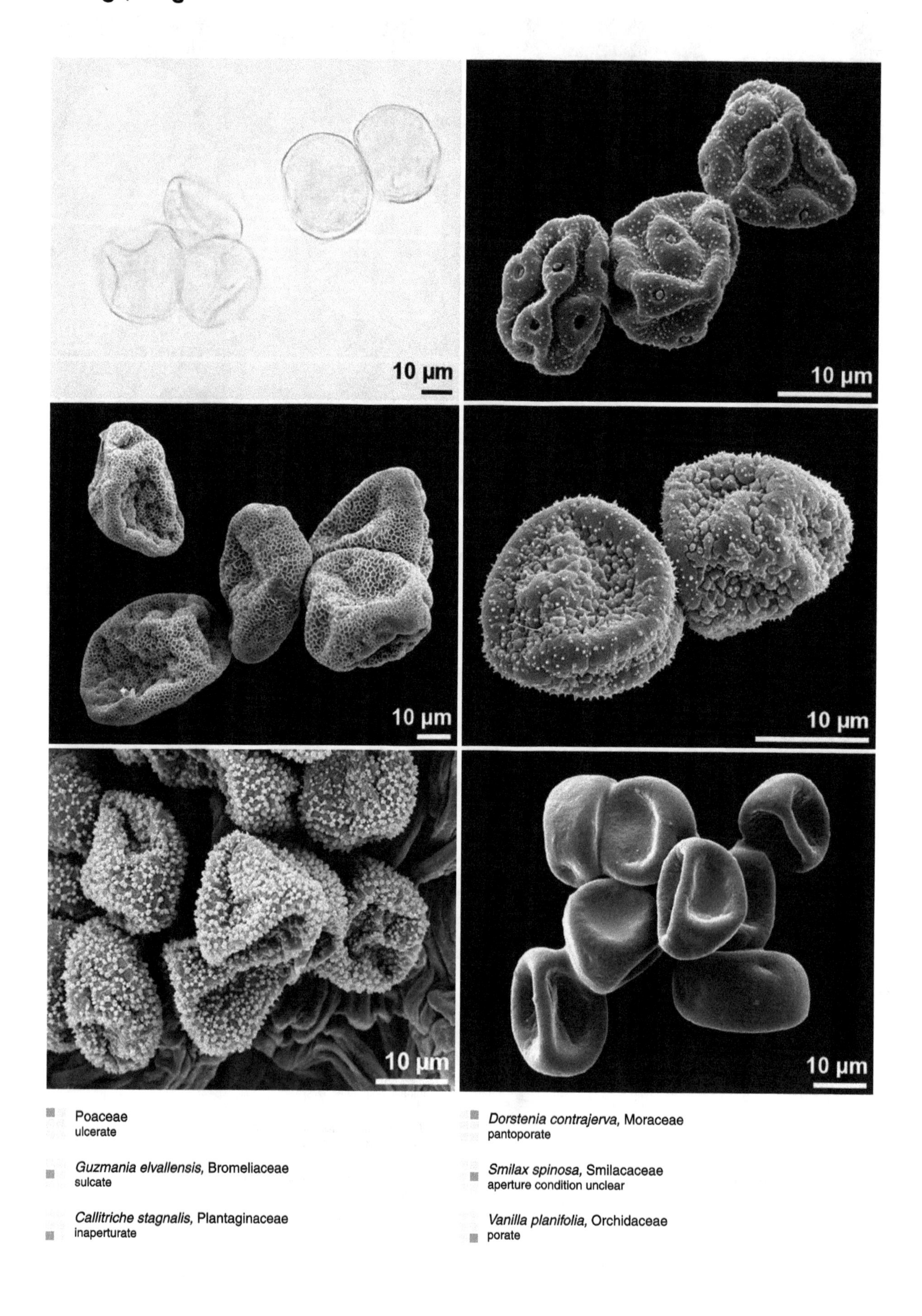

- Poaceae
  ulcerate
- *Guzmania elvallensis*, Bromeliaceae
  sulcate
- *Callitriche stagnalis*, Plantaginaceae
  inaperturate
- *Dorstenia contrajerva*, Moraceae
  pantoporate
- *Smilax spinosa*, Smilacaceae
  aperture condition unclear
- *Vanilla planifolia*, Orchidaceae
  porate

Urtica dioica, Urticaceae
porate

Populus alba, Salicaceae
inaperturate

Sesleria albicans, Poaceae
ulcerate

Anthurium radicans, Araceae
porate

Coriaria nepalensis, Coriariaceae
porate

Orobanche hederae, Orobanchaceae
inaperturate

Permissions

All chapters in this book were first published by Springer; hereby published with permission under the Creative Commons Attribution License or equivalent. Every chapter published in this book has been scrutinized by our experts. Their significance has been extensively debated. The topics covered herein carry significant information for a comprehensive understanding. They may even be implemented as practical applications or may be referred to as a beginning point for further studies.

The contributors of this book come from diverse backgrounds, making this book a truly international effort. We would like to thank all the contributing authors for lending their expertise to make the book truly unique. They have played a crucial role in the development of this book. Without their invaluable contributions this book wouldn't have been possible. They have made vital efforts to compile up to date information on the varied aspects of this subject to make this book a valuable addition to the collection of many professionals and students.

This book was conceptualized with the vision of imparting up-to-date and integrated information in this field. To ensure the same, a matchless editorial board was set up. Every individual on the board went through rigorous rounds of assessment to prove their worth. After which they invested a large part of their time researching and compiling the most relevant data for our readers.

The editorial board has been involved in producing this book since its inception. They have spent rigorous hours researching and exploring the diverse topics which have resulted in the successful publishing of this book. They have passed on their knowledge of decades through this book. To expedite this challenging task, the publisher supported the team at every step. A small team of assistant editors was also appointed to further simplify the editing procedure and attain best results for the readers.

Apart from the editorial board, the designing team has also invested a significant amount of their time in understanding the subject and creating the most relevant covers. They scrutinized every image to scout for the most suitable representation of the subject and create an appropriate cover for the book.

The publishing team has been an ardent support to the editorial, designing and production team. Their endless efforts to recruit the best for this project, has resulted in the accomplishment of this book. They are a veteran in the field of academics and their pool of knowledge is as vast as their experience in printing. Their expertise and guidance has proved useful at every step. Their uncompromising quality standards have made this book an exceptional effort. Their encouragement from time to time has been an inspiration for everyone.

The publisher and the editorial board hope that this book will prove to be a valuable piece of knowledge for students, practitioners and scholars across the globe.

Index